1981

High Performance Mass Spectrometry: Chemical Applications

Michael L. Gross, EDITOR
University of Nebraska

A symposium co-sponsored by the University of Nebraska—Lincoln, the National Science Foundation, AEI Scientific, and INCOS Corp., Lincoln, November 3–5, 1976.

ACS SYMPOSIUM SERIES **70**

AMERICAN CHEMICAL SOCIETY
WASHINGTON, D. C. 1978

Library of Congress CIP Data

Main entry under title:
High performance mass spectrometry.
(ACS symposium series; 70 ISSN 0097-6156)

Includes bibliographies and index.

1. Mass spectrometry—Congresses.
I. Gross, Michael L. II. University of Nebraska—Lincoln. III. Series: American Chemical Society. ACS symposium series; 70.

QD96.M3H53 543'.085 78-789
ISBN 0-8412-0422-5 ASCMC8 70 1-358 1978

PRINTED IN THE UNITED STATES OF AMERICA

ACS Symposium Series

Robert F. Gould, *Editor*

FOREWORD

The ACS Symposium Series was founded in 1974 to provide a medium for publishing symposia quickly in book form. The format of the Series parallels that of the continuing Advances in Chemistry Series except that in order to save time the papers are not typeset but are reproduced as they are submitted by the authors in camera-ready form. As a further means of saving time, the papers are not edited or reviewed except by the symposium chairman, who becomes editor of the book. Papers published in the ACS Symposium Series are original contributions not published elsewhere in whole or major part and include reports of research as well as reviews since symposia may embrace both types of presentation.

CONTENTS

PREFACE

Mass spectrometry has emerged as a mature scientific discipline in chemistry. Investigators no longer are limited to slow scans at low resolution of volatile samples ionized by electron impact. In recent years, a number of significant developments have occurred that are considered examples of "High Performance Mass Spectrometry." These include chemical and field ionization, ultra-high resolution, new methods of defocused metastable scans, collisional activation spectrometry, field ionization kinetics, and new techniques in gas chromatography/mass spectrometry and sample introduction.

To provide a forum for discussing the chemical applications of these techniques, a symposium was held by the Department of Chemistry, University of Nebraska, Lincoln, Nebraska, Nov. 3–5, 1976. The papers presented at that symposium have been updated during the past year (1977) and are collected in this volume.

The primary objective of this volume is to survey current research in high performance mass spectrometry. Solution to problems in chemical analysis is one major theme. Complete identification and quantitation of trace amounts of materials in biological and environmental systems requires many of the methods of high performance mass spectrometry: GC/MS, exact mass measurements, and improved methods of sample vaporization and ionization.

Another emphasis is the determination of structure and properties of gas-phase ions. This research is important in understanding the intrinsic properties of these species in the absence of solvent. Moreover, a thorough understanding of the mechanism of mass spectral fragmentations must rely on these fundamental studies.

Accordingly, the volume has been divided into two sections. The first includes chapters describing methods for acquiring metastable ion data that relate to stable and low-energy decomposing ions and the application of these methods for determining ion structures, properties, and potential energy surfaces. By way of contrast, the capability to investigate extremely rapid ionic reactions using field ionization kinetics is described.

The second section is a survey of analytical applications that can be explored with high performance mass spectrometry. General applications on qualitative and quantitative analyses using high resolution with direct inlet and GLC methods of sample introduction are described. Furthermore, selective methods of chemical ionization and the development of simultaneous recording of positive and negative ion chemical ionization are also reviewed.

The application of mass spectrometry to important problems in biochemistry and pharmaceutical chemistry are discussed in four of the chapters. In particular, methods of obtaining specificity in biological assays and the use of field desorption and an alternative to field desorption for handling nonvolatile materials are covered.

Petroleum and solid state analysis and the use of ultra-high resolution, metastable methods, and spark source ionization are discussed for the analysis of complex mixtures in three chapters.

The last two chapters are illustrative of two of the methods currently used in computer-assisted spectral interpretation. They are included because high performance techniques in mass spectrometry can produce tremendous quantities of data that place onerous interpretation demands on the user.

This volume and the original symposium are intended to introduce the various aspects of high performance mass spectrometry and to provide a review of the current state of knowledge. Our hope is that the book will be useful to scientists with a modest background in mass spectrometry who wish to make use of the technique in their own research and to the expert who is interested in a critical review.

At the opening remarks session of the symposium, G. G. Meisels, Chairman of the Chemistry Department at the University of Nebraska, offered an alternate definition of high performance mass spectrometry. He said that high performance mass spectrometry is a branch of research practiced by high performance scientists. Indeed, this volume is a collection of papers by many of the "highest performance" mass spectroscopists at work today. I must acknowledge their cooperation in assembling this book.

Moreover, I wish to thank Jan Deshayes for assisting in the organization of the symposium and overseeing the typing of the majority of manuscripts. The manuscripts were typed and assembled by Kathy Steen, Velaine Zbytniuk, and Sharon Coufal. Many of the conference details were handled by my graduate students: David Russell, David Hilker, and Stan Wojinski. I am also indebted to these persons for their efforts.

F. H. Field and T. L. Isenhour presented plenary lectures at the symposium, but the press of other duties did not allow them to contribute to this volume. However, I am indebted to them for their participation.

Finally, the symposium was made possible because of generous support from the National Science Foundation, the University of Nebraska Research Council, Kratos–AEI Scientific Apparatus, and the INCOS Corp. Their contributions are gratefully acknowledged.

University of Nebraska
Lincoln, Nebraska
December 30, 1977

MICHAEL L. GROSS

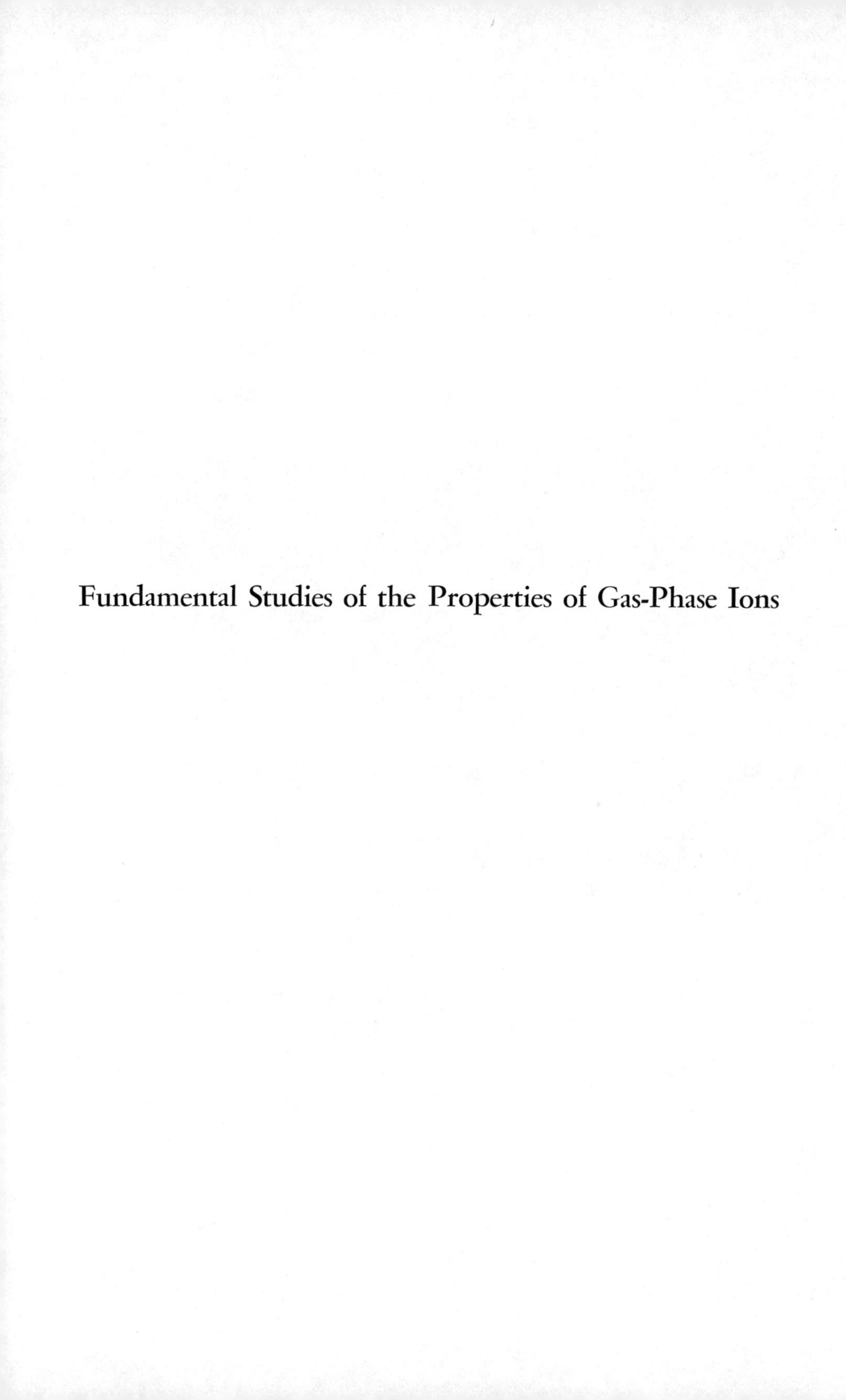

Fundamental Studies of the Properties of Gas-Phase Ions

1

Observation of Metastable Transitions in a High Performance Mass Spectrometer

K. R. JENNINGS

Department of Molecular Sciences, University of Warwick, Coventry, United Kingdom

A metastable transition is the unimolecular or collision induced decomposition of an ion after leaving the source exit slit and before reaching the collector of a mass spectrometer. The daughter ions produced in these decompositions give rise to metastable peaks in mass spectra produced in a magnetic deflection instrument. During the past ten years, various methods of observing metastable transitions in high performance double focussing instruments of various geometries have been introduced (1,2,3,4). This article aims to give a comparative account of these methods together with illustrations of the use of some of them in the author's laboratory.

No single method has established itself as ideal for all applications, and before considering the methods in detail, one must be clear what information is required or may be obtained from a study of metastable transitions. If one is concerned with obtaining detailed information about the potential energy surface over which a particular decomposition occurs, the method must be capable of yielding information on peak shapes, and if possible, the energy resolution should be high enough to reveal any structure in the peak. On the other hand, if one's prime aim is to establish whether or not there is a metastable peak arising from the process $m_1^+ \rightarrow m_2^+$ or alternatively to assign as precisely as possible the mass of either m_1^+ or m_2^+ when one of them is known, a method which produces a very narrow peak, the position of which on the mass or voltage scale can be measured with high precision, is to be preferred. The choice of method may also be influenced by a consideration of whether one is particularly interested in collision-induced decompositions or whether one particularly wishes to avoid them. Other factors such as sensitivity, rapidity of scan,

© 0-8412-0422-5/78/47-070-003$05.00/0

discrimination effects and the interpretation of relative peak intensities must also be considered.

Methods of Observing Metastable Transitions and their Applications.

In a double focussing instrument, the three possible variables are V, the accelerating voltage, E, the electric sector voltage and B, the magnetic field strength. These may be scanned in various ways to allow the collection of daughter ions formed in metastable transitions which occur in the field free regions between the source and the first sector and between the two sectors. The major characteristics of the five methods which have so far been used are summarized in Table I. In discussing these methods, it will be assumed, unless otherwise stated, that the mass spectrometer has a geometry in which the electric sector precedes the magnetic sector.

The first method simply makes use of the often diffuse peaks of low intensity which are observed in a normal mass spectrum in which B is scanned with V and E fixed. The metastable peaks may be masked by intense normal peaks, and assignment is usually not difficult for peaks given by a low molecular weight compound. However, it becomes increasingly difficult to assign low intensity metastable peaks in the mass spectrum of a compound of high molecular weight. If only the more intense metastable peaks are of interest, few problems are encountered, and the method has the advantage that the normal mass spectrum and metastable peaks are obtained in a single scan.

One of the most widely used methods is that in which the accelerating voltage V is scanned with E and B fixed. In this method the daughter ion m_2^+ is selected under normal operating conditions but with V and E set at typically half the maximum value. A scan of V is then initiated and a peak is obtained whenever different precursor ions m_1^+ fulfill the requirement that

$$V_1/V_0 = m_1/m_2 \quad (1)$$

where V_0 is the accelerating voltage required for the collection of m_2^+ ions formed in the source and V_1 is the value required to transmit m_2^+ ions formed from m_1^+ ions in the field-free region between the source and electric sector. Therefore, the method is well suited to give information on precursor ions of a given daughter ion. The signal is produced by the collection of daughter ions of a fixed mass and energy, defined by

Table I. Characteristics of Different Methods for Observation of Metastable Transitions.

	Scan	Fixed	Ease of Assignment	KE Release	Feature
1.	B	V, E	Often Difficult	Yes	Gives m_2^2/m_1 ratio. Low intensity relative to normal peaks. Overlap problems.
2.	V	E, B	Usually to nearest nominal mass	Yes	Gives all m_1^+ of selected m_2^+. Range limited by practical V_1/V_0 ratio. Source conditions vary during scan.
3.	E	V, B	i) Difficult ii) Usually to nearest nominal mass	Yes	i) E precedes B: gives all m_1/m_2 ratios without mass analysis. Requires EM between sectors. ii) B precedes E: gives all m_2^+ from selected m_1^+. Wide range; constant source conditions.
4.	$V^{\frac{1}{2}}/E$	B	Good; peaks narrow	No	Gives all m_2^+ from selected m_1^+. Range limited to $m_1/m_2 \simeq 3$. Source conditions vary during scan.
5.	B/E	V	Good; narrow peaks	No	Gives all m_2^+ from selected m_1^+. Wide range; constant source conditions.

the selected values of B and E. The spectrum is an energy spectrum of precursor ions which fragment to yield these daughter ions, and this is readily used to give the masses of precursor ions by means of equation (1).

If the fragmentation is accompanied by the release of translational energy, the metastable peak is broadened. If the β-slitwidth is reduced in an instrument of Nier-Johnson geometry, the increased discrimination at the β-slit causes the peak-shape to change and facilitates the evaluation of the energy release. Under such conditions, the intensity is reduced but signal averaging may be used to improve the signal-to-noise ratio, and, in certain cases, this has revealed fine structure in metastable peaks. Several fragmentations which lead to the production of $C_3H_3^+$ ions yield metastable peaks which possess fine structure (5). The signal given by the $C_4H_4^+$ ions formed in the process

$$C_7H_7^{++} \rightarrow C_4H_4^+ + C_3H_3^+$$

is reproduced in Figure 1. This peak structure is rationalized by assuming that the two processes yield linear and cyclic forms of the $C_3H_3^+$ ions, the enthalpies of formation of which differ by 1.09 eV. The difference in energies released in the two processes which give rise to the two components of the peak is 0.52 eV. This indicates that approximately half of the difference in enthalpies of formation appears as the difference in energies released. A similar conclusion is reached from the structure in the metastable peak arising from the process

$$C_5H_5^+ \rightarrow C_3H_3^+ + C_2H_2 \quad (3)$$

A further use of this type of scan has been in the investigation of the energy release which accompanies the fragmentation of an ion formed in an ion-molecule reaction in a high pressure source. For this type of experiment, the conditions require careful adjustment since if the pressure is too low, the yield of fragmenting ion is low and if the pressure is too high, collision-induced decompositions occur outside the source and mask the peak given by the unimolecular decomposition. In a study (6) of the fragmentation of the $CH_3OH_2^+$ ion, it was found that best results were obtained with a source pressure of hydrogen of 0.05 Torr with a trace of methanol. Metastable peaks arising from two fragmentations could be observed

$$H_3^+ + CH_3OH \rightarrow H_2 + [CH_3OH_2^+]^* \begin{cases} \rightarrow CH_2OH^+ + H_2 & (4a) \\ \rightarrow CH_3^+ + H_2O & (4b) \end{cases}$$

The broad metastable peak attributed to reaction (4a) indicates that it occurs with the release of 0.93 eV of translational energy, in excellent agreement with a value of 1.1 eV calculated from results based on ICR spectrometry and the quasi equilibrium theory (QET) of mass spectra (7). The metastable peak arising from reaction (4b) is very narrow and is consistent with the release of only 0.0016 eV of translational energy.

The effect of the β-slitwidth on sensitivity and on the ease of assignment of precursor ion masses has been studied on an AEI MS9 instrument, of Nier-Johnson geometry (8). The standard β-slitwidth in such an instrument is 0.508 cm. which has an energy bandpass of ± 0.66% of the nominal energy of the ion beam. If energy release in a metastable transition is sufficiently large to produce a beam with a spread in energy at the β-slit which is greater than this energy bandpass, some of the beam is lost with a consequent reduction in signal intensity. Any reduction in the β-slitwidth results in a further loss of signal. If on the other hand the energy spread within the beam is less than that passed by the β-slit, the slitwidth may be reduced without loss of sensitivity until the two become comparable after which any further reduction in slitwidth leads to a fall in sensitivity. Consequently, if one wishes to compare the relative intensities of the peaks given by two metastable transitions, it is usually necessary to specify the β-slitwidth at which the measurements are taken. For example, in Figure 2, the two signals are normalized to be equal at a slitwidth of 0.508 cm., but at any lower slitwidth they are clearly unequal.

Similar considerations apply when one reduces the β-slitwidth in an attempt to reduce the overlapping of metastable peaks in order to assign masses of precursor ions more readily. For example, if an intense, broad peak is observed for the transition $m_1^+ \rightarrow m_2^+$, it may be difficult to observe a low intensity signal for the transition $(m_1-1)^+ \rightarrow m_2^+$. If a metastable peak is intrinsically broad owing to the release of energy in the fragmentation, a reduction of the β-slitwidth reduces the intensity of the peak but not its width so that a "precursor mass resolving power" based on a 5% adjacent peak contribution rises very little as the slitwidth is decreased. If the energy release is small

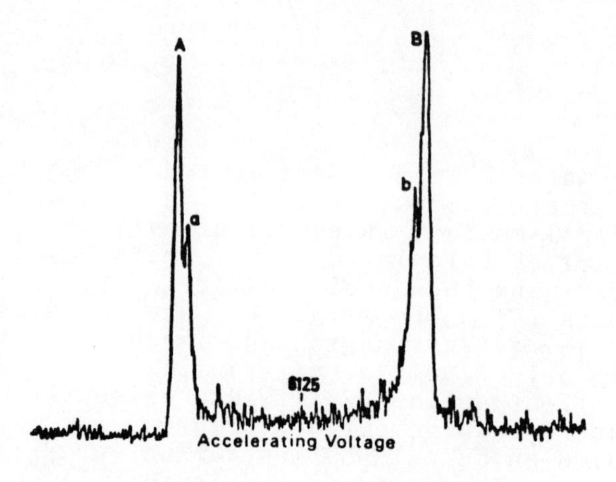

Figure 1. Typical peak shape for transition $C_7H_7^{++} \rightarrow C_4H_4^{+} (+ C_3H_3^{+})$

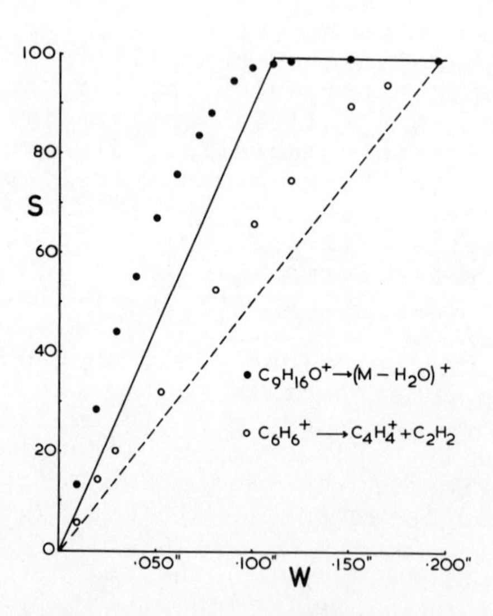

Figure 2. Effect of β-slitwidth W *on sensitivity* S *for two metastable transitions; the points are experimental data, and the lines are calculated assuming a rectangular beam profile.*

so that the energy spread in the beam is less than the energy bandpass of the β-slit, a substantial improvement in "precursor mass resolving power" is observed as the β-slitwidth is reduced. Results obtained for a number of metastable transitions are illustrated in Figure 3.

The main advantages of the accelerating voltage scan method of observing metastable transitions are its simplicity and sensitivity coupled with its ability to give detailed information on energy release in fragmentations. Its major disadvantages are that the practical range over which V_1/V_0 may be varied is limited to about four and that the source tuning characteristics may change during the scan thereby altering the efficiency of extraction of ions from the source. The latter problem can be overcome to a large extent by adjusting the tuning so that it is relatively insensitive to V_1 although this usually entails some sacrifice of sensitivity. In some applications, the broad peaks produced by this scan, even at reduced β-slitwidths, may cause some uncertainty in the assignment of precursor ion masses, particularly if the scan is used to study collision induced decompositions.

The third general method in which the electric sector voltage E is scanned at constant V and B shares many of the advantages and disadvantages of the previous method, especially when used with an instrument of reversed geometry (4). It makes use of the fact that if $m_1^+ \rightarrow m_2^+$ in the field-free region preceding the electric sector, m_2^+ are transmitted when

$$m_1/m_2 = E_1/E_2 \qquad (5)$$

where E_1 and E_2 are the electric sector voltages required to transmit ions m_1^+ (main beam) and m_2^+. In instruments of Nier-Johnson geometry, the method has not been widely applied since it requires the use of a sensitive detector between the two sectors. Even then, without mass analysis of the primary ion beam, peaks are observed for all transitions for which equation (5) is satisfied, i.e., without selection of either m_1^+ or m_2^+. Although assignment is not difficult for simple compounds, it becomes much more difficult as the complexity of the molecule increases. Nevertheless, such a display, known as an ion kinetic energy (IKE) spectrum, may have applications as a "fingerprint" spectrum.

In a reverse geometry instrument, however, the main beam is mass-analyzed by passing it through a magnetic sector before the electric sector so that m_1^+ is selected by adjusting the magnetic field. Decompositions of m_1^+ ions in the field-free region between the two sectors give rise to various daughter ions m_2^+ which are selected by scanning the electric sector voltage E. The peak-shape characteristics and other features are very similar to those of the V scan method except that one obtains information on the products given by the fragmentation of a particular ion m_1^+ rather than on all parent ions of a particular daughter ion m_2^+. The resulting spectrum is a mass-analyzed ion kinetic energy (MIKE) spectrum. Two advantages of this method over the V scan method are that the source conditions are constant throughout the scan so that it can be tuned for maximum sensitivity at the maximum practical value of V, and the range of E_1/E_2 which is practicable is substantially greater than the range of V_1/V_0 in the previous method. Problems associated with decomposition during acceleration are absent and collision-induced decompositions of products formed in ion-molecule reactions in a high pressure source are less likely to mask peaks arising from unimolecular decompositions. Set against these advantages, however, is the fact that since observations are made substantially later in the ion flight path than in the V scan method, a smaller number of ions, in general, will decompose in the field-free region before the electric sector of a reverse geometry instrument than in the same region of an instrument of Nier-Johnson geometry. Therefore, the sensitivity is lower. In addition, at high values of E_1/E_2, the energy of the transmitted m_2^+ ions is low, thereby leading to a reduced gain at the electron multiplier so that the wider range is obtained to some extent at the expense of sensitivity. Applications of this method are discussed by other authors elsewhere in this volume.

The last two methods differ from those described above in that two of the three parameters V, E and B are linked together and scanned while the third is held constant. Each has the feature of giving narrow metastable peaks comparable with peaks given by ions formed in the source in a normal mass spectrum so that, in effect, a high resolution spectrum of metastable peaks is obtained. In each scan, peaks are given by daughter ions formed from a chosen parent ion m_1^+ and the metastable transitions are readily assigned. Neither method gives any information on peak shape and hence no information on the release of translational energy in a fragmentation can be deduced.

In the first of the two linked scans B is held constant and the accelerating voltage V and the electric sector voltage E are linked such that $V^{1/2}/E$ is constant throughout the scan (9). Since E is directly proportional to $V^{\bullet}$, the energy of ions transmitted by the electric sector, $V^{\frac{1}{2}}/E$ is equivalent to maintaining $V^{\frac{1}{2}}/V^{\bullet}$ constant during the scan. At one point during the scan, at the cross-over point, $V = V^{\bullet}$ as shown in Figure 4, the main beam is transmitted and the parent ion m_1^+ is selected by adjusting the magnetic field B. If this ion fragments to give m_2^+ in the field-free region between the source and electric sector, m_2^+ ions are transmitted by the electric sector when

$$V_2^{\bullet} = (m_2/m_1)V_1 \qquad (6)$$

where V_1 is the accelerating voltage actually applied to the m_1^+ ions when m_2^+ ions are transmitted. In order for these m_2^+ ions to be collected at the fixed magnetic field strength B

$$V_1^{\bullet}m_1 = V_2^{\bullet}m_2 \qquad (7)$$

where $V_1^{\bullet} = V^{\bullet}$ at the crossover point. When both equations are satisfied

$$V_1^{\frac{1}{2}}/V_2^{\bullet} = (1/V_1^{\bullet})^{\frac{1}{2}} \qquad (8)$$

or

$$V_1^{\frac{1}{2}}/E = k(1/V_1^{\bullet})^{\frac{1}{2}}$$

In the scan, E and hence $V^{\bullet}$ is directly proportional to the time t which has elapsed since the initiation of the scan so that

$$m_2 = m_1t_1/t_2 = K/t_2 \qquad (9)$$

where t_1 and t_2 are the times after initiation of the scan at which m_1^+ and m_2^+ are collected.

This type of scan has been used on the AEI MS50 instrument with $V_1^{\bullet}$ = 2kV. When V reaches its instrumental maximum value of 8kV, $V^{\bullet}$ = 4kV so that the minimum value of $m_2/m_1 = 0.5$ for a metastable peak to be observable. If the crossover point is set at a lower value in the region of 1kV, the range can be extended so that the minimum value of m_2/m_1 is approximately 0.33 but practical problems may arise in maintaining source sensitivity at low accelerating voltages. This illustrates two of the main disadvantages of the method, limited range and variation of source tuning conditions during the scan. In addition, a narrow β-slitwidth is required to eliminate peaks arising from parent ions of mass $(m_1+1)^+$ or $(m_1-1)^+$,

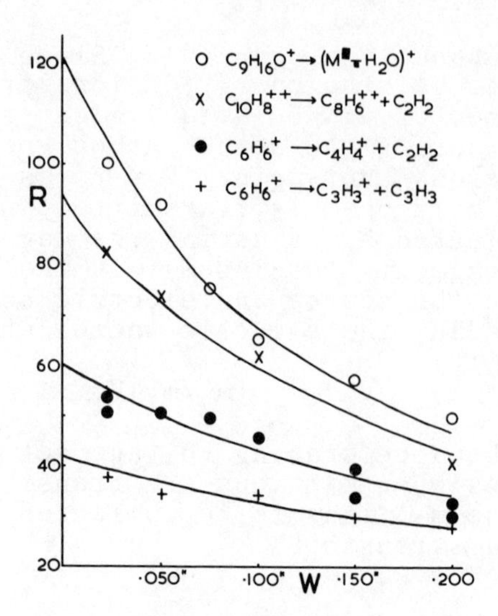

Figure 3. Effect of β-slitwidth W *on precursor mass resolution* R *for various metastable transitions*

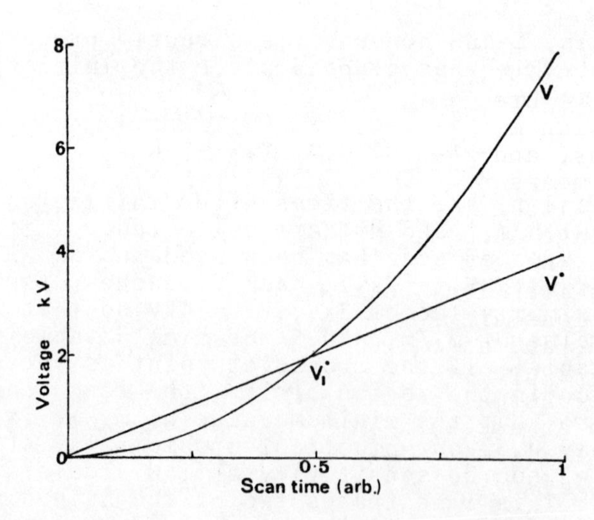

International Journal of
Mass Spectrometry and Ion Physics

Figure 4. Plot of voltage vs. scan-time showing the cross-over point at $V_1 \cdot$ *(2kV)*

but one cannot eliminate low intensity spurious peaks which occasionally arise from decompositions which occur in the accelerating region.

Despite these disadvantages, this linked scan has proved very useful in the investigation of the fragmentation of the molecular ions of a series of substituted tetrafluorocyclobutanes (10). The reactions of fluoroethylene and fluoropropene molecular ions with fluoroolefins studied using ICR spectroscopy have been interpreted as proceding through a reaction complex having a cyclobutane structure and the fragmentations of the molecular ions of the fluorocyclobutanes were shown to be consistent with this hypothesis. Metastable transitions of deuterated molecular ions showed that the specificity of the labelling is retained in their fragmentation so that, for example, peaks were observed for the fragmentations

$$\left[\text{cyclo-}C_4F_4H_2D_2\right]^{+\cdot} \longrightarrow C_2F_4^{+\cdot} + C_2H_2D_2 \qquad (10a)$$

$$\left[\text{cyclo-}C_4F_4H_2D_2\right]^{+\cdot} \longrightarrow CF_2CHD^{+\cdot} + CF_2CHD \qquad (10b)$$

In the second of the linked scans, the accelerating voltage V is held constant and the magnetic field B and electric sector voltage E are scanned such that B/E is constant throughout the scan (11). A prototype of this scan has recently been fitted to the Warwick MS50 instrument, and a schematic of the circuit is shown in Figure 5. The magnetic field is adjusted so that the ion m_1^+ is collected under normal operating conditions, and the electric sector voltage E is measured by means of a digital voltmeter. The linked scan is then switched in so that the electric sector reference voltage is controlled by the output of the field monitor. The variable gain amplifier is adjusted so that E is identical to that used in normal operation so that m_1^+ ions are again collected. The scan is then initiated, and as B and E fall, B/E remains constant as shown in Figure 6.

Under normal operating conditions, ions m_1^+ subjected to an accelerating voltage V will pass through an electric sector of radius R with velocity v_1 when the electric field E_0 is such that

$$m_1 v_1^2/R = eE_0 \quad \text{or} \quad 2V = RE_0 \qquad (11)$$

since $Ve = \frac{1}{2}m_1 v_1^2$. The m_1^+ ions are collected if they pass through a magnetic field of strength B_1 such that

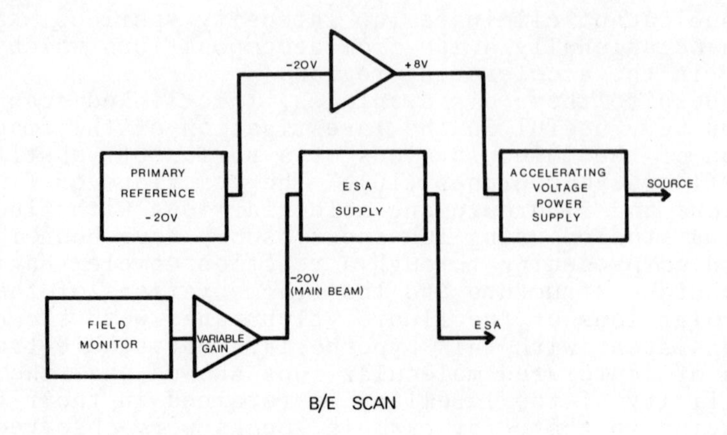

Figure 5. Circuit used for B/E *linked scan*

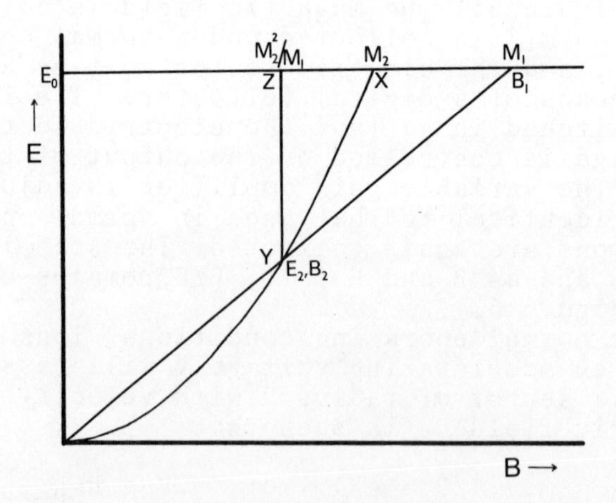

Figure 6. Plot of B/E *scan showing normal scan at* E_0 *and* B *and* E *values for collection of* m_2^+ *formed from various precursors*

$$m_1 v_1^2/r = B_1 e v_1 \quad \text{or} \quad m_1/e = r^2 B_1^2/2V = r^2 B_1^2/RE_0 \quad (12)$$

where r is the radius of the magnetic sector.

If $m_1^+ \rightarrow m_2^+$ occurs in the field-free region between the source and electric sector, m_2^+ ions of velocity v_1 are transmitted by the electric sector only if

$$E_2 = (m_2/m_1)E_0 \quad (13)$$

and by the magnetic sector only if

$$m_2 v_1^2/r = B_2 e v_1 \quad \text{or} \quad B_2 = (m_2/m_1)B_1 \quad (14)$$

Hence both E and B need to be reduced by the same fraction to (m_2/m_1) of their original value. Hence, the linked scan of B/E held constant allows all daughter ions m_2^+ from the selected m_1^+ ion to be collected in turn. The scan has the advantage that the source extraction efficiency is constant since V is not varied, and the range is limited only by the range over which the Hall probe is linear in B. The sensitivity is high since products of decompositions in the first field free region are collected, but, as in the scan of E in a reverse geometry instrument, the response of the electron multiplier will fall as the value of E falls. The relative intensities of peaks within a given scan will depend on the energy release for each decomposition, the linearity of the Hall probe, and the fall off in sensitivity. Nevertheless, the narrow peaks and ease of assignment make the scan ideal for applications in which one merely wants to ascertain whether or not there is a metastable transition for the process $m_1^+ \rightarrow m_2^+$.

If one considers the formation of m_2^+ ions from a series of parent ions, the particular values of B/E which are required for their collection are given by

$$m_2/e = (r^2 B_2^2)/E_2 \quad (15)$$

and so form a parabola as shown in Figure 6. The value required for the collection of m_2^+ from a particular m_1^+ is found at the intersection of the linear B/E plot and the parabola. One can show that the value of B is that at which the metastable peak would appear in a normal mass spectrum, i.e., at an apparent mass of m_2^2/m_1. When scanning to obtain daughter ions from a particular parent ion m_1^+, one may obtain spurious peaks of low intensity if there is an intense signal arising from $(m_1+1)^+ \rightarrow m_2^+$ or if there is significant release of internal energy. Such peaks can usually be recognized without difficulty and are, in any case, minimized by reducing the β-slitwidth.

When used to study collision-induced decompositions, either by raising the analyzer tube pressure or by using a collision chamber between the source and electric sector, this scan still gives sharp peaks and assignment is thereby simplified. For example, the collision-induced decomposition of the toluene molecular ion gave many sharp peaks down to masses as low as m/e = 25, C_2H^+. It has also been used in a study (12) of the reaction between the vinyl methyl ether molecular ion and 1,3-butadiene in which methanol is eliminated to yield what is considered to be the 1,3-cyclohexadiene molecular ion. The products obtained by the collision-induced decomposition of this ion produced by ion-molecule reactions in a high pressure source are very similar to those given by the 1,3-cyclohexadiene molecular ion when this compound is introduced directly in to the source, thereby supporting this suggestion.

Conclusion.

In the above account, we have attempted to point out the advantages and disadvantages of the various methods of observing the products of metastable transitions and it is clear that all have their uses in particular situations. For most applications, a V scan (Nier-Johnson geometry) or E scan (reverse geometry) together with a linked scan, preferably the B/E scan because of its greater range, will give the required information. The methods should be regarded as complementary rather than in competition.

Literature Cited.

1. Cooks, R.G., Beynon, J.H., Caprioli, R.M. and Lester, G.R., "Metastable Ions", (Elsevier Scientific Publishing Co.), Amsterdam, 1973.
2. Jennings, K.R. in "Mass Spectrometry, Techniques, and Applications", p. 419 (Ed. G. W. A. Milne, Wiley Interscience), New York, 1971.
3. Holmes, J.L. and Benoit, F.M. in MTP Int. Rev. of Sci. Physical Chemistry Series 1 (1972) 5, 259. (Ed. A. Maccoll, Butterworths), London.
4. Beynon, J.H. and Cooks, R.G. in TMP Int. Rev. of Sci. Physical Chemistry Series 2, (1975), 5, 159. (Ed. A. Maccoll, Butterworths), London.

5. Sen Sharma, D.K., Jennings, K.R. and Beynon, J.H., Organic Mass Spectrom., (1976), 11, 319.
6. Huntress, W.T., Jr., Sen Sharma, D.K., Jennings, K.R. and Bowers, M.T., Int. J. Mass Spectrom Ion Phys. (submitted for publication).
7. Bowers, M.T., Chesnavich, W.J. and Huntress, W.T., Jr., Int. J. Mass Spectrom Ion Phys., (1973), 12 357.
8. Jones, S., Elliott, R.M. and Jennings, K.R., unpublished work.
9. Weston, A.F., Jennings, K.R., Evans, S. and Elliott, R.M., Int. J. Mass Spectrom Ion Phys. (1976), 20, 317.
10. Derai, R., Ferrer-Correia, A.J.V., Mitchum, R.K., and Jennings, K.R., unpublished work.
11. Bruins, A.P., Jennings, K.R., Evans, S., Int. J. Mass Spectrom. Ion Phys. (in press).
12. van Doorn, R., Nibbering, N.M.M., Ferrer-Correia, A.J.V. and Jennings, K.R., unpublished work.

RECEIVED December 30, 1977

2

Potential Energy Surfaces for Unimolecular Reactions of Organic Ions

RICHARD D. BOWEN and DUDLEY H. WILLIAMS

University Chemical Laboratory, Cambridge, United Kingdom

In studying the reactions undergone by charged species it is often advantageous to examine the chemistry using ion beams. Such methods facilitate a tighter control of the chemistry than is possible in classical experiments in solution. For instance, complications caused by solvation are completely eliminated, and the occurrence of reactions is easily detected by the appearance of metastable peaks in the mass spectrum. The importance of metastable peaks in mass spectrometry has been discussed in detail (1-3); in the present context this importance arises because metastable peaks are usually produced by the slow, unimolecular decompositions of ions having a well-defined range of internal energy, just above the threshold for reaction. Although clearly defined examples of the intervention of isolated states are known (4), such cases are probably the exception rather than the rule. Indeed, the unimolecular chemistry of eleven members of the

$$C_nH_{2n+1}^+ \quad (n=2-12)$$

homologous series of ions can be either rationalized or predicted without invoking the intervention of isolated states (5). A consequence of the fact that metastable dissociations usually occur with little excess energy in the transition state is that the ability of possible decay channels to compete against one another is critically dependent on the activation energies for the processes concerned (6). This is elegantly demonstrated by the occurrence of isotope effects in the decomposition of suitably labelled ions. These isotope effects span the entire range of those known in solution chemistry, and in some cases are spectacularly large. For instance, the metastable decompositions of all the various D-labelled methanes have been documented (7), and, although calculations indicate that the threshold for loss of D• from the molecular ions is only 0.08 eV above that for H• loss, the energy discrimination is so strong that loss of D• is not observed except from $CD_4^{+\bullet}$. Studies of labelled ethane and propane have also been reported (8-10), and for $CD_3CH_2^{+\bullet}$ the ratio of H_2:D_2 loss is ca. 150:1 while that for H•:D• loss from $CD_3CH_3^{+\bullet}$ is > 600:1.

In addition to conclusions based on the relative abundances of metastable peaks, useful information concerning the transition states for reactions can sometimes be deduced from the shapes of metastable peaks. A necessary (but not sufficient) condition for decomposition of isomeric ions over the same potential surface, with closely similar internal energies, is that the metastable peaks for the dissociation(s) of such ions must be the same shape. For instance evidence has been presented (11) which strongly suggests that the $C_6H_6O^{+\bullet}$ ions in the mass spectra of phenol(1) and tropolone(2), which decompose by loss of CO in metastable transitions, have the same structure (or mixture of structures). In each case, the metastable peak for the reaction

$$C_6H_6O^{+\bullet} \rightarrow C_5H_6^{+\bullet} + CO$$

is the same shape (11).

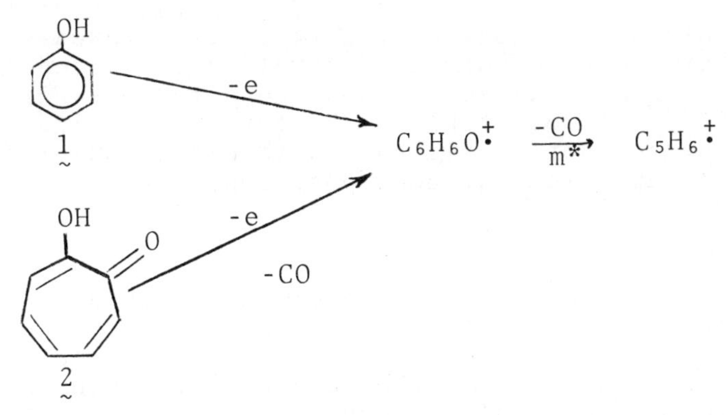

Conversely, if the metastable peaks for dissociation of isomeric ions are different, this constitutes strong evidence that the decay channels, through which reaction is occurring, are different. A classic case is the decomposition of the molecular ions of the isomeric nitrophenols via elimination of NO. In the cases of o- and p-nitrophenol molecular ions the metastable peak for NO loss is broad and flat-topped, whereas for m-nitrophenol the peak is gaussian (12).

NO_2 OH → (-NO, m*) → O^+ OH ⟷ O $\overset{+}{O}H$

3a 4a 5a

NO_2 OH → (-NO, m*) → O^+ OH

3b 4b

O_2N OH → (-NO, m*) → O^+ OH ⟷ O $\overset{+}{O}H$

3c 4c 5c

Clearly, there is a fundamental difference in the potential surface over which the molecular ion of *m*-nitrophenol (3b) decomposes. Consideration of the plausible product structures (4a, b, c) reveals that the charge can be more effectively delocalized in 4a and 4c *via* a "classical resonance effect". Furthermore, the different metastable peaks for NO loss indicate that 3b does not equilibrate with either 3a or 3c prior to unimolecular decomposition.

The Concept of the Potential Surface.

A relatively simple and elegant method of summarizing the chemistry of a given system is the construction of the potential surface over which the ions decompose (13). The method may be used when only one electronic state is involved (an adiabatic surface) and can be extended to include cases where more than one electronic state is involved (a diabatic surface). The latter must be employed when "crossings" occur (14).

The construction of the potential surface for an ionic system also enables the chemistry to be clearly understood. For instance, consider two isomeric ions A+ and B+, which can decompose into C^++D and E^++F, respectively. Clearly if the activation energies for isomerization of A^+ and B^+ are less than those needed to cause dissociation of either A^+ or B^+ then at energies appropriate to slow reactions, rapid equilibration of A^+ and B^+ will occur prior to metastable transitions (Fig. 1). Consequently, the slow reactions undergone by ions on such a surface will reflect the relative activation energies for dissociation to C^++D and E^++F, irrespective of whether the ions are initially formed as A^+ or B^+. An example of this situation is the potential surface for decomposition of $C_6H_{13}^+$ into $C_4H_9^+$ + C_2H_4 and $C_3H_7^+$ + C_3H_6. Irrespective of the precursor used to generate C_6H_{13}, such ions always undergo decomposition in approximately the same ratio (1.6:1 favoring C_3H_6 loss) (6) thus indicating that the activation energies for interconversion of all the accessible configurations of $C_6H_{13}^+$ are less than those for decomposition. A similar case occurs in the unimolecular decomposition of $C_4H_9^+$ where nearly all the metastable ion current is due to CH_4 loss (15) although a small and almost constant percentage is due to C_2H_4 loss (15). Rapid equilibrition of all the possible structures of $C_4H_9^+$ (Fig. 2) requires that complete loss of identity of all carbon and hydrogen atoms in labelled $C_4H_9^+$ ions must precede unimolecular decomposition. This is borne out by D and ^{13}C labelling studies, which establish statistical loss of CH_4 in metastable transitions, irrespective of the precursor structure (15).

This loss of identity of the atoms of labelled ions is sometimes referred to as "scrambling" or "randomization". Such expressions should be used with caution, for they may appear to imply that the chemistry of the system is not understood, whereas in fact, reference to a suitable potential surface (e.g. Fig. 2 above) readily explains the chemistry. Furthermore, it is important to grasp that it is the rapid equilibration of various isomeric structures which causes the loss of identity of atoms in labelled species and not the presence of a suitable symmetrical ion on the potential surface. That the presence of such a symmetrical ion is not a prerequisite is evident from the $C_4H_9^+$ case considered above, where loss of identity of all hydrogen and carbon atoms occurs prior to metastable transitions, even though no structure for $C_4H_9^+$ exists in which all the hydrogen and carbon atoms are equivalent.

In addition to the criterion of metastable abundances (16), and labelling studies, other important methods are available which assist in the construction of potential surfaces. These are Collisional Activation (CA), Ion Cyclotron Resonance (ICR) and energy measurements of various kinds. The first two techniques permit the detection of any ion which exists in a significant well on the potential surface. Thus, for instance, for $C_2H_5O^+$, CA studies indicate only two structures, presumably 6 and 7, which exist in relatively deep potential wells (17); this is in agreement with a study of the metastable decompositions of the ion (18) and with a previous ICR study (19).

$CH_3CH{=}\overset{+}{O}H$ **6** $CH_3\overset{+}{O}{=}CH_2$ **7**

Similar conclusions have also been reached in the $C_3H_7O^+$ system (20); and the $C_2H_6N^+$ and $C_3H_8N^+$ systems (21), where once again the onium type ions (8-11; 12-13 and 14-18) appear to exist in potential wells:

$CH_3CH_2CH{=}\overset{+}{O}H$ **8** $(CH_3)_2C{=}\overset{+}{O}H$ **9** $CH_3CH_2\overset{+}{O}{=}CH_2$ **10** $CH_3CH{=}\overset{\bullet +}{O}CH_3$ **11**

$CH_3CH{=}\overset{+}{N}H_2$ **12** $CH_3\overset{+}{N}H{=}CH_2$ **13**

$CH_3CH_2CH{=}\overset{+}{N}H_2$ **14** $(CH_3)_2C{=}\overset{+}{N}H_2$ **15** $CH_3CH_2\overset{+}{N}H{=}CH_2$ **16**

$CH_3CH{=}\overset{+}{N}HCH_3$ **17** $(CH_3)_2\overset{+}{N}{=}CH_2$ **18**

It should be noted, however, that the difference in CA spectra of two ions does not necessarily preclude isomerization prior to metastable dissociations. This is because the activation energies for interconversion of the ions may be less than those for dissociation but nevertheless large enough to cause considerable differences in the CA spectra. An example of this situation is found in the $C_3H_8N^+$ system where ions of structures 14 and 15 have different CA spectra (21) although an earlier metastable ion study reveals that these ions decompose over the same surface in metastable transitions (22).

CA spectroscopy has also been applied to the long-standing "benzyl versus tropylium" problem (23,24); several $C_7H_7^+$ species with different CA spectra are observed (23). Elegant results stem from a deuterium labelling study (24) where $C_7H_5D_2^+$ ions of nominal structure 19 are observed to lose $:CD_2$ in CA spectroscopy, thus proving that such low energy ions do not collapse to tropylium structures. These studies (23,24), together with ICR work (25) which reveals that two distinct $C_7H_7^+$ populations (presumably benzyl and tropylium), with different reactivities, exist (25), constitute strong evidence in favor of both tropylium and benzyl cations existing in potential wells in the gas phase.

$\overset{+}{C}D_2$

19

Nevertheless, studies on higher energy ions (metastable and source reactions) produced from 2,6-$^{13}C_2$-toluene (20) (26), the tropylium salt (21) (27) and doubly labelled cycloheptatriene (28) suggest isomerization of benzyl to tropylium (or equilibration of the two (26)) occurs. This is because 20, for instance, fragments to yield some unlabelled $C_5H_5^+$ and $^{13}C_2H_2^{+\cdot}$. This is not consistent with simple collapse to a

tropylium ion which then reacts without rearrangement since ions such as the cation of 21 can only lose C_2H_2 or $C^{13}CH_2$ without prior rearrangement.

The use of energy measurements, which yield heats of formation (ΔH_f) of species of interest, in constructing potential surfaces is in many ways self-evident. Where reliable values are available (for instance: several saturated alkyl ions (29,30), some unsaturated carbonium ions (31-34) and various other species (35-38)), the relative levels of reactants and product combinations on a potential surface can be accurately assigned. A case where this is possible is the C_4H_9+ surface referred to above, where the heats of formation of the isomers of $C_4H_9^+$ (29), product ions (29,31) and neutrals (38) are all available. In cases where there is incomplete data concerning the heats of formation of the species of interest, appearance potential (AP) measurements may be made. However, on conventional instruments there are systematic errors associated with AP determinations, and significant errors are likely to occur. Many of the difficulties associated with AP measurements are essentially instrumental in nature and have been discussed recently (39,40). AP measurements using conventional mass spectrometers may be useful in determining rough ΔH_f values, but in view of the difficulties attending such measurements (39,40), they should not be regarded as being accurate to better than (say) ±10 kcal mol^{-1}. Other ways around the problem of obtaining ΔH_f values for relevant species are calculations (41), estimation techniques (5,42,43), and proton affinity (PA) measurements (37,44). The first two are especially useful in cases where the species of interest are unlikely to be accessible *via* experimental AP or PA determination, e.g., the 3-hydroxy-1-propyl cation (**22**). The last method is often useful when the requisite cation cannot be generated *via* direct ionization of the corresponding radical and is capable of yielding accurate results. An ingenious application of the technique is the determination of the heat of formation of the 2-propenyl cation (**23**) by protonation of propyne (44).

$$\overset{+}{C}H_2CH_2CH_2OH \qquad CH_3\overset{+}{C}{=}CH_2$$

22 **23**

When the methods outlined above are applied, three basic types of potential surfaces are found to occur.

Complete Equilibration of Isomers Prior to Unimolecular Decomposition.

This corresponds to the general form of potential surface shown in Figure 1. Here the activation energies for isomerization processes are less than those for dissociation, and decomposition occurs over the same surface, from the same ion (or ions) irrespective of the origin of the ion under study. Examples of this general case are numerous, for instance the $C_4H_9^+$

(15) and $C_6H_{13}^+$ (6) isomers referred to above. Four general criteria of experimental measurements must be satisfied before this kind of surface may be safely assigned to a given ion.

(i) The criterion of metastable abundances (16) must hold good, i.e. the decomposition of ions generated initially as different structures must occur through the same channel(s) and in similar ratios. However, some changes in the ratios of ions decomposing through the various channels may be observed; these changes reflect the slight differences in internal energies of ions produced from different precursors (16).

(ii) The metastable peak shapes must be the same for each decomposition channel, irrespective of the precursor of the ion.

(iii) The AP's for each decomposition channel must be the same (after suitable corrections have been made to compensate for the different heats of formation of precursors) irrespective of the origin of the ions under study.

(iv) There must be loss of identity of some atoms in labelled ions, i.e., the decomposition of labelled ions must be statistical, or must be capable of being interpreted in terms of statistical selection of the necessary atoms together with an isotope effect. When isotope effects are in operation, they must be the same for ions generated from different precursors.

Whereas the third criterion is less reliable when the AP measurements are made using conventional mass spectrometers because of the systematic errors which may complicate the analysis (39,40), the other three are suitable for work using standard mass spectrometers.

An interesting example of the importance of considering isotope effects in conjunction with complete equilibration of the atoms in labelled ions is furnished by two independent studies (45,46) of the competitive loss of H• and D• from the molecular ions of labelled toluenes. For the three precursors **24**-**26**, loss of H• and D• in metastable transitions from the molecular ions can be accounted for if the loss of hydrogen radical is assumed to be statistical with an isotope effect of 2.8:1 (45) or 3.5:1 (46) in favor of H• loss. The slight discrepancy between the two values found in the two studies is presumably due to a population of ions with longer lifetimes (and hence lower internal energies) in the latter study (46). Similar results are found (45) for the labelled cycloheptatriene molecular ions **27** and **28**;

this is consistent with equilibration of the toluene and cycloheptatriene molecular ions prior to metastable decompositions (see criterion (iv) above).

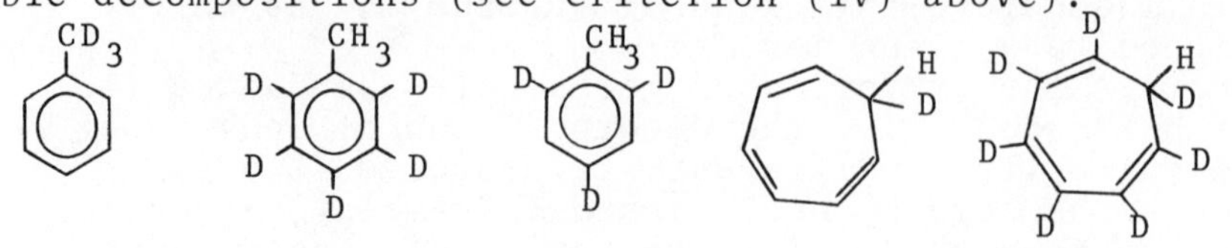

24 **25** **26** **27** **28**

Perhaps the best example of the joint application of the criteria is a definitive study of the $C_3H_6^{+\cdot}$ radical cations (47). In this study the metastable abundances, peak shapes and the appearance potentials for the various decomposition channels of $C_3H_6^{+\cdot}$ ions generated from propene and cyclopropane are shown to be consistent with complete equilibration of the two possible structures of the molecular ion prior to metastable transitions. Other members of the $C_nH_{2n}^{+\cdot}$ homologous series of radical cations behave similarly (48,49).

Other examples of the application of the criterion of metastable abundances (16) to show complete equilibration of all accessible reactant configurations prior to metastable decompositions include some $C_nH_{2n-1}^+$ and $C_nH_{2n-3}^+$ ions (50).

No Equilibration of Isomers Prior to Unimolecular Decomposition.

This corresponds to the general form of potential surface illustrated in Figure 3. Here the activation energies for isomerization processes are greater than those for dissociation, and decomposition occurs over two separate potential surfaces. Decomposition over this general potential surface can be detected by the following observations:

(i) The decomposition(s) of ions on the two separate "halves" of the surface will in general be different both in the nature of the reaction(s) undergone and in the extent to which any common reactions occur.

(ii) The shapes of the metastable peaks for any common reactions will not, in general, be the same. Indeed, in some cases (12,51), drastic differences are observed.

(iii) The AP's for any common decompositions will be different in general.

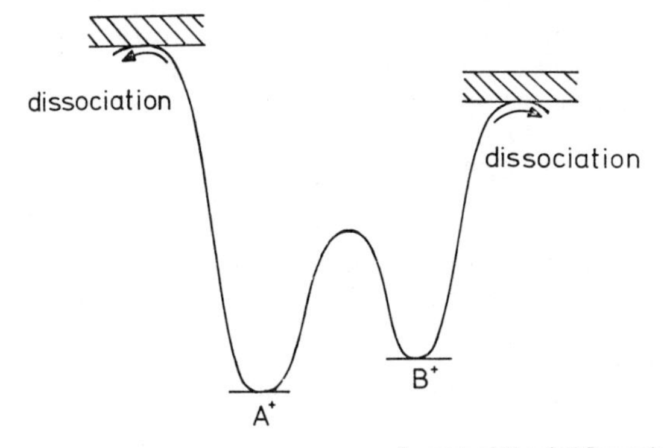

Figure 1. Potential energy surface corresponding to complete equilibration to two ions, A^+ and B^+, prior to slow unimolecular decomposition

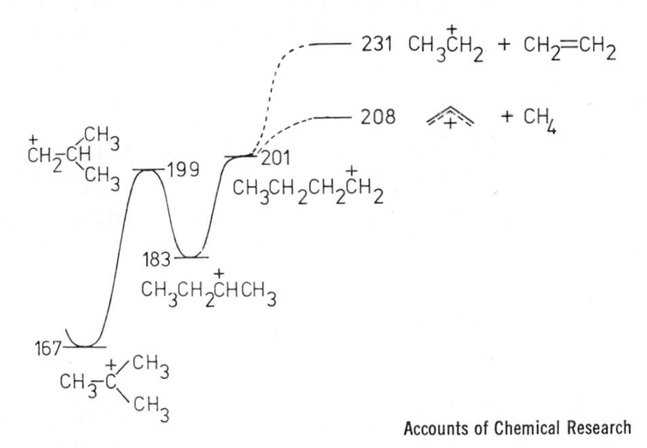

Figure 2. Potential energy surface for slow unimolecular decomposition of the isomeric cations $C_4H_9^+$

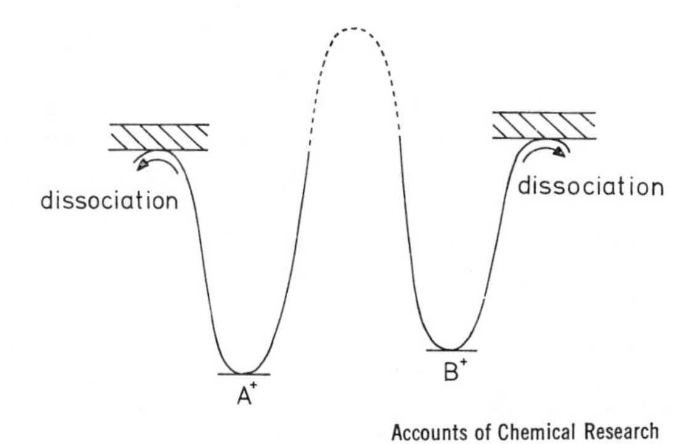

Figure 3. Potential energy surface corresponding to no interconversion of two ions, A^+ and B^+, prior to slow unimolecular decomposition

(iv) The behavior of labelled ions on the two "halves" of the surface will not usually be the same. In particular, isotope effects, when present, are unlikely to be identical.

An excellent example of this kind of potential surface is $C_2H_5O^+$. Ions generated from precursors having the ether moiety (structure **29**) behave in a completely different way to those generated from 1-alkanols (**30**), 2-alkanols (**31**) and compounds of formula (**32**), where Y is a variable group. Two

CH_3OCH_2Y	CH_3CHYOH	YCH_2CH_2OH	CH_3CH_2OY
29	**30**	**31**	**32**

metastable decompositions of $C_2H_5O^+$ ions are observed (51); C_2H_2 loss at m/e 8.02 and CH_4 loss at m/e 18.7. The abundance data, together with the kinetic energy release which accompanies CH_4 loss, are given in Table I (51).

Table I. Metastable Decompositions of $C_2H_5O^+$ ions.

Precursor structure	$\frac{m/e\ 8.02}{m/e\ 18.7}$	kinetic energy release ($kcal\ mol^{-1}$) of m/e 18.7
CH_3OCH_2Y	<0.01	<3
$HOCH_2CH_2Y$	1.8±0.2	11.5±0.9
CH_3CHYOH	1.9±0.1	10.1±0.5
CH_3CH_2OY	2.0±0.1	12.0±1.1

From the data of Table I it is evident that two distinct populations of ions, which do not interconvert appreciably at energies appropriate to metastable transitions, are being formed. Not only do $C_2H_5O^+$ ions formed from precursors of general formula **29** undergo exclusive methane loss, as opposed to ions generated from **30**, **31** or **32** which undergo loss of methane and acetylene, but also the metastable peak shapes for methane loss from the two types are different. That for methane loss from ions produced from **29** is gaussian, whereas methane loss from ions having **30**, **31** or **32** type precursors is flat-topped indicating kinetic energy release (52,53). Furthermore, the measured values for the AP's for CH_4 loss from the two kinds of ions, while only approximate, yield values for the transition state energies which are different (54). Finally, D labelling studies (55) on ions produced from 29 (presumably structure 7) reveal that CH_4 loss occurs with statistical selection of the hydrogen atoms while CH_4 loss from ions with precursor structures 30 or 31 (presumably structure 6 is the most stable form) involves specific elimination of the hydroxyl hydrogen, which does not lose its identity to a significant extent (56) (in the energy-releasing metastable transition).

The above data serve to illustrate the arguments in favor of 6 and 7 reacting over potential surfaces which are distinct. The overall case is overwhelmingly strong and further developments (57) of the potential surface approach can explain much of the detailed chemistry of the $C_2H_5O^+$ ion. The estimated potential surface is depicted in Figure 4, which schematically shows the different energy release profiles (metastable peaks) for the two independent dissociations leading to $HC{\equiv}O^+$ and CH_4 (see also Table I). It should be noted that protonated oxirane (33) and protonated acetaldehyde (6) are expected to interconvert prior to metastable dissociations. This explains why ^{13}C labelling studies reveal that both the carbon atoms of ions originally generated as 34 become equivalent prior to metastable transitions (58), and also why the deuterium atom attached to oxygen in 35 retains its identity (56) while all carbon-bound hydrogens in 6 become equivalent.

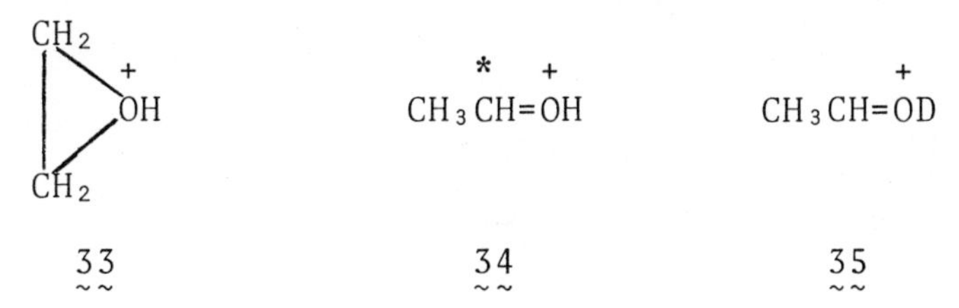

Other important examples of this general form of potential surface are $C_3H_8N^+$ (22), $C_3H_7O^+$ (20, 59-61) and $C_4H_9O^+$ (62), where the metastable abundance criterion can be used to classify the ions into 3, 4 and 5 groups respectively, which do not interconvert prior to unimolecular decompositions. The $C_3H_8N^+$ system is particularly instructive in that one ion (16) is unique in undergoing solely C_2H_4 loss. Deuterium labelling studies reveal that this is a specific process which may be rationalized in terms of a "four-center" mechanism (Scheme 1). The metastable peak for this reaction

SCHEME 1

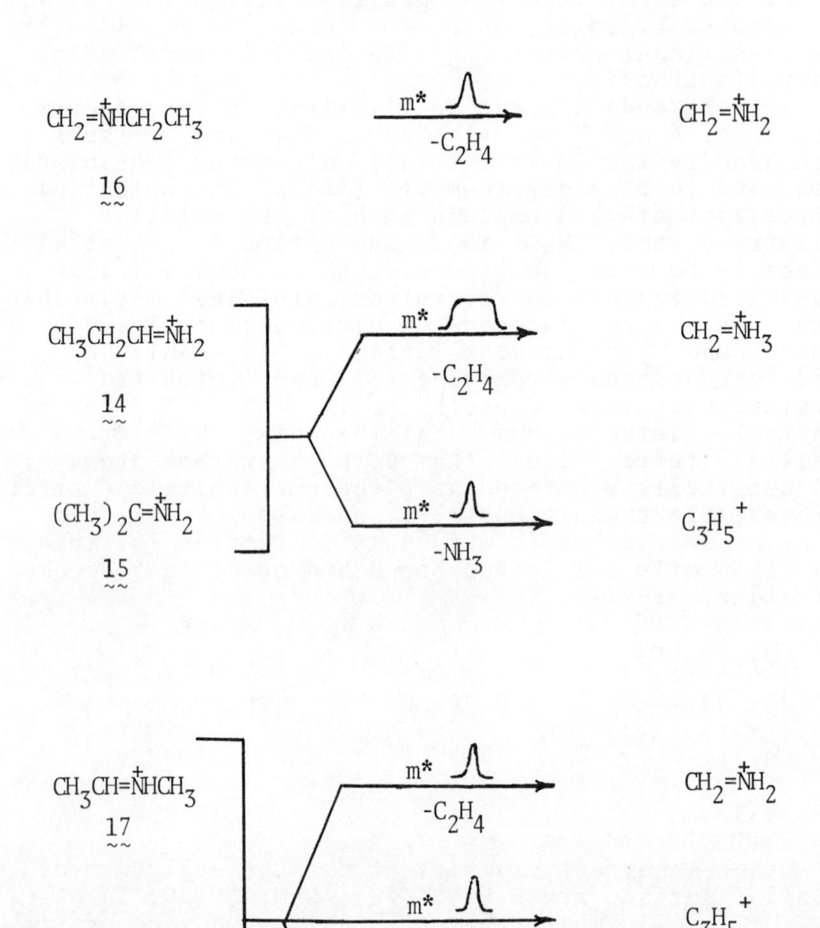
CH2=N+HCH2CH3
16
m*
-C2H4
CH2=N+H2
CH3CH2CH=N+H2
14
(CH3)2C=N+H2
15
m*
-C2H4
CH2=N+H3
m*
-NH3
C3H5+
CH3CH=N+HCH3
17
(CH3)2N+=CH2
18
m*
-C2H4
CH2=N+H2
m*
-NH3
C3H5+
m*
-H2
C3H6N+

is gaussian (22). Ions having structures 14 and 15 also undergo ethylene loss, but this process has a higher activation energy than isomerization reactions leading to loss of identity of all the carbon-bound hydrogens of 14 and 15 (22). The metastable peak for this reaction is flat-topped thus indicating that 14 and 15 decompose over a separate surface from 16. This conclusion is further strengthened by the occurrence of a new dissociation pathway, NH_3 loss, for ions of structure 14 and 15. The third class of ions, initially generated as 17 or 18, undergo both C_2H_4 and NH_3 loss together with a third reaction, H_2 loss, in metastable transitions. Furthermore, the abundances of C_2H_4, NH_3 and H_2 losses are independent, to a first approximation, of the precursor, thus indicating decomposition over a common surface. The metastable peak shapes for these reactions are gaussian for C_2H_4 and NH_3 loss and flat-topped for H_2 loss. Clearly, ions of initial structure 17 and 18 decompose over a common surface which is distinct† from those over which 16 or 14 and 15 dissociate (22). The unimolecular decomposition pathways and the corresponding metastable peak shapes are summarized in Scheme 1.

Another interesting example of the general form of potential energy surface depicted in Figure 3 is the unimolecular decompositions of the $C_7H_6NO_2^+$ ions in the mass spectra of some alkyl-nitrobenzenes (63). Starting from precursors of the general structure $O_2NC_6H_4CH_2R$ with R=Br, CH_3 or C_2H_5, the metastable decompositions of *m/e* 136 are found to be loss of NO for *para*-substituted precursors while for the corresponding *meta*-isomers only NO_2 loss is observed. This establishes that the ions formed by loss of R• from the molecular ions of *m*- and *p*-$O_2NC_6H_4CH_2R$ do not decompose over the same potential surface. This in turn leads to the conclusion that either or both of the benzylic ion structures (36 and 37) which might plausibly be formed at threshold do not undergo reversible ring expansion to the nitrotropylium ion 38.

---

† It is possible that 17 and 18 undergo a rate-determining isomerization onto the same potential surface as that over which 16 decomposes. However, the internal energy of such ions will be much higher than that of ions generated as 16 and so the new decomposition channels, for loss of H_2 and NH_3 are accessible. Nevertheless, this cannot be the case for 14 and 15 because the metastable peak for C_2H_4 loss is a different shape thus precluding decomposition over the same surface as 16 but with greatly different internal energies.

$$\underset{\mathbf{36},\ m/e\ 136}{p\text{-}O_2NC_6H_4CH_2^+} \xrightarrow[-NO]{m^*} C_7H_6O^{+\cdot} \xleftarrow[-NO]{m^*\ /\!/} \underset{\mathbf{37},\ m/e\ 136}{m\text{-}O_2NC_6H_4CH_2^+} \xrightarrow[-NO_2]{m^*} C_7H_6^{+\cdot}$$

$$\mathbf{36} \xrightarrow[-NO_2]{m^*\ /\!/} C_7H_6^{+\cdot}$$

$$\underset{\mathbf{38}}{C_7H_6\text{-}NO_2}$$

Rate-Determining Isomerization Prior to Unimolecular Decomposition.

This corresponds to the general form of potential surface illustrated in Figure 5. Here the energy needed to cause isomerization of structure B^+ to A^+ is more than that needed to cause unimolecular decomposition of B^+. Consequently, ion B^+ dissociates in preference to isomerizing to A^+. However, ion A^+ lies in a potential well which is so deep that the lowest activation energy process is isomerization to B^+, which is formed in a highly excited vibrational level. The ion B^+ thus formed, then dissociates more rapidly than it returns to structure A^+, and does so with a rate constant which is much greater than that ($k \simeq 10^5 sec^{-1}$) normally associated with metastable transitions. It is possible that A^+ and B^+ may undergo the same reactions in metastable transitions, but if the potential surface is of the general form depicted in Figure 5, then four conditions must be satisfied.

(i) The metastable peaks for any common processes must be similar. However, because of the larger amount of internal energy present in ions of structure B^+ which have their origin as isomerized "A^+" type ions, the metastable peaks for decomposition of such ions are likely to be broadened relative to those for dissociation of ions originally generated as B^+.

(ii) As dissociation of the rearranged ions of structure B^+ is fast relative to the normal time scale of metastable transitions, energy ceases to be the dominant factor and "entropy" effects become important. Thus, for rearranged ions, single bond-cleavage disso-

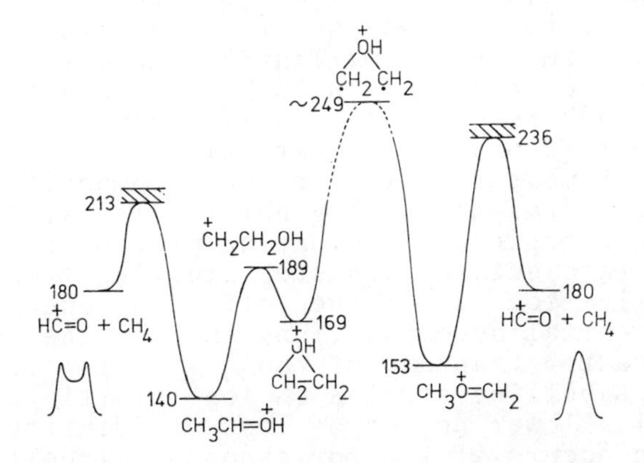

Figure 4. Potential energy surface and energy release profiles for slow unimolecular decomposition of the isomeric cations $C_2H_5O^+$

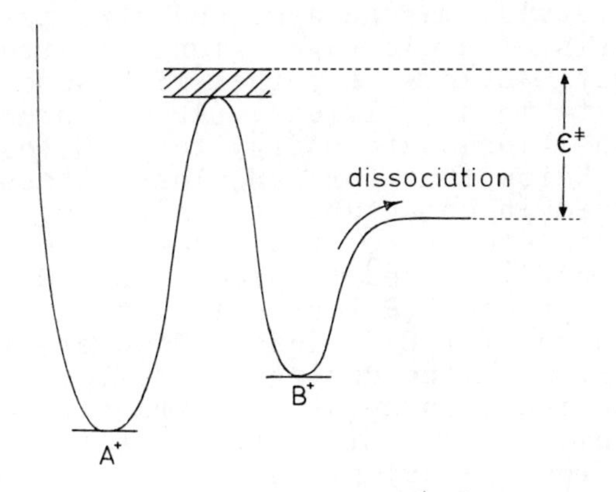

Figure 5. Potential energy surface corresponding to rate-determining isomerization of ion A^+ to B^+ prior to unimolecular decomposition

ciations are favored, whereas there is discrimination against those dissociations which have highly ordered transition states. Consequently, ions produced *via* the rate-determining isomerization are likely to show a marked preference for single bond-cleavage processes ("source type" reactions) even if such processes are not the most favorable energetically.

(iii) AP measurements for the decompositions of ions generated as B^+, will generally reveal different activation energies for competing processes. However, since the rate-determining step starting from A^+ is the isomerization to B^+, the activation energies for all the eventual decompositions must be the same and must be greater than any of those starting from B^+.

(iv) Labelling studies on ions starting from A^+ may reveal a lower degree of loss of identity of the label than occurs if the corresponding labelled B^+ ion is generated directly. This is because there is less time for isomerization processes to cause loss of integrity of labelled species of high internal energy (rearranged "A^+" type ions) which dissociate relatively rapidly.

Although allusions to the possibility of ions reacting over this kind of surface were made for $C_3H_8N^+$ (22) and $C_3H_5O^+$ (64) ions, it was not until 1975 that the first definite case, in the $C_3H_7O^+$ system (65), was documented. Here ions of *m/e* 59, generated from the fragmentation of the molecular ions of suitable ethers and acetals, may be assigned structures **10** and **11** on the basis of α-cleavage. Ions of structure **10** undergo mainly H_2O loss, together with some C_2H_4 loss in metastable transitions. AP measurements reveal that the transition state energy for H_2O loss is some 5 kcal mol^{-1} below that for C_2H_4 loss. Loss of $CH_2{=}O$ is not observed in metastable transitions, even though there is a complete $CH_2{=}O$ group present in **10**, because the minimum energy needed to produce $CH_2{=}O + C_2H_5^+$ is significantly greater (9 kcal mol^{-1}) than the measured activation energy for C_2H_4 loss. However, ions generated initially as **11** do undergo $CH_2{=}O$ loss, to an appreciable extent, in metastable transitions, and H_2O loss becomes a very minor (*ca* 1% total metastable ion current from *m/e* 59) process. One rationalization of this is given in Scheme 2, where **11** is considered to undergo a rate-determining isomerization to **10** prior to metastable dissociations. The rate-determining step is most simply formulated as a symmetry-forbidden (66) 1,3-hydride shift, though other mechanisms are possible.

Scheme 2

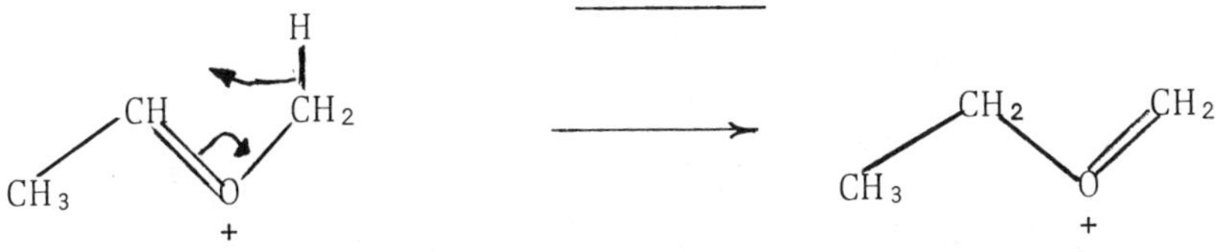

H
CH CH2
CH3 O
+
11

CH2 CH2
CH3 O
+
10

The resulting ion 10, formed by rate-determining isomerization of 11 now has sufficient internal energy to undergo $CH_2{=}O$ loss *via* simple bond cleavage of the C-O bond in 10. C_2H_4 loss is also possible, whereas H_2O loss, which requires considerable rearrangement of 10 prior to dissociation, is greatly disfavored by "entropy" factors.

Further evidence in favor of the isomerization is found in AP measurements for the decompositions of ions initially generated as 11, which are the same within experimental error, and correspond to an internal energy considerably greater than that needed to cause dissociation of 10. In addition, the metastable peaks for H_2O and C_2H_4 losses from ions generated as 10 and 11, respectively, are broader at half height in the latter case. This observation (65) together with D-labelling studies, which reveal a higher specificity for reactions starting from 11 than 10 (65), are consistent with the hypothesis that 11 undergoes a rate-determining isomerization to 10 prior to metastable transitions. The relevant portion of the potential surface is summarized in Figure 6.

The second reported case of rate-determining isomerization prior to unimolecular dissociations is also in the $C_3H_7O^+$ system, namely, protonated propionaldehyde and protonated acetone (67). The evidence presented is similar to that for 10 and 11 above, and the relevant part of the potential surface is depicted in Figure 7.

It is possible that this type of potential surface will prove to be more common than hitherto imagined as previous studies of the $C_3H_8N^+$ (22) and $C_3H_5O^+$ (64) systems reveal cases where isomerization may be the rate-determining step in dissociation of some ionic structures.

Having discussed the general shapes of potential surfaces over which ions may be considered to dissociate, it is instructive to examine the special effect which the nature of the final (dissociation) step has upon the shape of the metastable peak for that process.

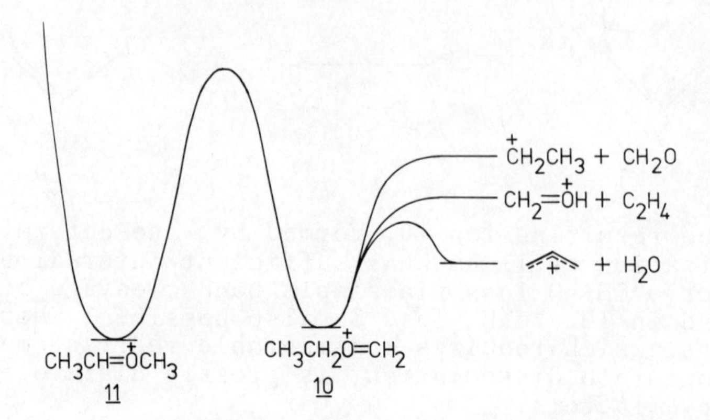

Figure 6. Potential energy surface for slow unimolecular decomposition of ions 10 *and* 11

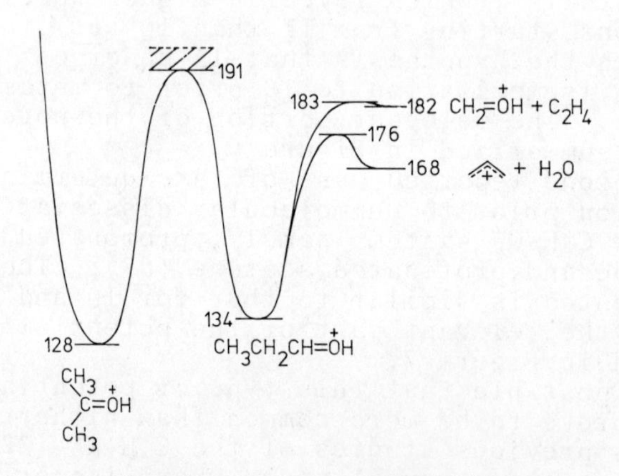

Figure 7. Potential energy surface for slow unimolecular decomposition of ions 8 *and* 9

Correlation of Metastable Peak Shape with Potential Energy Surface of the Dissociation Step.

Clearly there are two basic possibilities; namely that the reverse reaction will or will not involve activation energy. These correspond to the potential surfaces shown in Figures 8(a) and 8(b), respectively.

In the former case, it is clear that energy is released in the dissociation; and this energy may be absorbed into the vibrational modes of the products or may appear as translational energy. This kinetic energy release is evidenced by the broadening of the metastable peak, which becomes flat-topped in cases where a relatively *specific* and large amount of energy is involved. Conversely, in cases where there is no reverse activation energy, little or no kinetic energy release is expected and the metastable peak is gaussian in shape. Some cases are not so clear cut as to be bracketed into either category above; for instance, the transition state geometry may be fairly flexible, thus permitting the release of a *range* of kinetic energies which results in the broadening of the metastable peak for the process, although it frequently remains gaussian in shape. In such cases a value for the *average* kinetic energy release may be computed from the peak width at half height (68).

Alternatively, looking at the reaction in reverse, a gaussian metastable peak contains the information that the reverse reaction can be made to go by allowing the products (or possibly the products in a suitably excited vibrational level) to drift together. A flat-topped peak corresponds to a definite energy barrier to the reverse reaction, which cannot be overcome by merely exciting the products in the vibrational levels. Rather, the product ion and neutral must be given relative translational energy in order to make the reverse reaction go. Conceptually, a flat-topped metastable peak for a given reaction means that the products must be "pushed together" in order to make the reverse reaction go.

One situation in which the potential surface is of the general form likely to give rise to flat-topped metastable peaks is that of symmetry-forbidden dissociations (66). Here a "crossing" of two molecular orbital energy levels occurs, and upon attaining the transition state, which will generally be well defined geometrically, dissociation occurs over a repulsive potential surface.

For instance, consider the concerted 1,2-elimination of H_2 from ionized ethane (39) proceeding *via* a planar transition state. The relevant correlation diagram is given in Figure 9 (69). Analysis of this diagram, leads to the conclusion that the reaction is symmetry-forbidden. Kinetic energy release is expected since the high transition state energy is partially due to the occupancy of a molecular orbital characterized by mutual repulsion between the ionized ethylene and H_2 (see • in Figure 9). Experimentally, ionized ethane does indeed lose H_2 with kinetic energy release (70). A similar analysis for ionized methylamine (40), protonated formaldehyde (41), protonated thioformaldehyde (42) and protonated methylene imine (43) leads to the conclusion that these molecules, if they undergo concerted 1,2-eliminations of H_2, should do so with kinetic energy release. Experimentally, H_2 losses are observed to occur in metastable transitions and in each case the peak is of a flat-topped shape (70), thus indicating kinetic energy release (52,53). The fact that all these dissociations are 1,2-eliminations is established by deuterium labelling studies (8,10,70,71), which show that

$CH_3CH_3^{+\cdot}$	$CH_3NH_2^{+\cdot}$	$CH_2{=}\overset{+}{O}H$	$CH_2{=}\overset{+}{S}H$	$CH_2{=}\overset{+}{N}H_2$
39	40	41	42	43

$CH_3CD_3^{+\cdot}$ loses only HD in metastable transitions (8,10) and that the heteroatom deuterated analogues of 40 to 43 also eliminate only HD in unimolecular decompositions (70,71).

Further interesting applications of orbital symmetry to unimolecular decompositions exist (72,73) and one particularly relevant in the present context is the deduction that H_2 loss from ionized ethylene (44) occurs via a 1,1-elimination from the ionized carbene 45. Such a process is symmetry-allowed (66) and consequently no kinetic energy release is observed to accompany H_2 loss from $C_2H_4^{+\cdot}$. Were the reaction to occur via a concerted 1,2-elimination, it would be symmetry-forbidden (66) and a flat-topped metastable peak would be expected. A consequence of the proposed

$$\underset{44}{CH_2{=}CH_2^{+\cdot}} \underset{\text{shift}}{\overset{\text{1,2-H}}{\rightleftharpoons}} \underset{45}{{}^{+\cdot}CH{-}CH_3} \xrightarrow[\text{elimination}]{\text{1,1-}} CH{\equiv}CH^{+} + H_2$$

mechanism is that all the hydrogens in labelled ionized ethylene should become equivalent prior to metastable

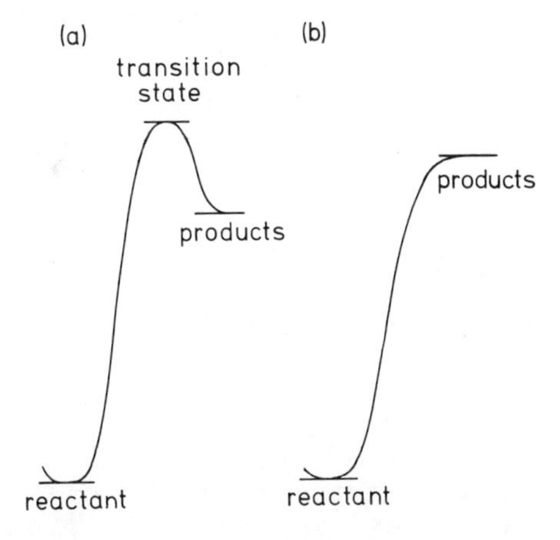

Figure 8. Potential energy surfaces corresponding to dissociation step; (a) with reverse activation energy, (b) with no reverse activation energy

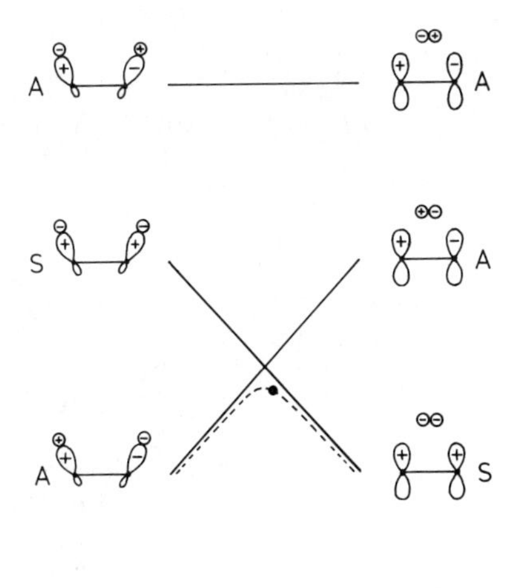

Figure 9. Correlation diagram for concerted, suprafacial 1,2-elimination of H_2 from ionized C_2H_6

transitions; hence, the decomposition of labelled forms of 44 must be explicable in terms of statistical selection of any hydrogen atoms together with an isotope effect favoring selection of H rather than D. Experimentally it is found (74) that the labelling results can be interpreted in this way. Moreover, the isotope effects observed are such as to suggest strongly that the elimination of H_2 is concerted. This is good evidence that the formulation of H_2 loss from $C_2H_4^{+\cdot}$ as a concerted 1,1-elimination from the ionized carbene 45 is correct.

The correlation of the metastable peak shape to the orbital symmetry of the reaction concerned is only one particular example of the potential surface approach. It should be noted that any dissociation which occurs through a channel in which the products are formed on a mutually repulsive potential surface can give rise to kinetic energy release.

An elegant example of this situation is found in the decomposition, *via* C_2H_4 loss, of the $C_3H_7O^+$ and $C_3H_8N^+$ species 8 and 14, which may be formulated as occurring *via* the mechanism of Scheme 3.

Scheme 3

$$\underset{\mathbf{8}}{CH_3CH_2CH{=}\overset{+}{O}H} \underset{\text{shift}}{\overset{\text{1,2-H}}{\rightleftharpoons}} CH_3\overset{+}{C}HCH_2OH \underset{\text{shift}}{\overset{\text{1,2-H}}{\rightleftharpoons}} \underset{\mathbf{22}}{\overset{+}{C}H_2CH_2CH_2OH} \xrightarrow[-C_2H_4]{} CH_2{=}\overset{+}{O}H$$

$$\underset{\mathbf{14}}{CH_3CH_2CH{=}\overset{+}{N}H_2} \rightleftharpoons CH_3\overset{+}{C}HCH_2NH_2 \rightleftharpoons \underset{\mathbf{46}}{\overset{+}{C}H_2CH_2CH_2NH_2} \xrightarrow[-C_2H_4]{} CH_2{=}\overset{+}{N}H_2$$

Using the isodesmic insertion method with a correction of 10 and 3 kcal mol^{-1} for the destabilizing effect of a β- and γ-OH (or $-NH_2$) group (41,43) yields the estimated potential surfaces shown in Figures 10 and 11. Also given in these figures are the metastable peak shapes for C_2H_4 loss from the ions in question. It is immediately apparent that in the oxygen system (Figure 10) energy must be continuously put into the system to attain and finally dissociate the reacting configuration 22. Consequently, the metastable peak for this process is expected to be gaussian and narrow, as is observed (67). In the corresponding nitrogen system (Figure 11), the reacting configuration 46 now

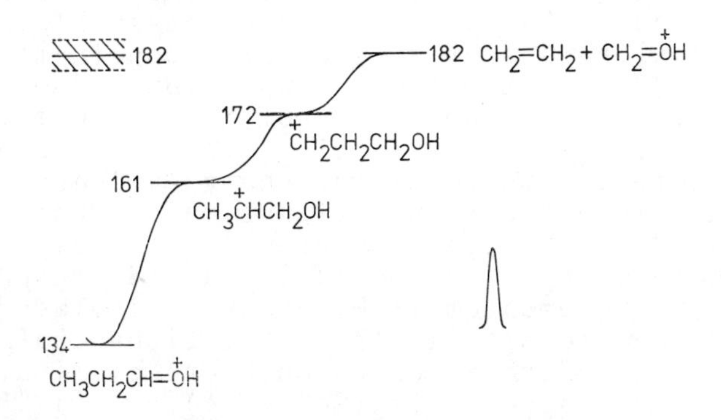

Journal of the American Chemical Society

Figure 10. Potential energy surface and energy release profile for C_2H_4 loss starting from 8

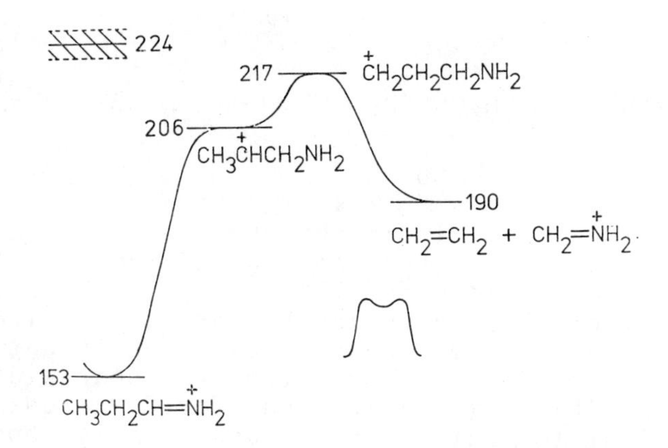

Journal of the American Chemical Society

Figure 11. Potential energy surface and energy release profile for C_2H_4 loss starting from 14

approximates the transition state and dissociates exothermically. Therefore, kinetic energy release is possible, and it is found that the metastable peak for this transition is flat-topped corresponding to a release of ~9 kcal mol^{-1} of translational energy (22). As the estimated exothermicity of the dissociation of 46 to products is ~27 kcal mol^{-1}, it can be seen that approximately one third of the energy released is partitioned into the translational mode. Using this figure, and constructing analogous potential surfaces for the homologues of 8 and 14 (47 and 48 respectively) in which the heteroatom [X-H] group is replaced by an [X-CH_3] group, results in the predictions that (i) both the oxygen and nitrogen cases ought to eliminate C_2H_4 via 49 and 50 respectively (Scheme 4). (ii) The oxygen case (47) is only relatively mildly exothermic (∿11 kcal mol^{-1}) and so only a small quantity of energy (∿4 kcal mol^{-1}) is expected to be released as translation. (iii) The nitrogen analogue (48) is estimated to dissociate via 50 with the release of ∿34 kcal mol^{-1} (i.e., ∿11 kcal mol^{-1} as translation).

Scheme 4

$$\underset{\mathbf{47}}{CH_3CH_2CH{=}\overset{+}{O}CH_3} \xrightleftharpoons{\text{1,2-H-shifts}} \underset{\mathbf{49}}{\overset{+}{C}H_2CH_2CH_2OCH_3} \xrightarrow[-C_2H_4]{} CH_2{=}\overset{+}{O}CH_3$$

$$\underset{\mathbf{48}}{CH_3CH_2CH{=}\overset{+}{N}HCH_3} \rightleftharpoons \underset{\mathbf{50}}{\overset{+}{C}H_2CH_2CH_2NHCH_3} \xrightarrow[-C_2H_4]{} CH_2{=}\overset{+}{N}HCH_3$$

Experimentally, it is observed that (i) both ions originally generated as 47 and 48 undergo C_2H_4 loss in metastable transitions (62,75). (ii) The peak for C_2H_4 loss from 47 is gaussian but broad; a value of ∿3 kcal mol^{-1} may be computed as the average kinetic energy release (75), measured from the peak width at half height (68). (iii) The peak for C_2H_4 loss from 48 is similar to that for C_2H_4 loss from 14; however, the kinetic energy release is somewhat greater (∿11 kcal mol^{-1} (75)), in good accord with the expected value.

Conclusion.

The concept that the metastable dissociations of organic ions may be considered as occurring over an appropriate potential surface is discussed. Its use assists in understanding phenomena such as "scrambling" and has a wider use in the systematic analysis of unimolecular ionic reactions. In particular, consideration of that portion of the potential surface which governs the final dissociation step often permits a better understanding of the shape of the metastable peak for that process. Conversely, consideration of the shape of the metastable peak for a unimolecular reaction may yield valuable mechanistic information concerning the reaction in question.

Abstract.

Results from many studies of metastable ion decompositions in double focussing mass spectrometers are interpreted in terms of the potential surfaces over which gas-phase organic ions decompose. The concept of the potential surface enables the chemistry of these ions to be clearly understood. Three general types of surfaces are proposed: (1) complete equilibration of isomers prior to unimolecular decomposition, (2) no equilibration of isomers prior to unimolecular decomposition, and (3) rate-determining isomerization prior to unimolecular decomposition. Examples of each are presented and discussed.

Literature Cited.

1. Cooks, R.G., Beynon, J.H., Caprioli, R.M., and Lester, G.R., "Metastable Ions", Elsevier, Amsterdam, (1973).
2. Williams, D.H. and Howe, I., "Principles of Organic Mass Spectrometry", McGraw-Hill, London, (1972).
3. See the relevant chapters in "Mass Spectrometry-Specialist Periodical Reports", Vols. 1-3, The Chemical Society, London.
4. Simm, I.G., Danby, C.J., and Eland, J.H.D., J.C.S. Chem. Comm., (1973), 832.
5. Bowen, R.D. and Williams, D.H., J.C.S. Perkin II, (1976), 1479.
6. Rosenstock, H.M., Dibeler, V.H., and Harlee, F.N., J. Chem. Phys., (1964), 40, 591.

7. Hills, L.P., Vestal, M.L. and Futrell, J.H., J. Chem. Phys., (1971), 54, 3834.
8. Löhle, U., and Ottinger, Ch., J. Chem. Phys., (1969), 51, 3097.
9. Vestal, M. and Futrell, J.H., J. Chem. Phys., (1970), 52, 978.
10. Lifshitz, C. and Sternberg, L., Int. J. Mass Spectrometry Ion Phys., (1969) 2, 303.
11. Williams, D.H., Cooks, R.G., and Howe, I., J. Am. Chem. Soc., (1968), 90, 6759.
12. Beynon, J.H., Saunders, R.A. and Williams, A.E., Ind. Chem. Belge., (1964), 29, 311.
13. Polanyi, J.H., Accounts Chem. Res., (1972), 5, 161.
14. Carrington, T., Accounts Chem. Res., (1974), 7, 20.
15. Davis, B., Williams, D.H., and Yeo, A.N.H., J. Chem. Soc. (B), (1970), 81.
16. Yeo, A.N.H. and Williams, D.H., J. Am. Chem. Soc., (1971), 93, 395.
17. McLafferty, F.W., Kornfeld, R., Haddon, W.F., Levsen, K., Sakai, I., Bente, P.F., Tsai, S.-C. and Schuddemage, H.D.R., J. Am. Chem. Soc., (1973), 95, 3886.
18. Keyes, B.G. and Harrison A.G., Org. Mass Spectrom., (1974), 9, 221.
19. Beauchamp, J.L. and Dunbar, R.C., J. Am. Chem. Soc., (1970), 92, 1477.
20. McLafferty, F.W. and Sakai, I., Org. Mass Spectrom., (1973), 7, 971.
21. Levsen, K. and McLafferty, F.W., J. Am. Chem. Soc., (1974), 96, 139.
22. Uccella, N.A., Howe, I., and Williams, D.H., J. Chem. Soc. (B), (1971), 1933.
23. Winkler, J. and McLafferty, F.W., J. Am. Chem. Soc., (1973), 95, 7533.
24. McLafferty, F.W. and Winkler, J., J. Am. Chem. Soc., (1974), 96, 5182.
25. Dunbar, R.C., J. Am. Chem. Soc., (1975) 97, 1382.
26. Bentley, T.W. and Johnstone, R.A.W., Adv. Phys. Org. Chem., (1970), 8, 197.
27. Siegel, A., J. Am. Chem. Soc., (1974), 96, 1251.
28. Davidson, R.A. and Skell, P.S., J. Am. Chem. Soc., (1973), 95, 6843.
29. Lossing, F.P. and Semeluk, G.P., Canad. J. Chem. (1970), 48, 955.
30. Lossing, F.P. and Maccoll, A., Canad. J. Chem. (1976), 54, 990.
31. Lossing, F.P., Canad. J. Chem., (1971), 49, 357.
32. Lossing, F.P., Canad. J. Chem., (1972), 50, 3973.

33. Lossing, F.P., reported at the American Society for Mass Spectrometry Conference, Houston, Texas, U.S.A., (1975).
34. Lossing, F.P. and Traeger, J.C., J. Am. Chem.Soc., (1975), 97, 1579.
35. e.g., Wanatabe, K., Nakayama, J., and Mottl, J., J. Quant. Spect. Rad. Transfer, (1962), 2, 369.
36. e.g. Refaey, K.M.A. and Chupka, W.A., J. Chem. Phys., (1968), 48, 5205.
37. e.g., Haney, M.A. and Franklin, J.L., J. Phys. Chem., (1969), 73, 4328.
38. For a compilation of ΔH_f values for various ions and neutrals see Franklin, J. L., Dillard, J.G., Rosenstock, H.M., Herron, J.L., Draxl, K., and Field, F.H., "Ionization Potentials, Appearance Potentials, and Heats of Formation of Gaseous Positive Ions", National Bureau of Standards, Washington, D.C., (1969).
39. Beynon, J.H., Cooks, R.G., Jennings, K.R. and Ferrer-Correia, A.J., Int. J. Mass Spectrometry Ion Phys., (1975), 18, 87.
40. Rosenstock, H.M., Int. J. Mass Spectrometry Ion Phys., (1976), 20, 139.
41. e.g., Radom, L., Pople, J.A., and Schleyer, P.von R., J. Am. Chem.Soc., (1972), 94, 5935.
42. Ref. 38, Appendix 2; see also Franklin, J.L., Ind. Eng. Chem., (1949), 41, 1070, and Franklin, J.L., J. Chem. Phys., (1953), 21, 2029.
43. Bowen, R.D. and Williams, D.H., Org. Mass Spectrom., 12, 475 (1977).
44. Aue, D.H. Davidson, W.R. and Bowers, M.T., J. Am. Chem. Soc., (1976), 98, 6700.
45. Howe, I. and McLafferty, F.W., J. Am. Chem. Soc., (1971), 93, 99.
46. Beynon, J.H., Corn, J.E., Battinger, W.E., Caprioli, R.M., and Benkeser, R.A., Org. Mass Spectrom., (1970), 3, 1371.
47. Holmes, J.L. and Terlouw, J.K., Org. Mass Spectrom., (1975), 10, 787.
48. Smith, G.A. and Williams, D.H., J. Chem. Soc. (B), (1970), 1529.
49. Bowen, R.C. and Williams, D.H., Org. Mass Spectrom, 12, 453 (1977).
50. Shaw, M.A., Westwood, R. and Williams, D.H., J. Chem. Soc. (B), (1970), 1773.
51. Shannon, T.W. and McLafferty, F.W., J. Am. Chem. Soc., (1966), 88, 5021.
52. Beynon, J.H. and Fontaine, A.E., Zeitschrift für Naturforschg, (1967), 22a, 334, and references cited therein.

53. Beynon, J.H., Saunders, R.A., and Williams, A.E., Zeitschrift für Naturforschg, (1965), 20a, 180.
54. Hvistendahl, G., Bowen, R.D. and Williams, D.H., J. Am. Chem. Soc., in press.
55. Hvistendahl, G. and Williams, D.H., J. Am. Chem. Soc., (1975) 97, 3097, (see pp. 3099 and 3100).
56. Van Raalte, D. and Harrison, A.G., Canad. J. Chem., (1963), 41, 3118.
57. Hvistendahl, G., Bowen, R.D. and Williams, D.H., unpublished results.
58. Harrison, A.G. and Keyes, B.G., J. Am. Chem. Soc., (1968), 90, 5046.
59. Tsang, C.W. and Harrison, A.G., Org. Mass Spectrom., (1970), 3, 647.
60. Tsang, C.W. and Harrison, A.G., Org. Mass Spectrom., (1971), 5, 877.
61. Tsang, C.W., and Harrison, A.G., Org. Mass Spectrom., (1973), 7, 1377.
62. Mead, T.J. and Williams, D.H., J.C.S. Perkin II, (1972), 876.
63. Westwood, R., Williams, D.H. and Yeo, A.N.H., Org. Mass Spectrom., (1970), 3, 1485.
64. Mead, T.J. and Williams, D.H., J. Chem. Soc. (B), (1971), 1654.
65. Hvistendahl, G. and Williams, D.H., J. Am. Chem. Soc., (1975), 97, 3097.
66. Woodward, R.B. and Hoffmann, R., "The Conservation of Orbital Symmetry", Verlag Chemie, Winheim/Bergstr., Germany, (1970).
67. Hvistendahl, G., Bowen, R.D. and Williams, D. H., J.C.S. Chem. Comm., (1976), 294.
68. Terwilliger, D.T., Beynon, J.H. and Cooks, R.G., Proc. Roy. Soc., (1974), 341, 135.
69. Smyth, K.C. and Shannon, T.W., J. Chem. Phys., (1969), 51, 4633.
70. Hvistendahl, G. and Williams, D.H., J. Am. Chem. Soc., (1974), 96, 6753.
71. Beynon, J.H., Fontaine, A.E., and Lester, G.R., Int. J. Mass Spectrometry Ion Phys., (1968) 1, 1.
72. Hvistendahl, G. and Williams, D.H., J.C.S. Perkin II, (1975), 881.
73. Williams, D. H. and Hvistendahl, G., J. Am. Chem. Soc., (1974), 96, 6755.
74. Hvistendahl, G. and Williams, D.H., J.C.S. Chem. Comm., (1975), 4.
75. Williams, D.H. and Bowen, R.D., J. Am. Chem. Soc., 99, 3192 (1977).

RECEIVED December 30, 1977

3

Structures of Gas-Phase Ions from Collisional Activation Spectra*

F. W. MC LAFFERTY

Department of Chemistry, Cornell University, Ithaca, NY 14853

"Double resonance" techniques have provided revolutionary new information for a variety of spectroscopic techniques in recent years. In mass spectrometry, double resonance has been utilized in the ion cyclotron resonance spectrometer to identify in a complex mixture of ions the precursor and product ions of a particular ion-molecule reaction. This paper concerns a technique in which a mass spectrum can be obtained of each peak in a mass spectrum; the ions representing that peak are activated by collision, causing their further decomposition to produce the "collisional activation" (CA) mass spectrum (1 - 3). The reverse-geometry double-focussing mass spectrometer (4) has special advantages for the measurement of such CA spectra. The sample is ionized and the resultant ions are mass analyzed in the magnetic analyzer; ions of a particular m/e value are brought into focus on a collision chamber containing a gas such as helium at $\sim 10^{-4}$ torr pressure. In some ion-atom interactions part of the ion's kinetic energy is converted into internal energy without appreciable scattering of the ion. The resulting decompositions yield product ions which are separated in the electrostatic analyzer to yield the CA mass spectrum. In both formation and utility the CA spectrum resembles a normal mass spectrum; the decompositions of the collisionally activated ion follow the predictions of the quasiequilibrium theory, so that the resulting product ions are indicative of the precursor ion's structure. In addition, in most of the observed decomposition processes sufficient energy has been added by collision to make the resulting ion abundances virtually independent of the original internal energy ("temperature") of the precursor ion; the only ion energy-dependent processes are those which are involved in the decomposition of metastable ions (the MI spectrum, measured in the same manner except in the absence of the collision gas). Thus

*Collisional Activation and Metastable Ion Characteristics. 55. For part 54, see reference 10.

both the masses and abundances of the CA spectrum, excluding peaks in the MI spectrum, are characteristic of the precursor ion structure on a quantitative basis, in the same way that a normal mass spectrum is a reproducible characteristic of the structure of the precursor molecule.

Areas in which CA-MS appears to be especially promising are fundamental studies of organic ion structures and reactions, MS structure determination of complex organic molecules through elucidation of the isomeric identity of fragment ions, and separation/identification of complex mixtures through MS separation of ionized molecular species which are then identified from their CA mass spectra. An excellent review of the field by Levsen and Schwarz has appeared (3), and Cooks reports on some related results (5); this article will use as examples more recent work of our laboratory.

CA-MS Instrumentation

Most recent CA-MS studies have utilized "direct analysis of daughter ions (DADI)" (4) (also called "mass analyzed ion kinetic energy spectroscopy [MIKES]") (6) with a reversed-geometry double-focussing mass spectrometer (7) although Jennings and coworkers have recently shown there are advantages in using a normal geometry double-focussing instrument with a "linked" scan of the electrostatic and magnetic analyzers (8). Instrumentation aspects of importance include achieving purity of the precursor ion, resolution of the CA spectrum, data acquisition and reduction, sensitivity, and quantitative reproducibility of spectra.

As is well know in normal mass spectrometry, the peaks in a unit resolution mass spectrum often contain isobaric multiplets; for example, $C_3H_7^+$ and $C_2H_3O^+$, both nominally *m/e* 43, actually differ in mass by 37 millimass units. "Pure" precursor ions of course are desirable for CA spectra, and this can be achieved with sufficient resolving power (9); a CA-MS instrument with a double-focussing mass spectrometer before the collision chamber is under construction at Cornell as a solution to this problem (10). Even if all the precursor ions have the same elemental composition, these still may contain isomeric impurities; formation of the ions using different ionizing electron energies, using different ionization methods (*e.g.*, chemical ionization), or from different precursor molecules can be used to improve isomeric purity.

A more serious problem is the resolution achievable in the CA spectrum. In the reversed-geometry instrument, the CA product ions are separated by the electrostatic analyzer, taking advantage of the fact that the loss in mass on decomposition results in a concommitant loss in kinetic energy. However, in the collision process a small fraction of the ion's kinetic energy is converted into internal energy and, more seriously, internal energy in excess of that required for the decomposition can be released in the reaction coordinate, appearing as a positive or negative velocity vector; this process has been extensively studied in connection with the "flat-top" peaks produced in metastable ion decompositions (6). This uncertainty in the kinetic energy of a particular product ion broadens the peak, the extent of this broadening depending on the amount of energy variation relative to the total kinetic energy. Increasing the ion accelerating voltage decreases this fraction and therefore increases the resolution of the CA spectrum (11). A similar approach is to use "post-acceleration" of the ions after the collision chamber (12); an instrument under construction at the FOM Institute in Amsterdam will float the electrostatic analyzer and collector at as much as 40 kV, which should greatly enhance the resolution of the CA spectrum. Double-focussing mass analysis is employed in instruments such as the spark source MS in order to bring ions of widely varying kinetic energies but the same mass into coincident focus, and this principle will be employed in the tandem double-focussing CA-MS instrument under construction at Cornell (10). The linked-scan, normal geometry double-focussing technique employed by Jennings and coworkers (8) apparently gives improved resolution for CA mass spectra; this arises because the ions produced by CA decomposition in the field-free region before the electrostatic analyzer thus traverse most of the double-focussing path of the instrument.

As already amply demonstrated in normal MS, utilization of a dedicated computer for automated data acquisition and reduction also is advantageous for CA-MS. The Cornell reversed-geometry CA-MS instrument was coupled to a DEC PDP-8 computer in 1972 (7), and in 1975 this was replaced by a PDP-11/10 dedicated computer which directs the multiple scanning of the CA spectrum through a D/A converter to the ESA (11). Special software and hardware give flexibility for selected mass scanning, averaging of the repeat scans, cathode ray tube display of the most recent and the averaged scans, comparison with previous spectra, multicomponent analysis of spectra, and so forth.

Of special importance for ion structure characterization is the ±2% precision in CA peak abundances which can be achieved with this new data system (11). In fact, this has been used to analyze ternary mixtures of isomeric $C_3H_7S^+$ ions with ±10% accuracy based on the peak abundances of their CA spectra (13). This is reminiscent of the quantitative analysis of light hydrocarbon mixtures using normal MS, the major incentive for the original development of the analytical mass spectrometer. In the Cornell tandem double-focussing CA-MS instrument under construction (10), we plan to have the PDP-11/45 computer also control the mass analysis of the precursor ion. The technique of "Signal Enhancement in Real-Time" (14) will be used to keep a particular precursor ion peak in focus on the collision chamber during the continuous scan of the first magnet by an offsetting increase in the ion accelerating voltage, during which time the CA-MS data (complete spectrum or selected ions) will be obtained; the computer will then direct the ion accelerating potential to focus the next precursor ion peak on the collision chamber while its CA spectrum is measured, and so forth (10). Such automated data acquisition and control for both MS sections in concert will be especially advantageous for its use as a separation/identification system for complex mixtures.

Structures of Gaseous Organic Ions

The structures of a wide variety of ions have been determined by CA-MS. Table I lists the elemental compositions of isomeric ions which have been studied by CA-MS at Cornell; additional studies by Levsen, Schwarz, Van de Sande, Beynon, Cooks, and their coworkers are listed in the Levsen-Schwarz review (3). Table II shows proposed structures for 9 isomeric $C_3H_5O^+$ ions which give distinguishable CA spectra (there is much better evidence for the first six than for the last three structures) (15). CA spectra, like normal mass spectra, can be used to characterize an ion's structure through rationalization of the decomposition products or through quantitative comparison with reference spectra of ions of established structures. This can be illustrated by studies of $C_8H_9^+$ isomers carried out by Dr. Claus Köppel at Cornell recently (16).

Table I. Collisional Activation Studies

CH_4^+, $C_3H_3^+$, $C_3H_7^+$, $C_4H_8^{+\cdot}$, $C_4H_9^+$, $C_5H_{10}^{+\cdot}$, $C_6H_{12}^{+\cdot}$

$C_7H_7^+$, $C_7H_8^{+\cdot}$, $C_8H_9^+$, $C_9H_{11}^+$, $C_{13}H_9^+$

$C_2H_4O^{+\cdot}$, $C_2H_5O^+$, $C_3H_5O^+$, $C_3H_6O^{+\cdot}$, $C_3H_7O^+$

$C_6H_6O^{+\cdot}$, $C_6H_{11}O^+$, $C_9H_{11}O^+$, $C_{12}H_9O^+$, $C_{12}H_{10}O^{+\cdot}$

$C_2H_6N^+$, $C_3H_8N^+$; $C_2H_5S^+$, $C_3H_7S^+$; $C_4H_8Cl^+$

Oligopeptides

Table II. Possible Isomeric Structures of $C_3H_5O^+$ Ions

$CH_2{=}CHCH{=}\overset{+}{O}H$	$CH_3CH_2C{\equiv}\overset{+}{O}$	$CH_2{=}CH\overset{+}{O}{=}CH_2$
$HC{\equiv}CCH_2\overset{+}{O}H_2$	$CH_3CH{=}\overset{+}{O}CH_2$ (cyclic)	$CH_3C{=}CH\overset{+}{O}H$ (cyclic)
$CH_2{=}CCH_2\overset{+}{O}H$ (cyclic)	$HC{=}CH\overset{+}{O}CH_3$ (cyclic)	$CH_2CH_2\overset{+}{O}{=}CH$ (cyclic)

A variety of molecules, including isotopically labeled derivatives, were used to form $C_8H_9^+$ ions; their CA spectra showed that at least 13 isomers have lifetimes $>10^{-5}$ seconds. Although many of the isomers do isomerize when formed with higher internal energies, these results contrast sharply with our previous conceptions of ready "scrambling" of hydrocarbon ions. Part of this misconception has arisen from the fact that most previous studies utilized the unimolecular decomposition products of an ion as evidence of its structure, and even the lowest activation energy for ion decomposition is still higher than that for an isomerization. For example, the scrambling of isotopic labels on decomposition of the $C_7H_7^+$ ion from toluene indicated that this ion had the symmetrical tropylium, not the benzyl, structure (17, 18), but CA spectra show conclusively that both ions are formed and are stable for $>10^{-5}$ seconds (19). Similarly, the $C_8H_9^+$ isomers methyltropylium and α, *o*-, *m*-, and *p*-methylbenzyl ions are found to be stable if formed with sufficiently low energy. However, as found for cycloheptatriene and toluene (19), isomerization does occur between the molecular ions of methylcycloheptatriene, ethylbenzene, and the xylenes (16).

A number of recent studies have increased our basic knowledge of both solution and gas phase ionic reactions by direct comparison of particular reactivities (*e.g.*, acidities, basicities) in the two media (20 - 22). A classic NMR study by Olah and Porter (23) of the ionization of β-phenylethyl chloride (*a*, X = Cl) in superacid medium gives detailed NMR evidence for the formation of the ethylenebenzenium (*c*) and α-phenylethyl (*d*) cations, with the β-ethylphenyl ion (*b*) as the proposed intermediate (see Scheme I). Although this medium should minimize solvent effects, we find that the unimolecular decomposition of *a*, with X as either Cl or CH_3, ions produces negligible amounts of either *c* or *d* as stable ionic products with the 10^{-5} sec lifetime requirements of the CA technique. Instead, at low electron energies these precursors yield ions whose CA spectra are identical to those formed by the protonation of benzocyclobutene, which thus presumably are formed originally as structure *e*. In contrast, ionization of β-phenylethyl bromide and iodide (*a*, X = Br or I) yields the ethylenebenzenium ion, *c*, at low

Scheme I

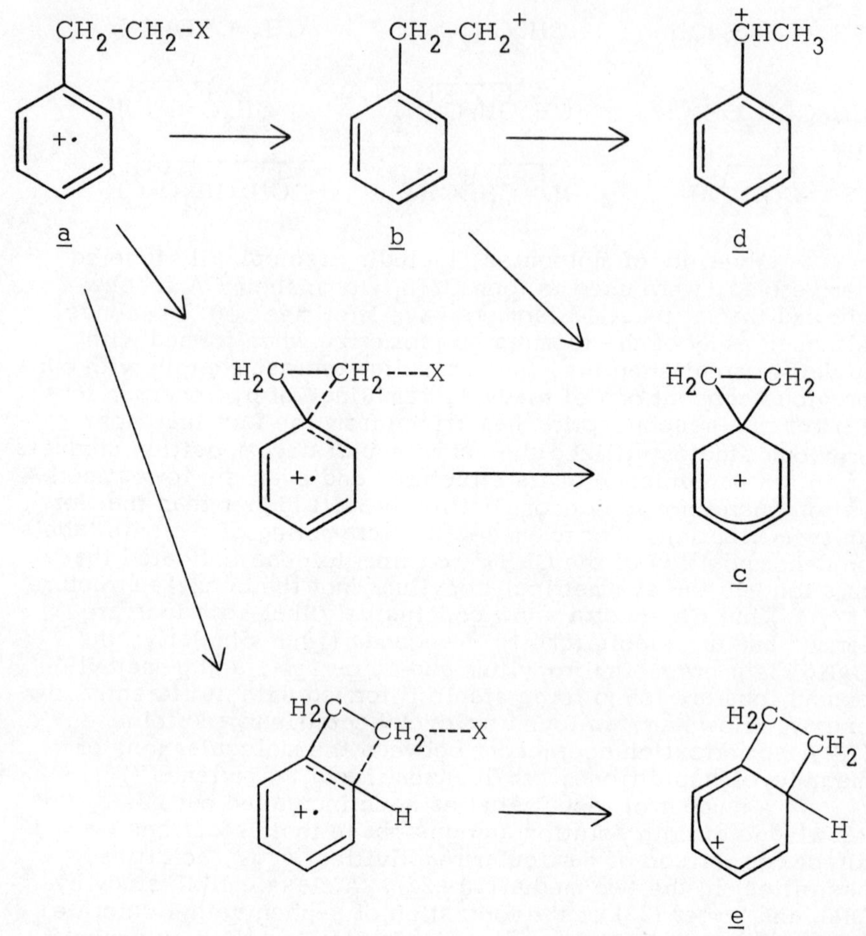

CH_2-CH_2-X (+·) a

$CH_2-CH_2^+$ b

$\overset{+}{C}HCH_3$ d

H_2C-CH_2----X (+·)

H_2C-CH_2 (+) c

H_2C CH_2----X H (+·)

H_2C CH_2 H (+) e

electron energies. This appears to indicate that the formation of e is favored energetically, possibly through an anchimeric-assisted reaction, but this pathway has more stringent steric requirements, so that the larger Br and I leaving groups cause the formation of c to be favored.

Another problem which has been studied extensively on a theoretical as well as experimental basis involves the relative energies of the possible $C_3H_7^+$ isomers. Equilibrium constant measurements on ion-molecule reactions by Chong and Franklin (24) indicate the isopropyl structure to be 8 kcal/mol more stable than the protonated cyclopropane, but ion cyclotron resonance measurements (25) are consistent with complete isomerization of the latter to the former structure in 10^{-3} seconds. Because the CA spectra of $C_3H_7^+$ ions formed by protonation of propene and of cyclopropane are identical (26), the isomerization also should be essentially complete in 10^{-5} seconds, indicating that the activation energy for the isomerization of protonated cyclopropane is very low. If the transition state for this isomerization is similar to the n-propyl structure, this also indicates that its heat of formation is surprisingly close to that of the protonated cyclopropane isomer (26).

The differences in the CA spectra of isomeric hydrocarbon ions are often relatively small, paralleling the behavior found for normal mass spectra of hydrocarbons. For example, the CA spectra of sec- and tert-butyl ions are nearly the same (27). One method to increase these differences is to form a derivative of the ions. In the case of the $C_4H_9^+$ ions, simple derivatives can be formed by the reaction of the newly formed ions directly in the MS ion source with acetyl chloride; the resulting $C_6H_{11}O^+$ ions show substantial differences in their CA spectra. The $C_4H_9^+$ ions formed from n-butyl bromide when derivatized in this fashion give a CA spectrum which is identical to that from sec-butyl ions, while isobutyl ions isomerize to give a 7:3 mixture of sec- and tert-butyl ions. Both gas phase and solution determinations show the heat of formation of tert-butyl ions to be 16 kcal/mol less than that of sec-butyl ions, so that conformational effects outweigh enthalpic factors. Obviously it is important to the understanding of solution reactions involving organic ions to be able to separate out completely the effect of the solvent on the reaction.

Applications of CA to Molecular Structure Determination

Although such applications have been restricted in the past because of instrumentation limitations, CA spectra should provide vital additional information for the structure elucidation of complex molecules. The interpretation of a normal, unit resolution mass spectrum utilizes the masses of the pieces of the molecule to reconstruct the molecule's structure. Exact mass measurement of the fragments by high-resolution mass spectrometry can give the elemental compositions of these fragments, an obvious aid to putting the pieces of the molecular puzzle together correctly. As described above, the CA spectrum can provide the isomeric structure of the fragment ion, obviously an even more valuable aid in the interpretation process. For example, if high-resolution measurements showed that an unknown molecule produced a significant $C_3H_5O^+$ ion, at least the nine isomeric structures of Table II should be considered as possible molecular fragments. Measuring the CA spectrum of the $C_3H_5O^+$ ions from the molecule should indicate the proper isomer. For example, a steroid containing a 17-α-hydroxyethyl group shows a significant $C_2H_5O^+$ peak in its normal mass spectrum; the CA spectrum of this peak identifies it as the $CH_3CH(OH)$ isomer eliminating the possibility, for example, that the steroid contains a $-CH_2OCH_3$ group (2).

Such applications of CA spectra have been especially useful for oligopeptide samples. The amino acid sequence in fragment ions can be deduced from their CA spectra, and the presence of leucine be distinguished from that of isoleucine (28).

MS/CAMS As a Separation/Identification System

Recognition of the usefulness of the gas chromatograph/mass spectrometer/computer (GC/MS/COM) system for the identification and quantitation of complex unknown mixtures has led to an amazing growth in its analytical applications in this decade. GC/MS/COM involves separation of the mixture and identification of the individual components, with automated data reduction; other such separation/identification systems include liquid chromatography/MS/COM. Separation can also be done mass spectrometrically, characterizing the products from the decompositions of metastable ions (29) or, as we suggested in a study on sequencing oligopeptides in mixtures in this fashion (30), the identification of the mass-separated ions can also be done by CAMS. Results from a surprising number of laboratories indicate a high potential for MS/CAMS as a separation/identification system. For example, Levsen and Schulten (31) identified major components of a complex mixture produced by the pyrolysis of deoxyridonucleic acid, and Cooks and coworkers (5) achieved accuracies of ±20% on the quantitative analysis of a five-component ketone mixture. Of special note is the 10^{-11} gram sensitivity achieved in the latter investigation. Recent studies at Cornell (9) have examined several aspects of MS/CAMS, including separation, sensitivity, mixture ionization techniques, computer identification of CA spectra, and automation. MS/CAMS appears to be especially promising in application to complex mixtures for the quantitative analysis for specific components (similar to the "selected ion monitoring" in GC/MS), and for the identification of individual components.

As a separation technique, mass spectrometry has the advantage over GC of submillisecond time requirements, which would be especially advantageous for continuous analyzers. Of special importance is the lower sample volatility requirement of MS, using direct probe introduction or special ionization methods such as field desorption or direct chemical ionization. Note that the lack of volatility restrictions on liquid chromatography is the chief incentive for the proposed LC/MS separation/identification systems. Resolution in MS is based on an entirely different principle than in chromatography, so that it should obviously be a complementary separation technique, and a possible solution for problems in which chromatographic resolution is inadequate. In normal mass spectrometry double-focussing instruments are employed to improve resolution, making possible the separation of isobaric multiplets. We have recently applied this (9) to improve the separation for MS/CAMS, obtaining characteristic CA spectra of each of the doublets $CH_2{}^{35}Cl_2$ and C_6H_{12} (nominal mass 84), and also of $CH_2{}^{35}Cl^{37}Cl$ and $C_5H_{10}O$ (nominal mass 86).

For MS/CAMS the ionization of the sample should give a minimum number (preferably 1) of peaks for each component of a complex mixture introduced into the MS ionization chamber. Lowering the ionizing electron energy reduces the number of peaks per component and increases the relative abundance of the molecular ion, but lowers the overall sensitivity. "Soft" ionization methods such as chemical ionization often give only one peak (plus its isotopic peaks) per compound, but different reagent gases can be necessary to obtain peaks for all components as the CI ionizing reagent gases (9).

Future Possibilities For Collisional Activation Mass Spectra

As with conventional mass spectrometry and GC/MS, the availability of CAMS systems which are convenient to use, have automated operation and data reduction, are reasonable in price, and which show improved sensitivity and resolution, should provide a powerful additional incentive to further applications in a variety of areas. Improved instrumentation such as the "ZAB" reverse-geometry double-focussing mass spectrometer of VG Micromass, the "post-acceleration" reverse-geometry spectrometer at the FOM Institute (12), as well as the tandem double-focussing instrument under construction at Cornell (10) should aid this problem, but smaller inexpensive instruments must also become available commercially. In this regard, it should be possible to utilize a quadrupole MS as the mass analyzer, accelerate the ions into a collision region, possibly using a molecular beam of collision gas as demonstrated by Moran and collaborators (32), followed by post-acceleration and separation of the CA spectrum in an electrically isolated electrostatic analyzer (12). With no magnet, complete computer control with rapid mass response in both the mass analysis steps should be possible. The analysis of highly complex mixtures such as biological fluids, pollutants, insect pheromones, food aromas and flavors, and pyrolysis products is becoming of increasing importance. The sensitivity, specificity, and speed of analysis which should be possible with MS/CAMS should make it applicable to mixtures of organic compounds on a routine basis, even to on-line control, for many important problems which can only be analyzed with difficulty at present.

Literature Cited

1. McLafferty, F. W., Bente, III, P. F., Kornfeld, R., Tsai, R.-C., and Howe, I., J. Am. Chem. Soc., (1973), 95, 2120.
2. McLafferty, F. W., Kornfeld, R., Haddon, W. F., Levsen, K., Sakai, I., Bente, III, P. F., Tsai, S.-C., and Schuddemage, H. D. R., J. Am. Chem. Soc., (1973), 95, 3886.
3. Levsen, K. and Schwarz, H., Angew. Chem., Internat. Edit., (1976), 15, 509.
4. Maurer, J. H., Brunee, C., Kappus, G., Habfast, K., Schröder, U., and Schulze, P., 19th Annual Conference on Mass Spectrometry, Atlanta, 1971, paper K-9.
5. Kruger, T. L., Litton, J. F., Kondrat, R. W., and Cooks, R. G., Anal. Chem., (1976), 48, 2113.
6. Cooks, R. G., Beynon, J. H., Caprioli, R. M., and Lester, G. R., "Metastable Ions", Elsevier, Amsterdam, 1973.
7. Wachs, T., Bente, III, P. F., and McLafferty, F. W., Int. J. Mass Spectrom. Ion Phys., (1972), 9, 333.
8. Weston, A. F., Jennings, K. R., Evans, S., and Elliot, R. M., Int. J. Mass Spectrom. Ion Phys., (1976), 20, 317.
9. McLafferty, F. W. and Bockhoff, F. M., Anal. Chem., submitted.
10. McLafferty, F. W. in "Analytical Pyrolysis", H. L. C. Meuzelaar, Ed., Elsevier, Amsterdam, 1977.
11. Wachs, T., Van de Sande, C. C., Bente, III, P. F., Dymerski, P. P., and McLafferty, F. W., Int. J. Mass Spectrom. Ion Phys., accepted.
12. Tuithof, H. H., Int. J. Mass Spectrom. Ion Phys., submitted.
13. Van de Graaf, B. and McLafferty, F. W., in preparation.
14. McLafferty, F. W., Michnowicz, J. A., Venkataraghavan, R., Rogerson, P., and Giessner, B. G., Anal. Chem., (1972), 44, 2282.
15. Dymerski, P. P. and McLafferty, F. W., in preparation.
16. Köppel, C., Van de Sande, C. C., Nibbering, N. M. M., Nishishita, T., and McLafferty, F. W., J. Am. Chem. Soc., in press.

17. Rylander, P. N., Meyerson, S., Grubb, H. M., J. Am. Chem. Soc., (1957), 79, 842.
18. Bursey, J. T., Bursey, M. M., and Kingston, D. G. I., Chem. Rev., (1973), 73, 191.
19. McLafferty, F. W. and Winkler, J., J. Am. Chem. Soc., (1974), 96, 5182.
20. Arnett, E. M., Acc. Chem. Res., (1973), 6, 404.
21. Taagepera, M., Hehre, W. J., Topsom, R. D., and Taft, R. W., J. Am. Chem. Soc., (1976), 98, 7438.
22. Farneth, W. E. and Brauman, J. I., J. Am. Chem. Soc., (1976), 98, 7891.
23. Olah, G. A. and Porter, R. D., J. Am. Chem. Soc., (1971), 93, 6877.
24. Chong, S.-L. and Franklin, J. L., J. Am. Chem. Soc., (1972), 94, 6347.
25. McAdoo, D. J., McLafferty, F. W., and Bente, III, P. F., J. Am. Chem. Soc., (1972), 94, 2027.
26. Dymerski, P. P., Prinstein, R. M., Bente, III, P. F., and McLafferty, F. W., J. Am. Chem. Soc., (1976), 98, 6834.
27. Dymerski, P. P. and McLafferty, F. W., J. Am. Chem. Soc., (1976), 98, 6070.
28. Levsen, K., Wipf, H.-K., and McLafferty, F. W., Org. Mass Spectrom., (1974), 8, 117.
29. McLafferty, F. W. and Bryce, T. A., Chem. Commun., (1967), 1215.
30. Wipf, H.-K., Irving, P., McCamish, M., Venkataraghavan, R., and McLafferty, F. W., J. Am. Chem. Soc., (1973), 95, 3369.
31. Levsen, K. and Schulten, H.-R., Biomed. Mass Spectrom., (1976), 3, 137.
32. Moran, T. F., Wilcox, J. B., and Abbey, L. E., J. Chem. Phys., (1976), 65, 4540.

Acknowledgment

We are indebted to the National Institutes of Health (grant 16609) and the Army Research Office, Durham, (grant 13136-C) for generous financial support of research described in this review.

RECEIVED December 30, 1977

4

Ion Chemistry via Kinetic Energy Spectrometry

R. G. COOKS

Department of Chemistry, Purdue University, W. Lafayette, IN 47907

This paper describes studies on ion structure, reaction mechanism and thermochemistry all of which rely on the measurement of ion kinetic energy. In all experiments the ion beam undergoes modification after it leaves the source and it is subjected to sequential analysis by magnetic and electric fields. This ensures that a particular reaction occurring in the analyzer can be selected for study. The methodology is derived from that developed for the study of metastable ions, i.e., the spontaneous delayed fragmentations of ionized molecules. In this work, measurements on metastable ions are made in conjunction with ion/molecule interactions at high relative kinetic energy occurring after ion formation and acceleration.

The reactions of interest are classified as secondary processes in contrast to primary processes which occur in the ion source. According to this view, primary processes may involve complex reaction sequences, and chemical and photochemical processes may follow the primary ionization steps, but the beam is not perturbed after extraction from the ion source. Simple mass spectra are recorded. This has been the traditional source of information on ion chemistry as well as the most important route to the applications of mass spectrometry. Secondary processes on the other hand include all those in which the ions issuing from the ion source subsequently undergo measurable alterations in charge state, mass, gross structure or electronic state. Adequate characterization of these processes usually requires that two independent measurements be performed on the ion beam after it leaves the source. Secondary processes may be spontaneous and involve emission of a photon, an electron, an atom or a molecule. Of these processes, only the latter is well-characterized, and it represents a metastable ion dissociation. Secondary processes which are not spontaneous may occur when the ion interacts with a gas atom or molecule, an electron, a surface, a photon beam or an electromagnetic field. Reactive collisions are of no importance in the kilovolt energy range so the inelastic reactions of interest are confined to electronic transitions, charge stripping, charge exchange and photoemission and the dissociations which can follow these processes. Thus, our study of chemistry will be based on measurements on the processes shown in Table I.

The structure of the reactant ion is frequently of major interest when secondary processes are studied. This structural information is often derived by monitoring the dissociations, either spontaneous or collision-induced, of the ion in question. This often provides direct access to molecular structure and constitutes a method of product analysis for ion source reactions.

Other methods of characterizing an ion through its charge changing collisions or via simple excitation reactions will also be noted below. Thermochemical measurements accomplish this purpose, and ion enthalpy measurements made in the ion source by varying the energy deposited upon ionization represent the traditional source of this data. Using fast ion beams kinetic energy loss measurements associated with ion/molecule reactions provide analogous data. It is a fortunate fact that for conditions commonly fulfilled in mass spectrometers, the heat of reaction for ion/molecule reactions is measured to a good approximation as the kinetic energy change (loss if endothermic, gain if exothermic) of the fast species.* The value of kinetic energy measurements in studying secondary processes is further enhanced by the kinetic energy amplification associated with the disposal of internal energy of an excited species as relative translational energy of its fragments. The range of kinetic energies of the fragment ions can be 10^5 the actual center of mass energy release. This makes possible sophisticated measurements on energy partitioning as well as the characterization of the reacting ion and in favorable cases its fragmentation mechanism. This measurement is applicable to both spontaneous and collision-induced dissociations.

In summary, the chief concern of this paper is the use of secondary processes to characterize ions formed in a chemical ionization source. To fully exploit the rich chemistry associated with low energy, gas-phase ion/molecule reactions, direct methods for characterizing the structures of product ions are needed. For this purpose, we use the conjunction of a chemical ionization source as reaction vessel with the methods of ion kinetic energy spectrometry for analysis.

*The target recoil energy is given by $\frac{1}{4}\frac{m}{N}\frac{Q^2}{E}$ where m is the mass of the ion, N that of the target, Q is the heat of reaction and E is the reactant ion translational energy.

Table I. Reactions Studied Via Translational Energy Measurements

			Reaction	Equation
SECONDARY PROCESSES	SPONTANEOUS PROCESSES		Metastable Fragmentation	$m_1^+ \rightarrow \underline{m_2^+} + m_3$
			Emission	$m_1^{+*} \rightarrow \underline{m_1^+} + h\nu$
			Autoionization	$m_1^+ \rightarrow \underline{m_1^{++}} + e^-$
	INDUCED PROCESSES	Collision-induced (Surface-induced)	Dissociation	$m_1^+ + N \rightarrow \underline{m_2^+} + m_3 + N$
			Stripping	$m_1^+ + N \rightarrow \underline{m_1^{++}} + N + e^-$
			Charge exchange	$m_1^{++} + N \rightarrow m_1^+ + N^+$
			Excitation	$m_1^+ + N \rightarrow m_1^{+*} + N$
		Photon-induced	Photo Dissociation	$m_1^+ + h\nu \rightarrow \underline{m_2^+} + m_3^+$
			Photoionization	$m_1^+ + h\nu \rightarrow \underline{m_1^{++}} + e^-$
			Photoexcitation	$m_1^+ + h\nu \rightarrow \underline{m_1^{+*}}$

The fast analyzed species is underlined.

Analysis of Mixtures by MIKES/CID

By choosing an appropriate configuration for a double focusing mass spectrometer and by studying secondary fragmentations, the analysis of mixtures can readily be achieved.(1-4) The mass-analyzed ion kinetic energy (MIKE) arrangement allows for ion selection as a prelude to dissociation. Kinetic energy analysis of the fragments provides a fragmentation pattern from which the structure of the reactant ion, and hence that of the corresponding neutral molecule, can be inferred. Figure 1 summarizes the principles of this method of mixture analysis. The method is an alternative to gas chromatography/mass spectrometry and it has the advantage that it employs electromagnetic rather than chromatographic separation. Difficulties of volatility, lengthy development times and chemical contamination or degradation associated with chromatography are avoided. This is accomplished at the expense of a loss in sensitivity associated with measurements on secondary rather than primary ions.

Although both electron impact and chemical ionization have been employed in these analyses, chemical ionization is preferred because it secures the structural relationship between the neutral molecule of interest and the corresponding ion which is actually separated and analyzed. The secondary fragmentation reaction may be spontaneous or collision-induced. The latter are much preferred since (i) more reactions are observed, (ii) the total signal due to secondary ions is higher, typically by an order of magnitude, (iii) there is a greater tendency for simple bond cleavages in CID than in metastable ion reactions.

To illustrate the sensitivity of this technique consider the scan shown in Figure 2. This peak is a part of the MIKE spectrum taken in the presence of collision gas, of the $(M + H)^+$ ion of p-nitrophenol which was present in a mixture of nitrobenzenes. Of special note in these spectra (compare ref. 1c) is the cleanliness of the background as contrasted to gc/ms spectra. This is a consequence of the double filter which operates in obtaining these spectra. The signal to noise characteristics of the peak shown in Figure 2 are consistent with a detectability limit of ca. 10^{-11}g for this experiment.

Quantitation of the procedure has not been investigated in detail but the general considerations should be similar to those involved in quantitation of gc/ms data. For highest accuracy this demands use of an internal standard, preferably an isotopically labeled variant of the compound of interest. An additional consideration, if collision-induced dissociation is employed, is that the pressure of the target gas must be held constant. Figure 3 demonstrates the type of accuracy which is readily achieved, variations in collision gas pressure being the major source of error in this experiment.

One of the more attractive features of this method is its insensitivity to the presence of large amounts of impurities. Solvent

M1 M2 M3 Ionize CI (M1+H)+ (M2+H)+ (M3+H)+ Mass Analyze (M3+H)+ Excite [(M3+H)+*] m31+ m32+ m33+ Energy Analyze

Separation Characterization

Collision Spectroscopy

Figure 1. Principle of mixture analysis using the reversed sector mass-analyzed ion kinetic energy (MIKE) spectrometer

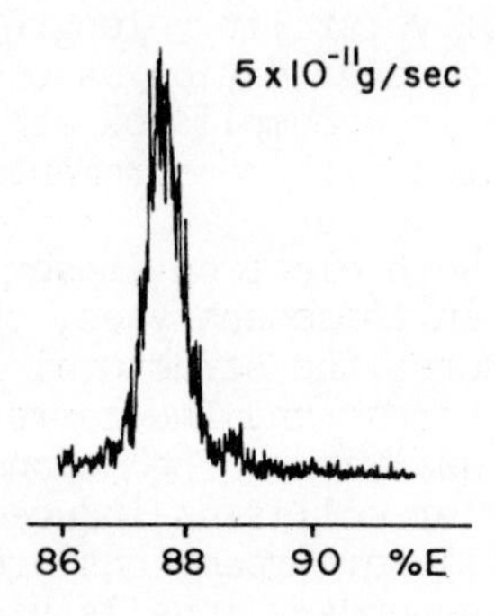

Figure 2. Illustration of sensitivity of the MIKES mixture analysis method. The peak shown, scanned in 3 min, is a characteristic fragmentation (loss of OH·) of the protonated molecular ion of p-nitrophenol.

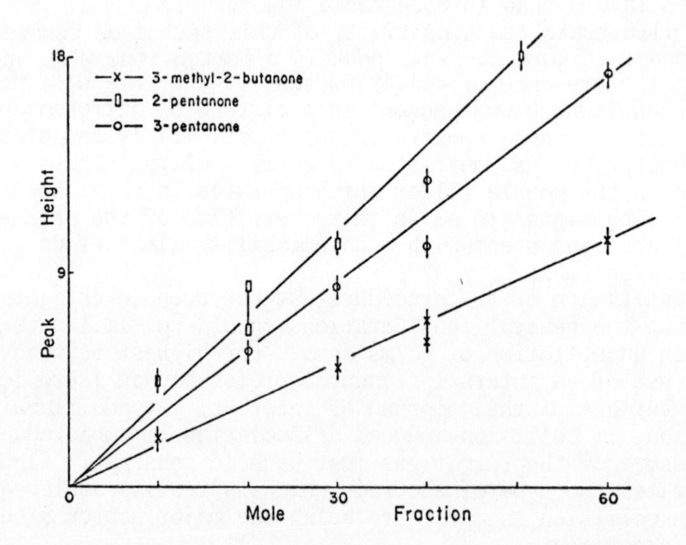

Analytical Letters

Figure 3. Analysis of ketones in mixtures of varying composition by electron impact ionization followed by MIKES

may constitute the major portion of the sample without deleterious results. Indeed, the solvent could probably be used to advantage as the CI reagent gas in some cases. These considerations suggest that the method might find appropriate application in the continuous analysis of organic reaction mixtures or effluents. Cases in which workup must be minimized, e.g., where chromatography alters the composition of the mixture are of special interest. Results on the analysis of the mixture resulting from the methylation of 2-butanone by methyl iodide in the presence of base are given in Table II. Each compound was quantitated using characteristic peaks in the MIKE spectrum of its molecular ion (electron impact). It is noteworthy that the mixture included isomers which are not separated in the mass analysis stage of the procedure. The results given are for an early and a late aliquot. They indicate the progress of methylation and the steric control to which this reaction is subject. The agreement between results employing metastable ion dissociations and CID is of the order expected in the absence of more sophisticated internal standards.

Currently, mixture analysis by CI/MIKES is being extended to more complex compounds including natural products and pharmaceuticals. Classes of compounds which, because of volatility or thermal instability, cannot readily be separated by gas chromatography are of special interest. For this reason the MIKE spectra of barbiturates have been examined. (2) There have been continued difficulties in the analysis of these compounds, even using the newer mass spectrometric techniques. The ethyl- and allyl-substituted barbiturates give most difficulty in analysis and we therefore chose to examine several of the most refractory isomeric and homologous barbiturates. The MIKE spectra of the $(M + H)^+$ and $(M + C_2H_5)^+$ ions were found to clearly distinguish them. Figure 4 compares the MIKE spectra taken in the presence of collision gas of four protonated barbiturates. Two sets of isomers and two sets of homologs are represented. The sec-butyl group of butabarbital and the sec-amyl group of pentobarbital are associated with intense fragmentation by alkene elimination. The substituents which bear fewer hydrogens on the α-carbon show this reaction to a lesser extent. The similarities in the spectra shown are typical of those observed for homologous barbiturates.

The current situation regarding mixture analysis by MIKES is that the technique appears to be complementary to gc/ms, being applicable in situations in which gc is difficult. The spectra are readily interpreted and quantitation seems not to present any new problems. As occurred in the development of gc/ms, it seems likely that automation of the instrumentation will be required to achieve its full potential which should include continuous monitoring capabilities. The somewhat lower sensitivity compared with gc/ms should not often preclude its use, and the potential to analyze involatile and thermally unstable biological compounds is a most attractive compensatory feature.

Table II. Analysis of Reaction Mixtures from the Methylation of Butanone

Compound	$M^{+\cdot}$	Early		Late	
		Unimol	Unimol + CID	Unimol	Unimol + CID
	72	0.15	0.15	--	--
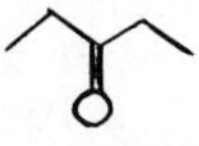	86	0.18	0.20	--	--
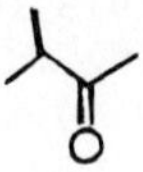	86	0.17	0.16	--	--
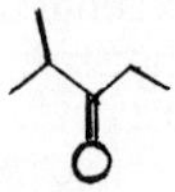	100	0.35	0.35	0.04	0.04
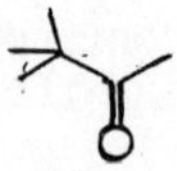	100	0.02	0.02	--	--
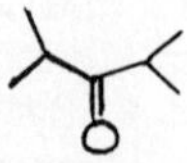	114	0.09	0.10	0.88	0.92
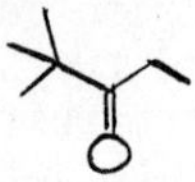	114	0.03	0.02	0.08	0.02
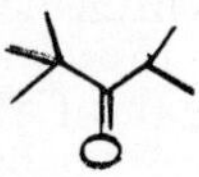	128	--	--	0.01	0.01

Fragmentation Mechanisms and Metastable Ions in Chemical Ionization

Chemical ionization allows ready access to types of ions which cannot be generated by electron impact. There has, however, been relatively little attention given the fragmentation mechanisms of these ions. This has been a consequence of the emphasis accorded the determination of molecular weights by CI and the resulting efforts to minimize fragmentation. When secondary reactions are studied, fragmentations are naturally highlighted and it has been observed that metastable ions in CI often have intensities comparable to those observed in EI.

The study of metastable ions in CI reveals the existence of a fragmentation mechanism in which a neutral alkane molecule is eliminated from onium type ions. For example, protonated t-butyl ethyl ether fragments by eliminating a butane molecule (1) while

$$(CH_3)_3C\text{-}\overset{+}{O}(H)\text{-}CH_2CH_3 \longrightarrow CH_3CH{=}\overset{+}{O}H + C_4H_{10} \qquad (1)$$

the trimethyloxonium ion losses methane. The reaction has also been observed in halonium and ammonium ions. In all cases it probably occurs *via* a four-centered mechanism (see below). It is sometimes observed in competition with alkyl radical loss and alkene molecule elimination. The reaction is quite general, since it also occurs in odd-electron ions. (3) Thus it has been observed from the molecular ions of ketones and amines. Figure 5 compares the fragmentations of the molecular ion and the $(M + H)^+$ ion of diisopropyl ketone. The similarity in behavior of the odd and even electron ions in undergoing loss of an alkane molecule is remarkable.

Alkane loss has long been known in the mass spectra of alkanes themselves, but, except for an isolated observation on ketones (4), the reaction has until recently gone unnoticed. It now seems that it constitutes an important general fragmentation mechanism which parallels the well-known alkene elimination (McLafferty rearrangement) in the types of ions in which it occurs and in its importance as a favored process at low internal energy. The four centered mechanism has been explicitly demonstrated for diisopropyl ketone by deuterium labeling.

$$(CH_3)_2CH\text{-}C({=}O^{+\cdot})\text{-}CD(CH_3)_2 \longrightarrow H\text{-}C(CH_3)_2\text{-}D + O{=}C{=}C(CH_3)_2^{+\cdot}$$

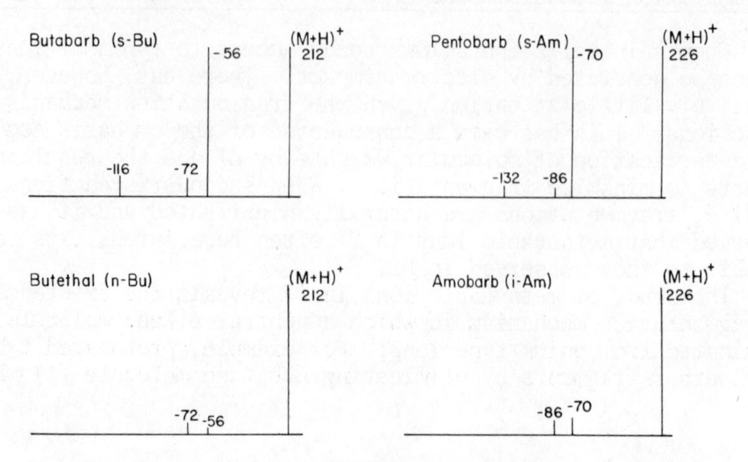

Figure 4. MIKE spectra taken in presence of target gas of the protonated barbiturates indicated. The stable ion beam, $(M + H)^+$ is recorded at an attenuation of 10^3.

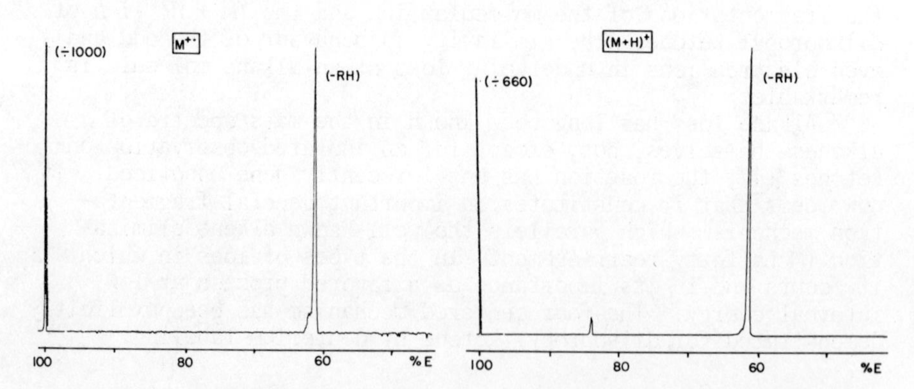

Figure 5. MIKE spectra showing spontaneous alkane elimination from both the molecular ion and the protonated molecule for diisopropyl ketone

Studies in other cases are in hand. A tentative generalization is that the reaction is facilitated by positive charge density on the heteroatom or functionalized carbon.

The study of metastable ions in CI not only provides information on fragmentation mechanisms but it is a valuable source of information on the internal energies of ions extracted from the source. If collisions relax the ions below the activation energy for the lowest energy fragmentation then no metastable ions will be observed. Experiments in which an inert bath gas is admitted to the CI source and metastable ion abundances are followed are particularly attractive as a new method of studying energy transfer in ion/molecule collisions. In a related type of experiment a correlation has been observed between the presence of metastable ions in CI and the activation energy for the lowest energy fragmentation.(5) Thus, $C_3H_5^+$ ions which require 2-3 eV of internal energy to undergo H_2 loss show essentially no metastable peak even when generated by a variety of ion/molecule reactions. The process can, however, be induced by collision and the kinetic energy release is then 0.47 eV in good agreement with the value of 0.43 eV found for collision-induced dissociation of $C_3H_5^+$ ions generated by electron impact. Figure 6 which is adapted from ref. 6 shows these results as compared to the metastable dissociations of the same ions. The latter occur via two competitive reactions to give a composite metastable peak. On collision, only the higher activation energy, higher frequency factor reaction is observed.

In contrast to the behavior of $C_3H_5^+$, the molecular ions of substituted nitrobenzenes which can undergo NO$^\cdot$ loss at considerably lower internal energies, show abundant metastable peaks. These reactions correspond entirely to those observed for the same ions generated by electron impact, both sets of ions showing composite metastable peaks with the same kinetic energy releases. Figure 7 illustrates the composite metastable peaks associated with NO$^\cdot$ loss from the molecular ions of p-chloro and p-hydroxy nitrobenzene. The contrasting behavior of the nitrocompounds and $C_3H_5^+$ can be interpreted in terms of the efficiencies of quenching of the states responsible for fragmentation on the time scale associated with metastable ions.

Isotope Labeling and the MIKE Spectrometer

It would be difficult to overestimate the importance that labeled compounds have had in the elucidation of gas-phase ion chemistry. Frequently the synthesis of pure compounds, highly incorporated at specific sites, has been the challenging and time consuming step in such work. The MIKE spectrometer relaxes the requirements for both chemical and isotopic purity and so facilitates this type of endeavor.(7) Quite simply, mass selection acts as a final stage of chemical and isotopic purification in the synthesis. Subsequent to mass-analysis, either metastable or collision-

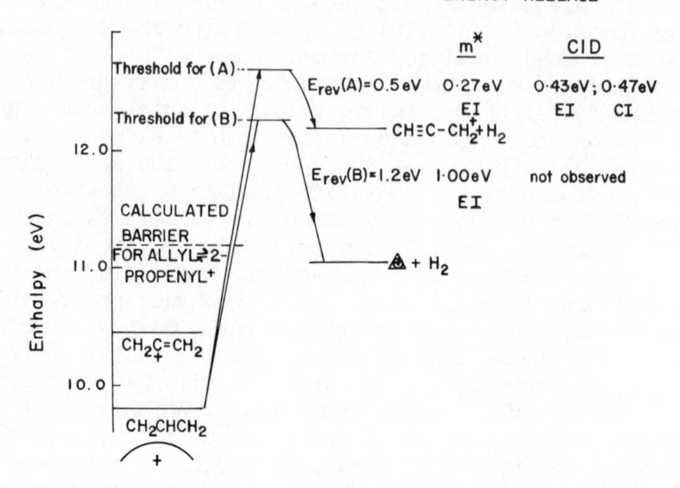

Figure 6. Thermochemistry of the reactions $C_3H_5^+ \rightarrow C_3H_3^+ + H_2$ (6). Also shown are the measured kinetic energy releases for the unimolecular (m) and collision-induced dissociations (CID) when the ion is generated by electron impact (EI) and chemical ionization (CI).*

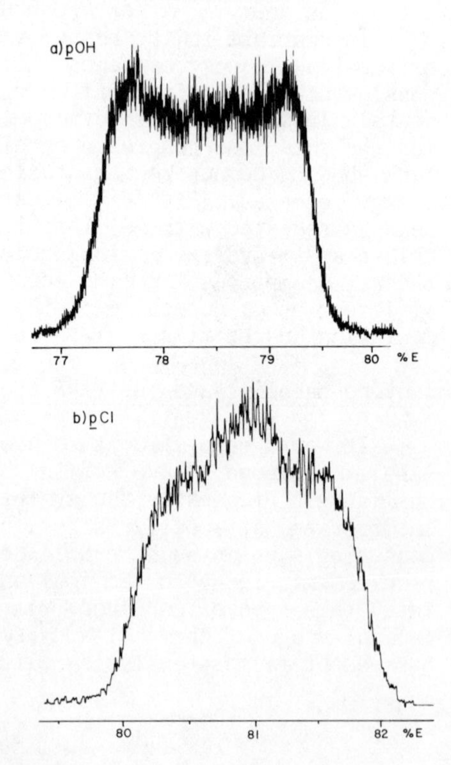

Figure 7. Composite metastable peaks for NO· loss from the molecular ions of the substituted nitrobenzenes indicated. The molecular ions were formed in the CI source by charge exchange with argon ions, were mass selected, and their energy spectra were plotted.

induced dissociations may be studied depending upon whether reacting ions of higher or lower internal energy are of interest.

These points are illustrated by the elimination of propane from the diisopropyl ketone molecular ion.(1d) The acidic hydrogens on the α-carbons are readily exchanged with D_2O in base. The partially incorporated compound obtained by a single exchange step serves for analysis by the MIKES technique as illustrated in Figure 8. This shows part of the MIKE spectrum of the d_2-compound (40% of the ketone sample) compared to that of the unlabeled compound. Alkane elimination is seen to involve loss of both labeled hydrogen atoms. Indeed, informative results are obtained on both the d_2-ketone and the d_1-ketone which can be selected independently for study. Experiments run in the presence of target gas show, in addition, collision-induced dissociation by loss of an alkyl radical. The MIKE spectra of the labeled compounds confirm this and show that no hydrogen isomerization involving the methyl groups occurs.

As a variation on the above approach in which the MIKES is used to purify a crude labeled compound, the synthesis itself can be carried out in the CI source. Thus, using simple labeled reagents the ion of interest can be synthesized, purified and its reactions characterized, all within the mass spectrometer.

Proton Affinities using Metastable Ions

Metastable ions, because they refer to species which have just enough energy to react at a rate of approximately 10^5 sec^{-1}, provide a source of thermochemical data on reaction thresholds. A long standing generalization is that if competitive fragmentations from the same ion both give rise to metastable peaks, the reactions must have similar activation energies. If in addition, the reactions are of similar type so that their frequency factors are comparable, then the relative magnitudes of their activation energies can be expected to determine the relative metastable peak heights. Thus, comparison of a kinetic quantity, the relative rates of two reactions from a given ion, allows the determination of a thermochemical quantity, the relative free energies of activation. In the cases of interest here, a direct correlation with enthalpy of activation is expected because of the detailed similarity in the competitive reactions examined. Furthermore, this correlation should extend to the enthalpies of reaction provided reverse activation energies are negligible or similar. Thus, the relative metastable ion abundance should correlate with the relative product enthalpies. Figure 9 illustrates the basis of the method.

We have applied (8) these concepts to the competitive metastable ion dissociations of proton bound dimers, $B_1HB_2^+$. These reactions

$$B_1HB_2^+ \rightarrow B_1H^+ + B_2 \quad (2)$$

$$B_1HB_2^+ \rightarrow B_1 + B_2H^+ \quad (3)$$

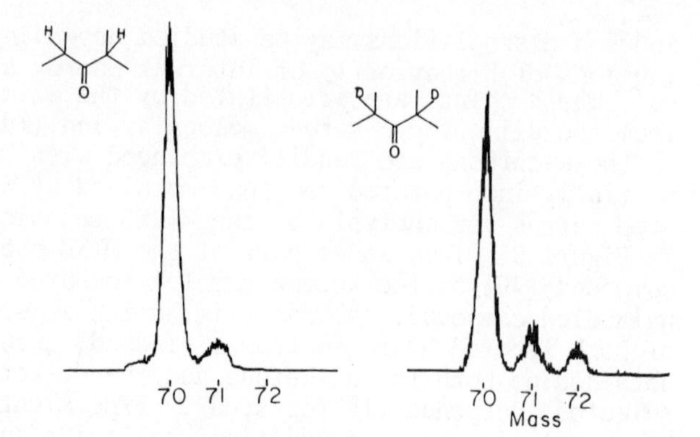

Analytical Chemistry

Figure 8. Partial MIKE spectra (no collision gas) of the molecular ions of unlabeled and labeled ketones mass selected from a partially incoroprated sample

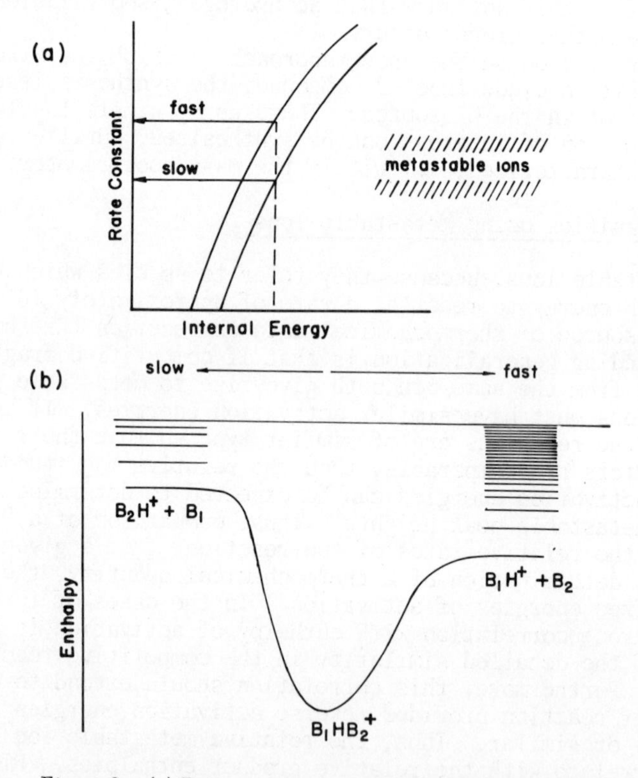

Figure 9. (a) Rate constant vs. internal energy curves for competitive reactions of different activation energy but similar frequency factor. When the internal energy is high enough for the higher energy process to occur fast enough to be observed as a metastable ion, the lower energy process is going faster still, and it dominates the spectrum at all internal energies; (b) the same situation illustrated by the very different state densities for the competitive fragmentations at an internal energy in excess of both activation energies.

are particularly appropriate for this treatment since they are simple bond cleavages, presumably involving steeply rising rate constant vs internal energy curves. Furthermore reverse activation energies are likely to be small. The difference in product enthalpies for the competitive reactions (2) and (3) is

$$\Delta(\Delta H_f) = \Delta H_f(B_1H^+) + \Delta H_f(B_2) - \Delta H_f(B_1) - \Delta H_f(B_2H^+) \quad (4)$$

but the proton affinity PA is the negative of the enthalpy change on addition of a proton,

$$PA(B_1) = \Delta H_f(B_1) + \Delta H_f(H^+) - \Delta H_f(B_1H^+) \quad (5)$$

hence substituting in (4) we have,

$$\Delta(\Delta H_f) = PA(B_1) - PA(B_2) \quad (6)$$

Thus, the more abundant metastable peak for reactions (2) and (3) should indicate the base of greater proton affinity. The ordering of proton affinities should be independent of the internal energy distribution of the reactant ion $B_1HB_2^+$ although the metastable ion abundance ratio will depend on this factor. We leave for elsewhere an attempt to develop this quantitative relationship further; here we present results on the relative ordering of proton affinities by this new method.

A mixture of 3-aminopentane and 2-aminobutane introduced into the CI source via a methane gas stream gave the mass spectrum shown as Figure 10. The ions at m/e 88 and 74 are due to protonated amines, those at 114, 128 and 142 appear to be immonium ions formed by transamination of the $(M - H)^+$ ions and those at 147, 161 and 175 are the proton bound dimers. The MIKE spectrum of the asymmetrical dimer m/e 161 is shown in Figure 11. As expected from data taken by equilibrium studies (9), 3-aminopentane has the larger proton affinity (the difference is 1 kcal $mole^{-1}$).

The method is expected to be sensitive to small differences in proton affinity as the schematic illustration of Figure 9 shows. This sensitivity is illustrated by a comparison of s-butylamine and pyridine which have been found by icr (10) to have virtually identical proton affinities. Figure 12 is a slow scan through part of the MIKE spectrum of the proton bound dimer of these two bases. Protonated pyridine is formed in higher abundance suggesting that pyridine has the greater proton affinity, probably by 0.1 to 0.2 kcal $mole^{-1}$. These results were obtained without correcting for collection and detector efficiencies. This can be done by studying labeled ions if isotope effects can be ignored. The dimer of pyridine and d_5-pyridine provides a case in point, results for it being shown in Figure 13. This type of experiment can also be designed for the determination of isotope effects on proton affinities.

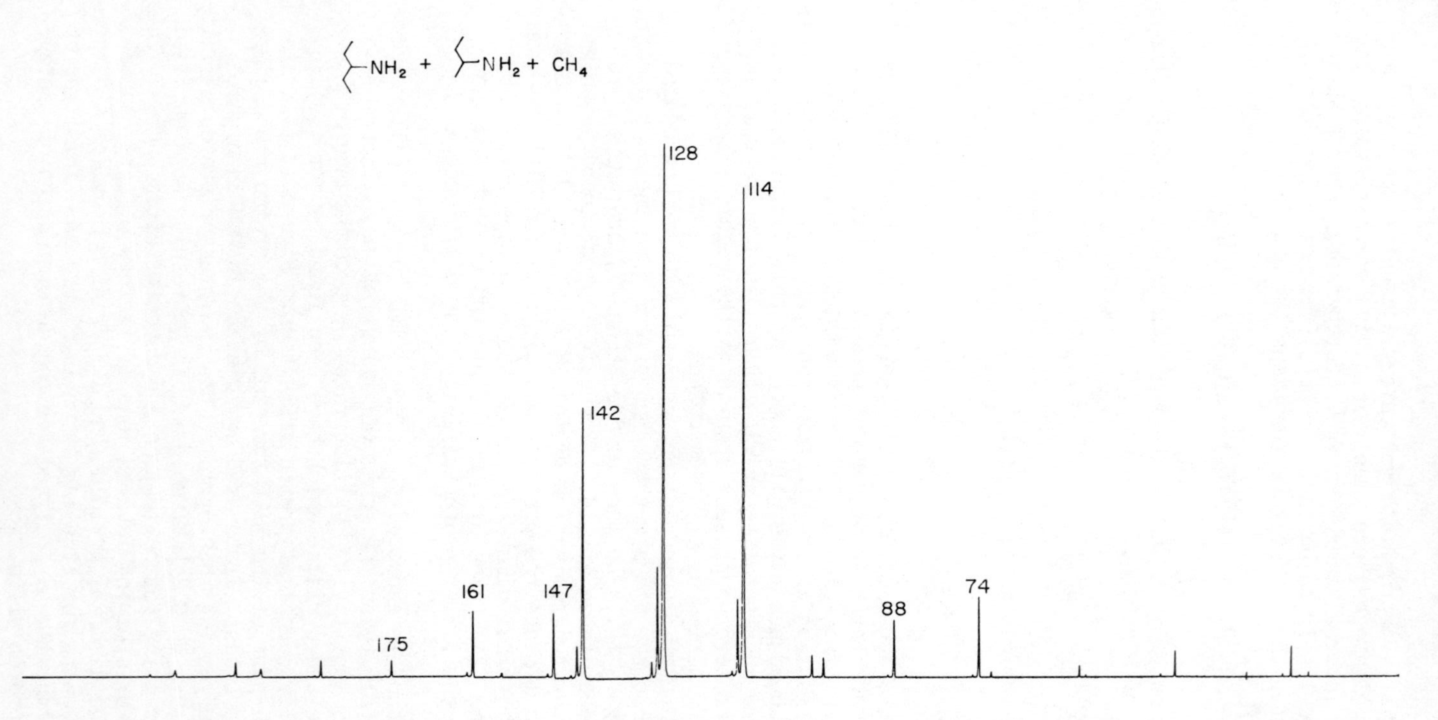

Figure 10. CI (CH_4) mass spectrum of a mixture of 3-aminopentane and 2-aminobutane showing formation of proton bound dimers

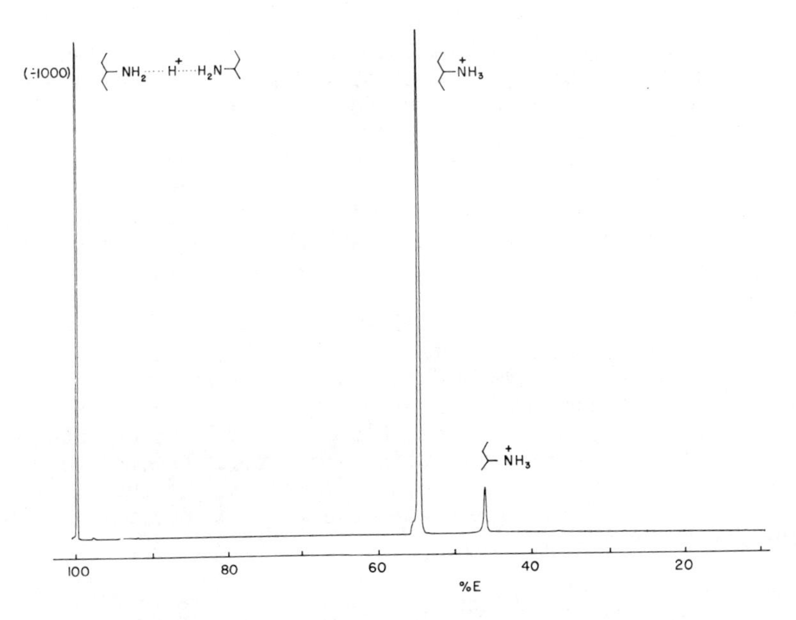

Figure 11. MIKE spectrum (no collision gas) of the proton bound dimer formed from 2-aminobutane and 3-aminopentane. The stronger base gives the large metastable peak for formation of the BH^+ ion.

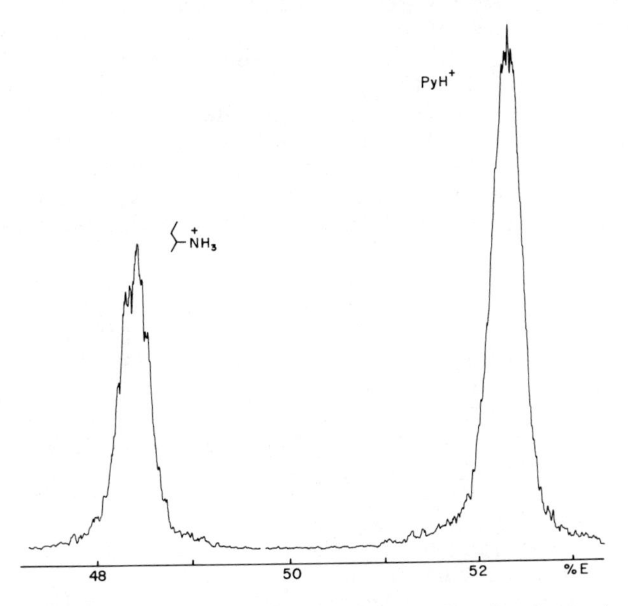

Figure 12. Part of the MIKE spectrum of the s-butylamine/pyridine proton bound dimer illustrating a case in which the metastable peaks for Reactions 2 and 3 differ in height by less than a factor of two. The proton affinity difference in this case is estimated to be 0.1 to 0.2 kcal mol^{-1}.

As a final illustration of the sensitivity of the MIKES approach in investigating ion chemistry, Figure 14 shows a partial MIKE spectrum of a ^{13}C-labeled pyridine/s-$BuNH_2$ dimer formed using only unenriched compounds. These species were selected from the complicated mixture of ions present in the ion source. The doubly labeled ion represents only 1% of the protonated dimer ion current. Nevertheless, its metastable reactions can readily be studied the label pattern shown in part (b) being that associated with pyridine and agreeing with the statistically required ratio of 3:10:5 for 80^+:81^+:82^+.

Work currently in hand has shown that the method can be extended to bases belonging to other functional groups, and a ladder of basicities is being established. Collision-induced dissociation, although less sensitive to small differences in proton affinities than are metastable ion reactions, has been found to give corresponding results. The kinetic energy releases for some of these reactions have been measured, and the small values recorded are consistent with the proposed negligible effect of reverse activation energy.

Ion Structure Determinations

In keeping with the importance of determining ion structures, there has been a great deal of emphasis placed on this subject in recent years. Measurements of metastable ion characteristics, especially kinetic energy release, ion enthalpies, and collisional activation (collision-induced dissociation) spectra have been most valuable in this regard. Our recent work in this area has fallen into two categories, (a) the application of established methodology to investigate the structures of ions generated by chemical ionization, (b) the testing of new methods for characterizing ions. Examples from each topic are covered in turn below.

The combination of a chemical ionization source with a MIKE spectrometer allows ions formed by CI to be independently characterized. This is readily done by mass-selection, followed by collision-induced dissociation. The resulting MIKE spectrum will contain metastable peaks as well as those due to CID. Figure 15 shows a set of these spectra for the ethylated nitrobenzenes indicated.(11) A striking feature of the spectra is the occurrence of major reactions common to all four $(M + C_2H_5)^+$ ions. The reactions correspond to loss of C_2H_4, $C_2H_5O^\cdot$ and C_2H_5ONO. The latter two processes can occur as simple bond cleavages only if the nitro group is ethylated. As a further test of this structure, the corresponding protonated molecules were examined in the same way. Major reactions now observed were loss of $HO^\cdot$ and $NO_2H^\cdot$, supporting the conclusion that the nitro group is more basic than the ring or the para substituent.

A characteristic of ring alkylation or protonation is the loss of the substituent as a free radical. Thus, one can contrast the absence of $Cl^\cdot$ loss in the spectrum of ethylated

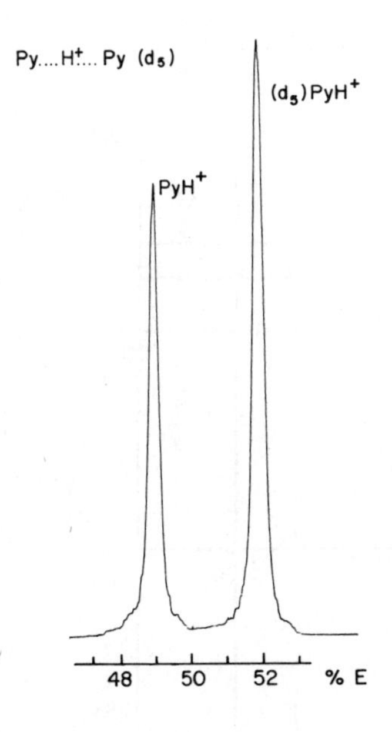

Figure 13. Metastable peaks for the formation of labeled and unlabeled pyridinium ions from a partially labeled dimer. Collection and detection efficiencies can be estimated from this data if secondary isotope affects are assumed negligible.

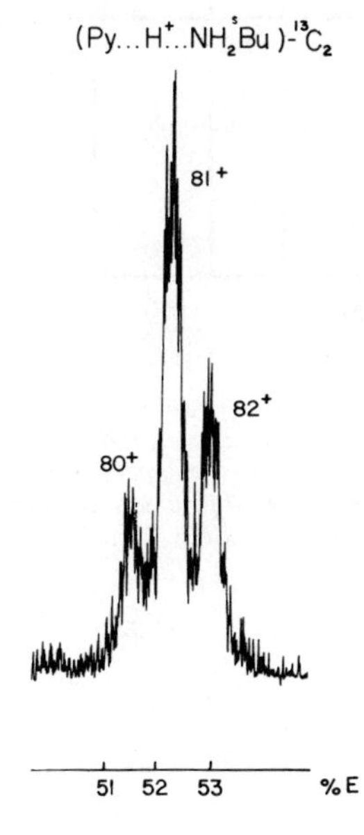

Figure 14. Illustration of the selectivity and sensitivity of the MIKES technique for studying ion thermochemistry. The doubly labeled ion studied was selected from the naturally occurring mixture.

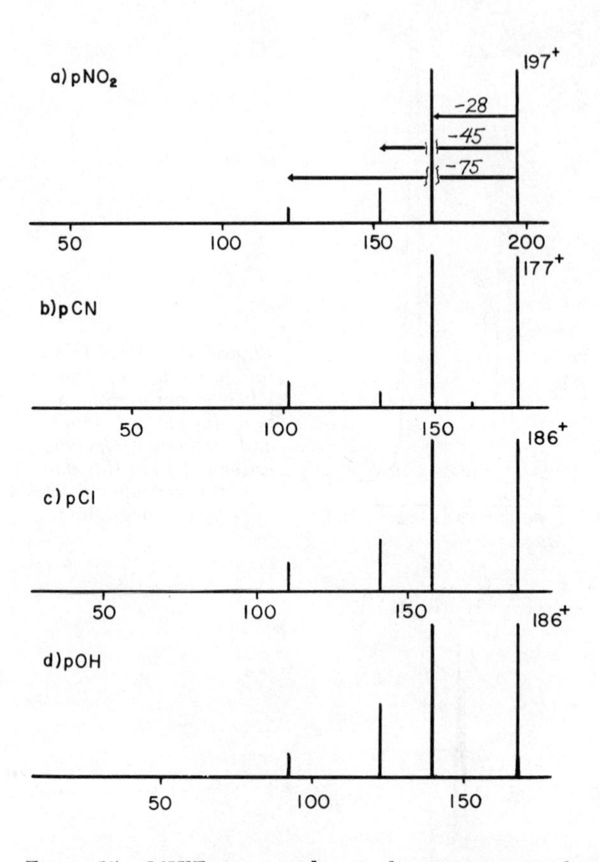

Figure 15. MIKE spectra taken in the presence of collision gas of the $(M + Et)^+$ ions of the substituted nitrobenzenes indicated. The abscissae of the spectra are calibrated in terms of ion mass. The stable ion beams have been divided by approximately 10^3.

p-chloronitrobenzene (Figure 15c) with the spectra of protonated p-chlorophenol, ethylated p-chlorobenzoic acid and ethylated p-chlorobenzonitrile. Loss of Cl• in these compounds gives peaks which are 53%, 50%, and 3% of the most abundant fragment ions, respectively. In these and other cases there is also evidence for a mixture of ring and substituent alkylated ions.

We turn now to the testing of newer methods of characterizing ions. Those which employ the same instrumentation and methods as CID and metastable ion reactions are most attractive. Thus, charge stripping (7) provides a case in point, the cross section

$$m_1^{+} + N \rightarrow m_1^{++} + N + e^- \qquad (7)$$

for the reaction being readily measured relative to that for an ion chosen to serve as an internal standard. This procedure was adopted in comparing the isomeric $C_3H_8N^+$ ions generated from t-butylamine and diethylamine. The latter showed a peak due to charge stripping which was more than an order of magnitude more intense than the former. It seems probable that this difference is due less to differences in the cross section for stripping (a process with a large energy loss which should be relatively insensitive to structural variation) than it is due to differences in the stability of the nascent doubly charged ions. Stripping at kilovolt energies is a vertical process, the doubly charged ion being generated with the same structure as its singly charged precursor. Thus, the greater extent of fragmentation of the more highly branched $C_3H_8N^{++}$ ion generated from t-butylamine is the expected result. Although this method of characterizing ions relies on a single parameter, it is of some interest as a means of sampling singly charged ions of lower average internal energy than is the case for collision-induced dissociation. (12) A consequence is that comparisons between stripping and CID should aid in uncovering isomerizations occurring below the threshold for fragmentation.

As a second example of newer methods of characterizing ions, we shall consider measurements of kinetic energy loss for collision-induced dissociations. Kinetic energy releases can be measured from the same data so we consider them together. As an example, H_2 loss from alkyl cations can be considered. Analogies in structure and fragmentation mechanism of $C_2H_3^+$, $C_2H_2Cl^+$, $C_3H_3^+$, $C_3H_5^+$, and $C_4H_3^+$ are inferred from the fact that these ions when generated from a variety of precursors give the energy losses and kinetic energy releases shown in Table III. The methyl cation shows contrasting behavior in both its kinetic energy loss and its energy release.

Table III. Energy Releases and Losses for Alkyl Ions

Ion	Energy Loss (eV)	Energy Release (meV)
$C_2H_3^+$	10.3±1.6	330±15
$C_2H_2Cl^+$	9.4±1.4	339±31
$C_3H_3^+$	11.3±0.5	334±4
$C_3H_5^+$	11.0	334
$C_4H_3^+$	10±1	296±17
CH_3^+	6.4±0.9	83±4

Conclusion

Ion chemistry and its analytical applications have been treated here with a focus on particular instrumentation and types of measurements. The combination of chemical ionization with the mass-analyzed ion kinetic energy (MIKE) technique is valuable in determining the structures of ions generated by low energy ion/molecule reactions, in the analysis of complex mixtures without prior separation and in mechanistic studies employing labeled compounds.

The usefulness of kilovolt energy ion/molecule reactions as a surprisingly subtle probe of ion properties is shown. The compatibility of these and metastable ion reactions in terms of instrumentation and the complementary information they supply is noteworthy. Wider use of charge changing collisions as well as collision-induced dissociation for organic ions seems appropriate.

The range of inquiry which can be based on the rather simple measurement of ion kinetic energy remains a constant source of pleasure to this research group.

Acknowledgement

This work has been supported by the National Science Foundation and done in collaboration with T. L. Kruger, V. M. Franchetti, J. F. Litton, R. W. Kondrat, C. S. Hsu, R. Flammang, E. Soltero-Rigau and D. Cameron.

Abstract

Secondary reactions including spontaneous and collision-induced dissociation and charge changing collisions, have been studied by the methods of ion kinetic energy spectrometry. Emphasis is on ions generated by chemical ionization which are mass-selected before reaction and energy analyzed after. Applications to the analysis of mixtures without chromatographic separation are demonstrated. Ion structure determinations are made on CI generated ions. Kinetic energy release and loss measurements are used to characterize ions. Studies on labeled compounds employing partially labeled material are shown to be possible so facilitating mechanistic investigations. Alkane elimination is shown to be a general low energy fragmentation occurring in both odd- and even-electron ions. The occurrence of metastable ion dissociations in chemical ionization experiments has been explored and these reactions form the basis for a new and sensitive method of ordering proton affinities in the gas phase.

Literature Cited

1. Smith, D.H., Djerassi, C., Maurer, K.H., and Rapp, U., J. Am. Chem. Soc., (1974), 96, 3482.
2. Schlunegger, U.P., Angew. Chem., Int. Ed. Engl., (1975), 14, 679.
3. Kruger, T.L., Litton, J.F., and Cooks, R.G., Anal. Chem., (1976), 9, 533.
4. Kruger, T.L., Litton, J.F., Kondrat, R.W., and Cooks, R.G., Anal. Chem., (1976), 48, 2113.
5. Levsen, K., and Schulten, H.R., Biomed. Mass Spectrom., (1976), 3, 137.
6. Soltero-Rigau, E., Kruger, T.L., and Cooks, R.G., Anal. Chem., (1977), 49, 435.
7. Litton, J.F., Kruger, T.L., and Cooks, R.G., J. Am. Chem. Soc., (1976), 98, 2011.
8. Cooks, R.G., Yeo, A.N.H., and Williams, D.H., Org. Mass Spectrom., (1969), 2, 985.
9. Cameron, D., Clark, J.E., Kruger, T.L., and Cooks, R.G., Org. Mass Spectrom., in press.
10. Holmes, J.L., Osborne, A.D., and Weese, G.M., Org. Mass Spectrom., (1975), 10, 867.
11. Beynon, J.H., Brothers, D.F., and Cooks, R.G., Anal. Chem., (1974), 46, 1299.
12. Cooks, R.G., and Kruger, T.L., J. Am. Chem. Soc., (1977), 99, 1279.
13. Aue, D.H., Webb, H.M., and Bowers, M.T., J. Am. Chem. Soc., (1972), 94, 4726.
14. Kruger, T.L., Flammang, R., Litton, J.F., and Cooks, R.G., Tetrahedron Letters, (1976), 4555.
15. Cooks, R.G., Beynon, J.H., and Litton, J.F., Org. Mass Spectrom., (1975), 10, 503.

RECEIVED December 30, 1977

5

Field Ionization Kinetic Studies of Gas-Phase Ion Chemistry

N. M. M. NIBBERING

Laboratory for Organic Chemistry, University of Amsterdam, Amsterdam, The Netherlands

In the last few years the method of Field Ionization Kinetics (FIK) has been developed rapidly by several groups (1). The technique has proved to be successful in providing detailed information on the mechanisms of decompositions of ions in the gas phase (1,2). This information has been difficult or even impossible to obtain from conventional electron impact (EI) studies.

Although the underlying theory of the FIK method has been described extensively in detail (1,3), it will be repeated here in a very simple way.

If, for example, a 10 μm conditioned wire emitter (4) at potential V_0 (~ 8 kV) is positioned approximately 1.5 mm from a grounded and slotted cathode, gas phase molecules may be ionized in the strong electric field at a very narrow region close to the emitter. To a good approximation, the potential is equal to the emitter potential V_0 at this point (Figure 1). Suppose that the ions, formed by either direct ionization or unimolecular decomposition close to the emitter, are stable for $\geq 10^{-5}$ sec. They will acquire a kinetic energy eV_0 during acceleration from the emitter to the cathode and will pass the electric sector of a double focussing mass spectrometer, set to transmit ions with a kinetic energy eV_0. Consequently, they will be mass analyzed at their correct *m/e*-values. In this way a field ionization (FI) spectrum is obtained, containing peaks corresponding to the ions generated very near to the emitter, *i.e.*, within approximately 10^{-11} sec.

Fragment ions, m^+, generated by expulsion of (M-m) from M^+ ions between the emitter and cathode at potential V_x (Figure 1) are not transmitted through the electric sector, because they have insufficient kinetic energy (em/M (V_O-V_X) + eV_X) to be focussed by this sector. However, these ions can be transmitted through the electric sector by increasing the emitter potential, V_O, by an amount ΔV, such that the kinetic energy of the fragment ions, m^+, is given by equation (1) (3a).

$$eV_O = \frac{em}{M}(V_O + \Delta V - V_X) + eV_X \quad (1)$$

Thus it is possible to monitor at a fixed electric sector voltage the abundance of m^+ ions as a function of the emitter potential $V_O + \Delta V$ (3a). A scan of the emitter potential, achieved by increasing ΔV, allows one to view the decompositions of ions M^+ occurring at ever-increasing distances from the emitter. Of course, these increased distances correspond to longer lifetimes (5) of the parent ion, M^+. Expressing the abundance of m^+ ions due to loss of (M-m) as a function of time following FI of M, the FIK curve of m^+ is obtained (Figure 2). The time range, covered continuously by the FIK method, is approximately 10^{-11} to 10^{-9} sec. Additional points at longer times can be obtained from decompositions of M^+ ions in the first and second field free regions, which are also accessible by the conventional EI technique (Figure 1).

The strength of the FIK method is shown to full advantage, if it is used in combination with a double focussing mass spectrometer and with isotopic labelling, for the elucidation of mechanisms of gas-phase ion decomposition processes at 10^{-11} to 10^{-9} sec. (3a). The first and already classical example of such mechanistic studies has been the FIK study on cyclohexene-3,3,6,6-d_4 (6). Since then many mechanistic FIK studies have appeared in the literature, to which the reader is referred (1c,2,7).

In our laboratory, the FIK method in combination with deuterium labelling has been applied to gas phase decompositions of the molecular ions of 2-phenoxyethyl chloride and of 3-phenylpropanal, as will be discussed below. In this way, we intend to show the application of this powerful technique for investigation of gas-phase ionic decompositions.

FIK Study on 2-Phenoxyethyl Chloride (8).

A previous EI study of 2-phenoxyethyl chloride, $C_6H_5OC^2H_2C^1H_2Cl$, in combination with deuterium labelling has shown, that the molecular ions lose in approximately equal amounts $\cdot C^1H_2Cl$ and $\cdot C^2H_2Cl$ (9). This must

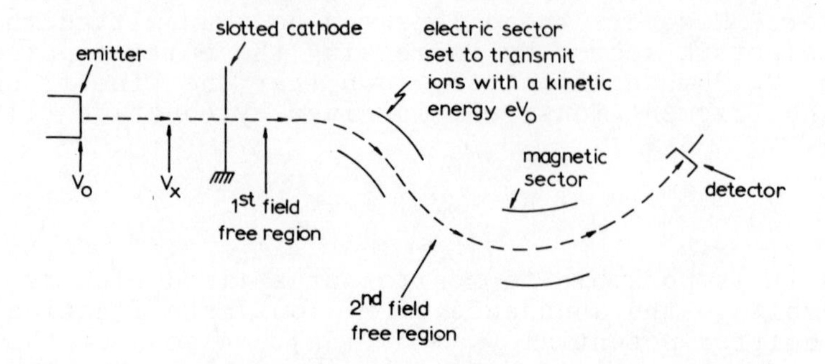

Figure 1. Double-focusing field ionization mass spectrometer (not drawn to scale)

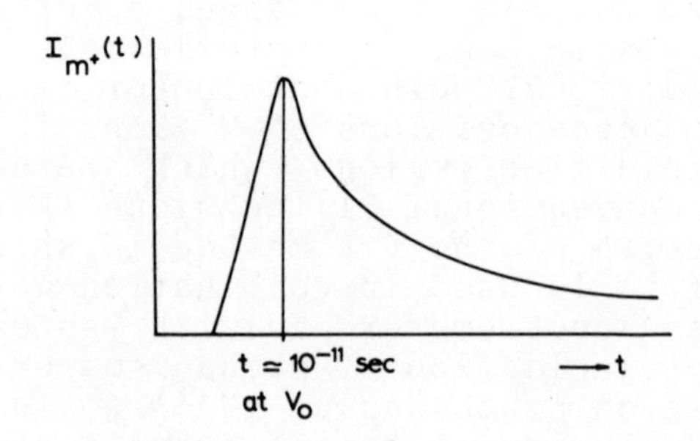

Figure 2. FIK curve of ions m⁺

(right) ⟶

Advances in Mass Spectrometry

Figure 3. FI spectra of 2-phenoxyethyl chloride and of its 1,1-d_2- and 2,2,-d_2 analogs. The left ordinate scale refers to the intensities of ions with m/e < 130; the right ordinate scale to those of ions with m/e > 130.

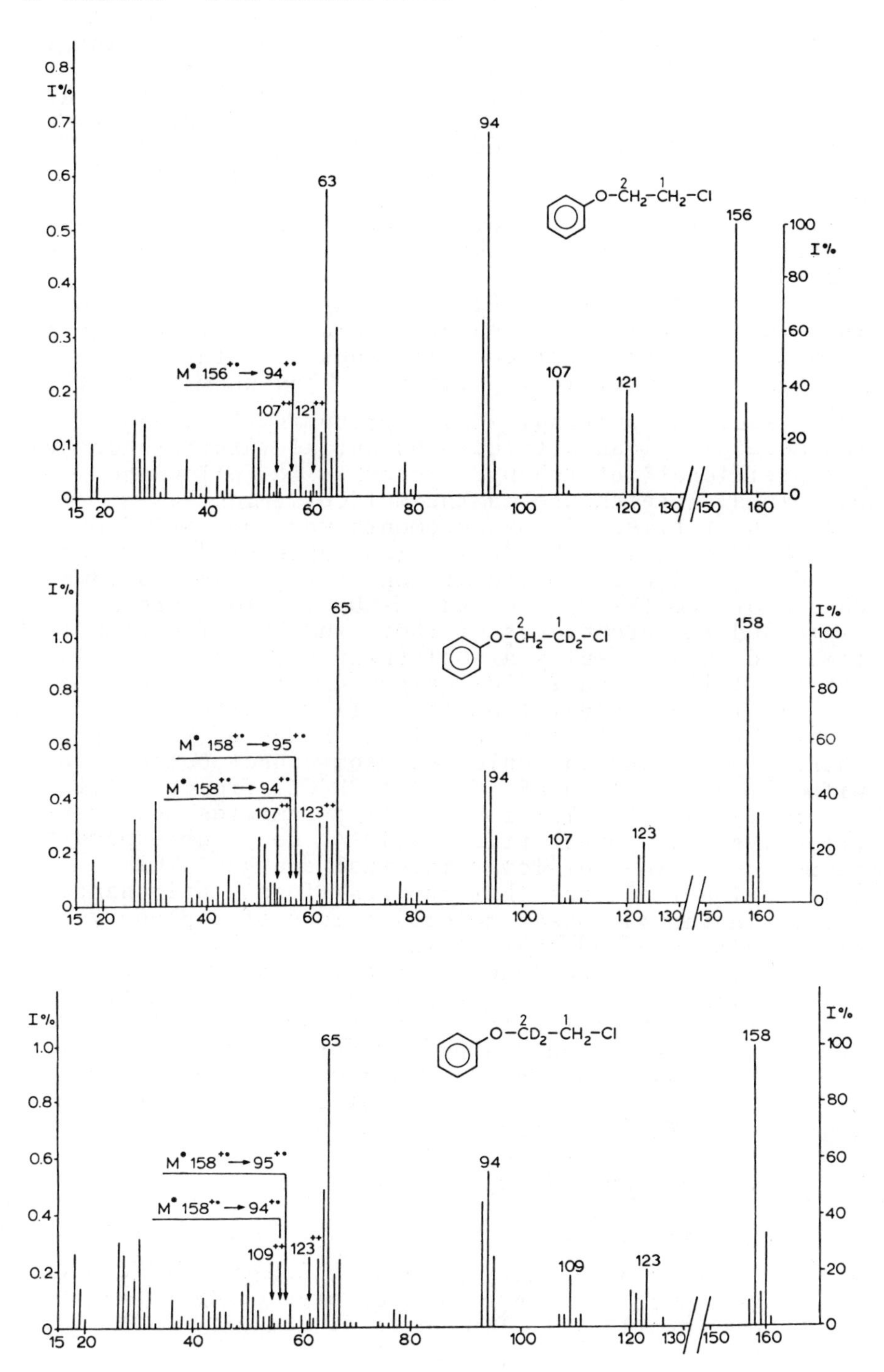

O–CH2–CH2–Cl
M* 156+• → 94+•
107++ 121++
63
94
107
121
156
O–CH2–CD2–Cl
M* 158+• → 95+•
M* 158+• → 94+•
107++ 123++
65
94
107
123
158
O–CD2–CH2–Cl
M* 158+• → 95+•
M* 158+• → 94+•
109++ 123++
65
94
109
123
158
I%

arise from a nearly completely equilibrated positional interchange of the phenoxy group and the chlorine atom in the molecular ions prior to loss of a CH_2Cl radical. We have addressed ourselves to the question of whether FIK could show the time dependence of this positional interchange. Of course, normal electron-impact (EI) mass spectrometers present an integrated view of ion chemistry from the time of formation to $\sim 10^{-6}$ sec, and, therefore, this information is not obtainable on these instruments.

The FI spectra of 2-phenoxyethyl chloride and of the 1,1-d_2-and 2,2,-d_2 analogues are given in Figure 3 without correction for contributions of natural isotopes and of isotopic impurities. They show that at 10^{-11} sec, the original $\cdot\overset{1}{C}H_2Cl$ group is eliminated from the molecular ions without substantial interference by the positional interchange process. To follow the positional interchange in the molecular ions as a function of time, FIK measurements were performed on the $(M-CH_2Cl)^+$ and $(M-CD_2Cl)^+$ ions generated from the 1,1-d_2- and 2,2-d_2 molecular ions. The measured abundances of the $(M-CH_2Cl)^+$ and $(M-CD_2Cl)^+$ ions are expressed in percentages of their sum as a function of time and these results are given in Figures 4a and 4b for the 1,1-d_2- and 2,2-d_2 compounds, respectively.

It is very clear from these figures, that at shorter times the positional interchange of the phenoxy group and the chlorine atom cannot compete with the elimination of $\cdot CD_2Cl$ ($\cdot CH_2Cl$) from the 1,1-d_2- (2,2-d_2) molecular ions, but it competes much more effectively at longer times ($\sim 10^{-10}$ sec) corresponding to molecular ions of lower internal energy. This observation indicates that the positional interchange occurs in the gas phase molecular ions of 2-phenoxyethyl chloride, and that it is a process having a lower frequency factor and lower activation energy than the elimination of $\cdot CH_2Cl$ via a simple cleavage. Note also from Figure 4 that the time-resolved positional interchange reaches at least a higher percentage of equilibration (87-93%) than can be deduced from previous EI results (85%) (9), which represent only an integrated value over the time range up to $\sim 10^{-6}$ sec.

FIK Study on 3-Phenylpropanal (10).

The eliminations of C_2H_2O and of C_3H_4O from the molecular ions of 3-phenylpropanal, $C_6H_5\overset{3}{C}H_2\overset{2}{C}H_2\overset{1}{C}HO$ upon electron impact have been studied previously by D- and C-13 labeling (11,12). In spite of the extensive

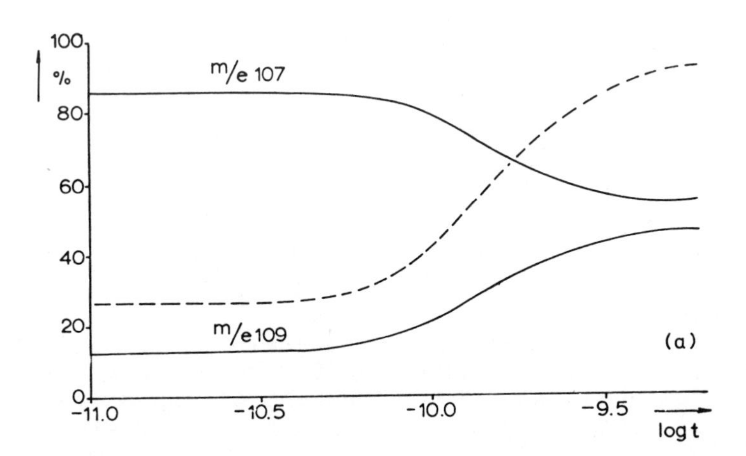

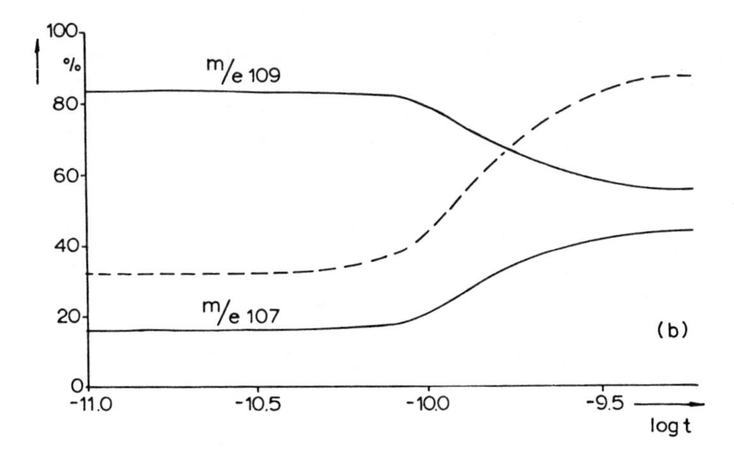

Advances in Mass Spectrometry

Figure 4. Intensities of m/e *107 and* m/e *109 expressed in percentage of their sum as a function of time for the 1,1-*d_2*- and 2,2-*d_2*-2-phenoxyethyl chlorides [(a) and (b), respectively]. The dashed lines refer to the percentage of equilibrated positional interchange of the phenoxy group and the chlorine atom.*

hydrogen randomization in the molecular ions prior to or during fragmentation, it has been suggested that a hydrogen atom from position 2 migrates to the phenyl ring during the eliminations of C_2H_2O and of C_3H_4O (11). This hypothesis can be tested with the FIK method.

In Figure 5 the EI spectrum and the FI spectra of 3-phenylpropanal at an unheated and heated emitter have been reproduced. It is seen, that the ions *m/e* 92 and *m/e* 78, generated by eliminations of C_2H_2O and C_3H_4O from the molecular ions respectively, are present in the FI spectra. The FI spectra of 3-phenylpropanals, labelled with deuterium in the 1 position (1-d_1), the 2 position, (2,2-d_2), the 3 position (3,3-d_2) and in the phenyl ring (Ø-d_5) show that the ions *m/e* 92 and *m/e* 78 retain the hydrogen atom from position 1 to the extent of approximately 90% (10). Contrary to conclusions from previous EI results (11) we now conclude that a hydrogen atom from position 1 and *not* from position 2 migrates to the phenyl ring during the eliminations of C_2H_2O and of C_3H_4O, at least within 10^{-11} sec. The elimination of C_2H_2O occurs by a 1,5 hydrogen migration, *i.e.*, the well-known McLafferty rearrangement, whereas the second elimination might proceed *via* a 1,5- and/or a 1,4-hydrogen shift (Scheme 1).

Scheme 1

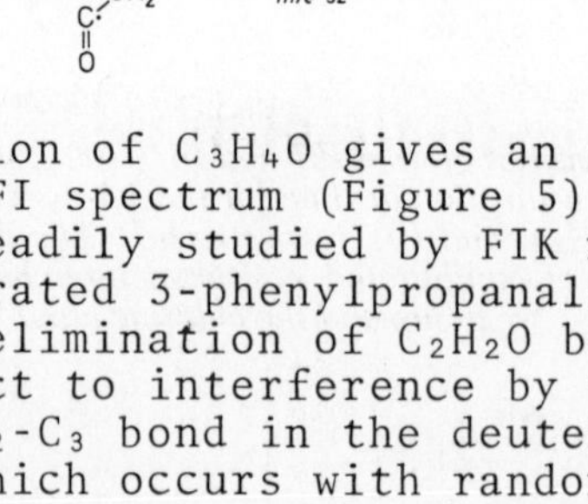

$$\longrightarrow C_6H_6^{+\cdot} + (CO + C_2H_4(2,3)) \quad m/e\ 78$$

$$\longrightarrow C_7H_8^{+\cdot} + C_2H_2(2)O \quad m/e\ 92$$

The elimination of C_3H_4O gives an isolated peak at *m/e* 78 in the FI spectrum (Figure 5), so that this reaction can be readily studied by FIK making use of the various deuterated 3-phenylpropanals. This is not true for the elimination of C_2H_2O because this reaction is subject to interference by the simple cleavage of the C_2-C_3 bond in the deuterated 3-phenylpropanals which occurs with randomization of hydrogen atoms (*cf.* peaks at *m/e* 91 and 92 in the FI spectra, given in Figure 5).

FIK curves have been measured for the $[M-C_3H_xD_yO\ (x+y=4)]^{+\cdot}$ ions from the various deuterated

3-phenylpropanals. Expressing the measured abundances of these ions from deuterated 3-phenylpropanal in percentages of their sum as a function of time in the graphs, presented in Figure 6, the following interesting results are obtained.

i) The hydrogen atom originally at position 1 is predominantly transferred to give $C_6H_6^{+\bullet}$ ions at short ion lifetimes.

ii) Transfer of the hydrogen atoms from position 3 to yield the $C_6H_6^{+\bullet}$ ions becomes more important at longer times (note the increase of *m/e* 80, *i.e.* $C_6H_4D_2^{+\bullet}$, with increasing time).

iii) The hydrogen atoms from position 2 are not involved in the production of $C_6H_6^{+\bullet}$ ions in the measured time range.

iv) One of the original phenyl ring hydrogen atoms is increasingly retained in the neutral eliminated C_3H_4O moiety at longer times, albeit to a small extent.

The following explanation is proposed for these results. In the measured time range of $10^{-10.3}$ to $10^{-9.5}$ sec, two specific hydrogen exchange processes in the molecular ions of 3-phenylpropanal occur prior to or during the formation of the $C_6H_6^{+\bullet}$ ions. The most important hydrogen exchange process takes place between the hydrogen atoms from positions 1 and 3, as depicted in Scheme 2.

Scheme 2

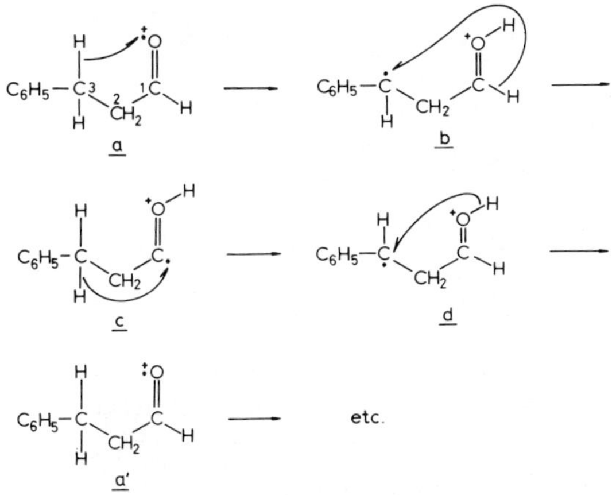

C_6H_5—C(H)(H)—CH_2—C(H)=$O^{+\bullet}$ (a) ⟶ C_6H_5—$\dot{C}$(H)—CH_2—C(H)=O^{+}H (b) ⟶

C_6H_5—C(H)(H)—CH_2—$\dot{C}$=O^{+}H (c) ⟶ C_6H_5—$\dot{C}$(H)—CH_2—C(H)=O^{+}H (d) ⟶

C_6H_5—C(H)(H)—CH_2—C(H)=$O^{+\bullet}$ (a′) ⟶ etc.

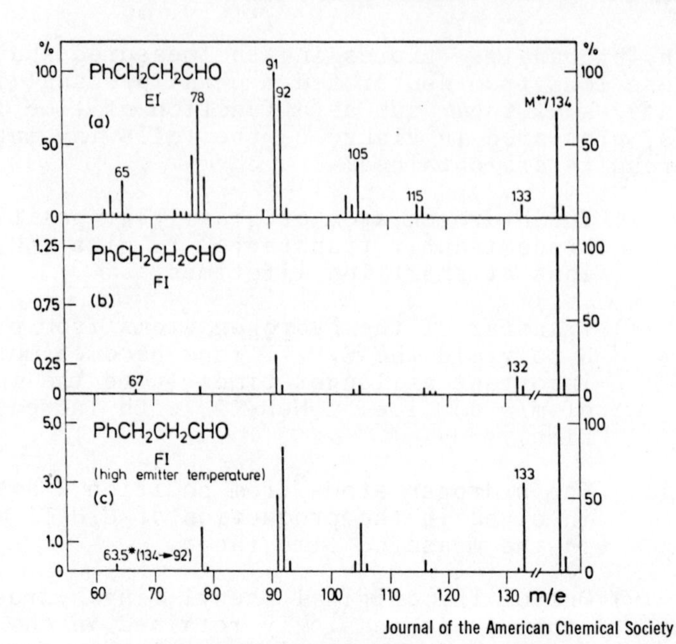

Figure 5. EI and FI spectra of 3-phenylpropanal [(b) unheated emitter, (c) heated emitter]. In the FI spectra the left ordinate scale refers to the intensities of ions with m/e < *134; the right ordinate scale to those of ions with* m/e ≥ *134.*

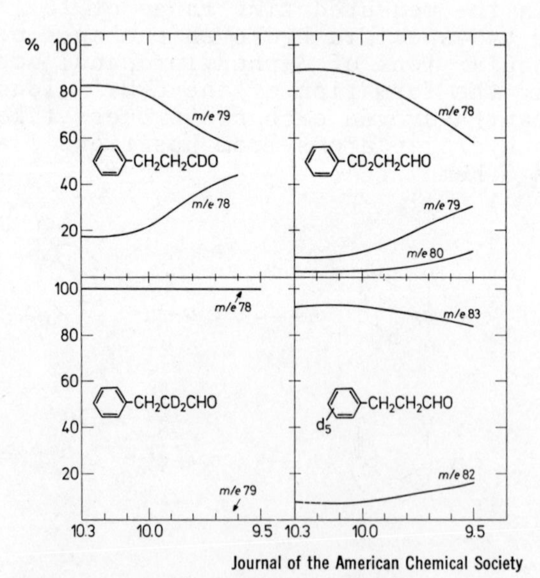

Figure 6. Intensities of the $(M\text{-}C_3H_xD_yO\ (x + y = 4))^{+\cdot}$ *ions expressed in percentage of their sum as a function of time for 3-phenylpropanals, deuterated in the 1, 2, and 3 positions and in the phenyl ring as indicated*

It should be noted that the hydrogen atom abstraction from position 3 by the radical site at the oxygen atom in the reaction a → b (or the equivalent proton abstraction from position 3 by the lone pair of electrons at the oxygen atom, if the charge resides initially in the phenyl ring) has been observed previously in the molecular ions of ethyl 3-phenylpropionate (13).

The other hydrogen exchange process at short times is best explained by an exchange between the hydrogen atoms from the phenyl ring and position 1, as rationalized in Scheme 3.

Scheme 3

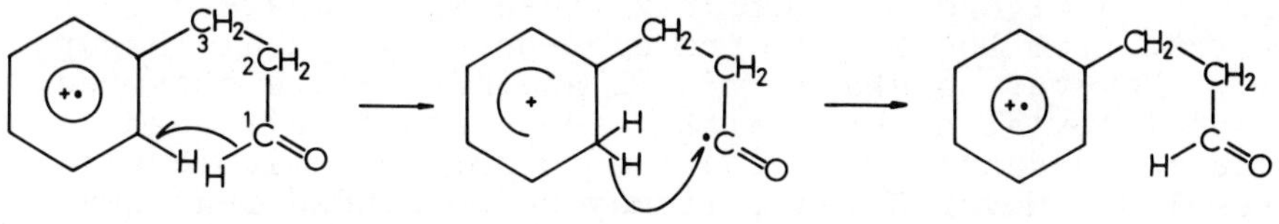

a″ e a‴

A similar exchange process, as presented in Scheme 3, has been observed in the molecular ions of 3-phenylpropyl bromide (14). It further supports the idea that the elimination of C_2H_2O *via* a McLafferty rearrangement (*vide supra*) occurs stepwise, which is now well accepted (15). In any event, the operation of both schemes 2 and 3 accounts for the incorporation of the two hydrogen atoms from position 3 in the $C_6H_6^{+\bullet}$ ions at longer times (*vide supra*). This FIK study does not show any active participation of the hydrogen atoms from position 2 in the hydrogen exchange processes prior to or during the formation of the $C_6H_6^{+\bullet}$ ions. It is of interest to note that such exchanges have been observed for metastable decompositions in the first field free region for ions generated by EI (11). This suggests that an exchange between the hydrogen atoms at position 2 and, for example, the phenyl ring can only compete with the exchange processes given in schemes 2 and 3 for molecular ions of lower internal energy, *i.e.* those decomposing in the first field free region. Unfortunately, this proposal could not be tested because of the absence of a metastable peak for the elimination of C_3H_4O from the molecular ions of 3-phenylpropanal under FI conditions.

The mechanism for the C_3H_4O elimination requires further comment. The $C_6H_6^{+\bullet}$ ions, resulting from this elimination, appear to originate from the molecular ions and from the $(M-28)^{+\bullet}$ ions, as shown by appropriate metastable decompositions in the first field free region under EI conditions. Therefore, it may be concluded that the $C_6H_6^{+\bullet}$ ions are generated by successive hydrogen migration from position 1 to the phenyl ring, carbon monoxide loss in a slow and rate determining step and a fast elimination of ethylene, as indicated in scheme 1.

Finally, some comments should be made on cinnamyl alcohol, which has been studied extensively by EI in combination with D- and ^{13}C-labelling (16,17). This compound, an isomer of 3-phenylpropanal, behaves quite differently upon EI, although its molecular ions eliminate C_2H_2O and C_3H_4O as well (16,17). FI spectra of cinnamylalcohol at an unheated and heated emitter are given in Figure 7. Clearly these spectra are different from the FI spectra of 3-phenylpropanal (Figure 5). However, peaks at m/e 92 and m/e 78 are observed, especially at higher emitter temperature, although their abundances were too low to obtain reliable FIK results. Nevertheless, it may be concluded that small fraction of the molecular ions of cinnamyl alcohol may have isomerised to 3-phenylpropanal, possibly as follows:

$$\left[C_6H_5CH{=}CHCH_2OH\right]^{+\bullet} \rightarrow C_6H_5\dot{C}H{-}CH_2CH{=}\overset{+}{O}H \rightarrow \left[C_6H_5CH_2CH_2CHO\right]^{+\bullet}$$

f g h

Note, that ion g is identical to ion b in Scheme 2. It is interesting in this respect that the ratio of intensities of the metastable peaks in the first field free region for the loss of C_2H_2O and C_3H_4O upon EI are $\sim$150 for cinnamyl alcohol and $\sim$200 for 3-phenylpropanal. This would mean that in cinnamyl alcohol the elimination of C_3H_4O which requires a higher activation energy, can compete more effectively with the loss of C_2H_2O than in 3-phenylpropanal, if their decompositions in the same time window are considered. Ions h seem, therefore to be slightly more excited than those directly generated from 3-phenylpropanal. This can be explained, if the isomerization of cinnamyl alcohol proceeds through a slow step, presumably f $\rightarrow$ g, to eventually give a relatively higher excited 3-phenylpropanal ion. The eliminations of C_2H_2O and of C_3H_4O can then compete more effectively with each other in the first field free region. A similar phenomenon has been described recently. Fast unimolecular dis-

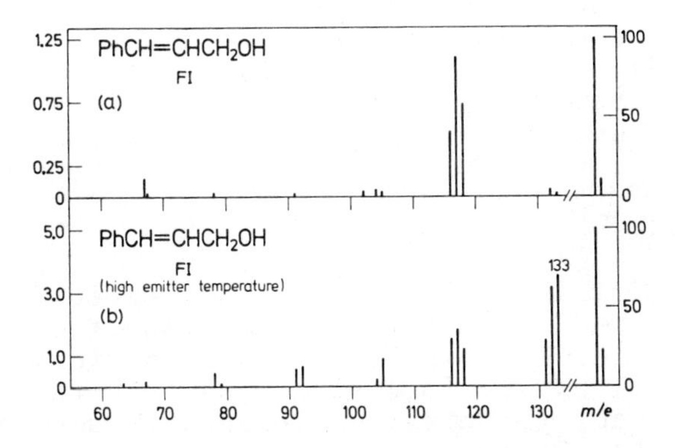

Journal of the American Chemical Society

Figure 7. FI spectra of cinnamylalcohol at an unheated (a) and heated (b) emitter. The left ordinate scale refers to the intensities of ions with m/e $<$ *134; the right ordinate scale to those of ions with* m/e $\geq$ *134.*

sociations of the (M-methyl) ion from methylisopropyl ether in the metastable region are attributed to a rate determining isomerization of this ion to the (M-methyl) structure from diethyl ether (18).

Conclusions.

The results, obtained from FIK for the systems described, show that this method is extremely valuable for obtaining a deeper insight in true gas-phase ion decompositions and in the long-standing problem of scrambling of hydrogen atoms. This latter phenomenon appears to consist of a series of specific hydrogen exchange processes, as noted previously by other authors (2a).

Experimental.

Experimental details will be described fully elsewhere (8,10) and the most important conditions are only given here. The FIK experiments were performed on a Varian MAT 711 double focussing mass spectrometer (Mattauch-Herzog geometry) equipped with a commercially combined EI/FI/FD source. Data acquisition and data processing were achieved with the Varian Spectro System 100 on line and special computer programmes, written for the FIK experiments, were used (19). Further experimental conditions for 2-phenoxyethyl chloride were:

Inlet system: reference inlet system at 130°C
Ion source temperature: 100°C
Emitter Current: 5 mA
Analyzer pressure: $6x10^{-9}$ Torr
Source pressure: $5x10^{-6}$ Torr
Mass resolution (10% valley difinition): 1200
Energy resolution (at an accelerating voltage of 8453 Volts): 1.4%
Voltage at focussing electrode: ±6 kV.

The experimental conditions employed for the 3-phenylpropanal studies were:
Inlet system: cooled direct insertion probe at -5 to -10°C
Ion source temperature: 98°C
Emitter current: 0 mA
Analyzer pressure: $6x10^{-9}$ Torr
Source pressure: $7x10^{-6}$ Torr
Mass resolution (10% valley definition): 500
Energy resolution (at an accelerating voltage of 8400 Volts): 1%
Voltage at focussing electrode: ±6 kV.

For both studies activated tungsten wire emitters of 10 µm with an average needle length of 10 µm were used and positioned 1.45 mm from the grounded slotted cathode.

Acknowledgement.

The author is extremely grateful to Drs. J. van der Greef and Mr. F. A. Pinkse, who, with great enthusiasm, have developed the FIK method in his laboratory. He further acknowledges the contributions of Drs. C. B. Theissling and Dr. P. Wolkoff Christensen in the chemical field and the contribution of Dr. C. W. F. Kort who developed the software for FIK. Finally, he would like to thank Prof. H. D. Beckey and his group of the University of Bonn for stimulating discussions and the gift of some well-activated emitters.

Abstract.

Field Ionization Kinetics has provided a clear insight into the positional interchange of the phenoxy group and halogen atom in the molecular ions of 2-phenoxyethyl chloride and into the various hydrogen exchange processes in the molecular ions of 3-phenylpropanal prior to or during fragmentation. This information could not be obtained from previous electron impact studies.

Literature Cited.

1. a. Beckey, H. D., Field Ionisation Mass Spectroetry, (1971), Akademie-Verlag, Berlin and Pergamon Press, Oxford. b. Robertson, A. J. B., in "Mass Spectrometry", Maccoll, A. (Ed.), Int. Rev. Sci., Physical Chemistry Ser. One, Vol. 5, Butterworths, London and University Park Press, Baltimore, (1972), p. 103. c) Derrick, P. J., in "Mass Spectrometry", Maccoll, A. (Ed.), Int. Rev. Sci., Physical Chemistry Ser. Two, Vol. 5, Butterworths, London and Boston, Mass., (1975), p. 1.
2. a. Derrick, P. J. and Burlingame, A. L., Acc. Chem. Res., (1974) 7, 328. b. McMaster, B. N., Johnstone, R. A. W. (Ed.), in Mass Spectrometry (Specialist Periodical Reports), The Chemical Society, London, (1975), Vol. 3, p. 33. c. Burlingame, A. L., Kimble, B. J., and Derrick, P. J., Anal. Chem., (1976) 48, 368R.

3. a. Falick, A. M., Derrick, P. J., and Burlingame, A. L., Int. J. Mass Spectrom. Ion Phys., (1973) 12, 101. b. Beckey, H. D. Hey, H., Levsen, K. and Tenschert, G., Int. J. Mass Spectrom. Ion Phys., (1969) 2, 101.
4. Schulten, H.-R. and Beckey, H. D., Org. Mass Spectrom., (1972), 6, 885.
5. Falick, A. M., Int. J. Mass Spectrom. Ion Phys., (1974) 14, 313.
6. Derrick, P. J., Falick, A. M. and Burlingame, A. L., J. Am. Chem. Soc., (1972) 94, 6794.
7. Borchers, F., Levsen, K. and Beckey, H. D., Int. J. Mass Spectrom. Ion Phys., (1976) 21, 125.
8. Greef, J. van der, Theissling, C. B., and Nibbering, N. M. M., 7th International Mass Spectrometry Conference, Florence, Italy, 1976, Paper No. 168; will appear in Advan. Mass Spectrom. 7 Daly, N. (Ed.).
9. Theissling, C. B., Nibbering, N. M. M., and Boer, Th.J. de, Advan. Mass Spectrom., (1971) 5, 642.
10. Christensen, P. Wolkoff, Greef, J. van der, and Nibbering, N. M. M. J. Am. Chem. Soc. (submitted)
11. Venema, A., Nibbering, N. M. M., and Boer, Th.J. de, Org. Mass Spectrom., (1970) 3, 583.
12. Venema, A. and Nibbering, N. M. M., Org. Mass Spectrom., (1974) 9, 628.
13. Resink, J. J., Venema, A., and Nibbering, N. M. M. Org. Mass Spectrom., (1974) 9, 1055.
14. Nibbering, N. M. M. and Boer, Th.J. de, Tetrahedron, (1968) 24, 1427.
15. Kingston, D. G. I., Bursey, J. T., and Bursey, M. M., Chem. Rev., (1974) 74, 215.
16. Schwarz, H., Köppel, C. and Bohlmann, F., Org. Mass Spectrom., (1973) 7, 881.
17. Köppel, C. and Schwarz, H., Org. Mass Spectrom., (1976) 11, 101.
18. Hvistendahl, G. and Williams, D. H., J. Am. Chem. Soc., (1975) 97, 3097.
19. Greef, J. van der, Pinkse, F. A., Kort, C. W. F. and Nibbering, N. M. M., Int. J. Mass Spectrom. Ion Phys., in press.

RECEIVED December 30, 1977

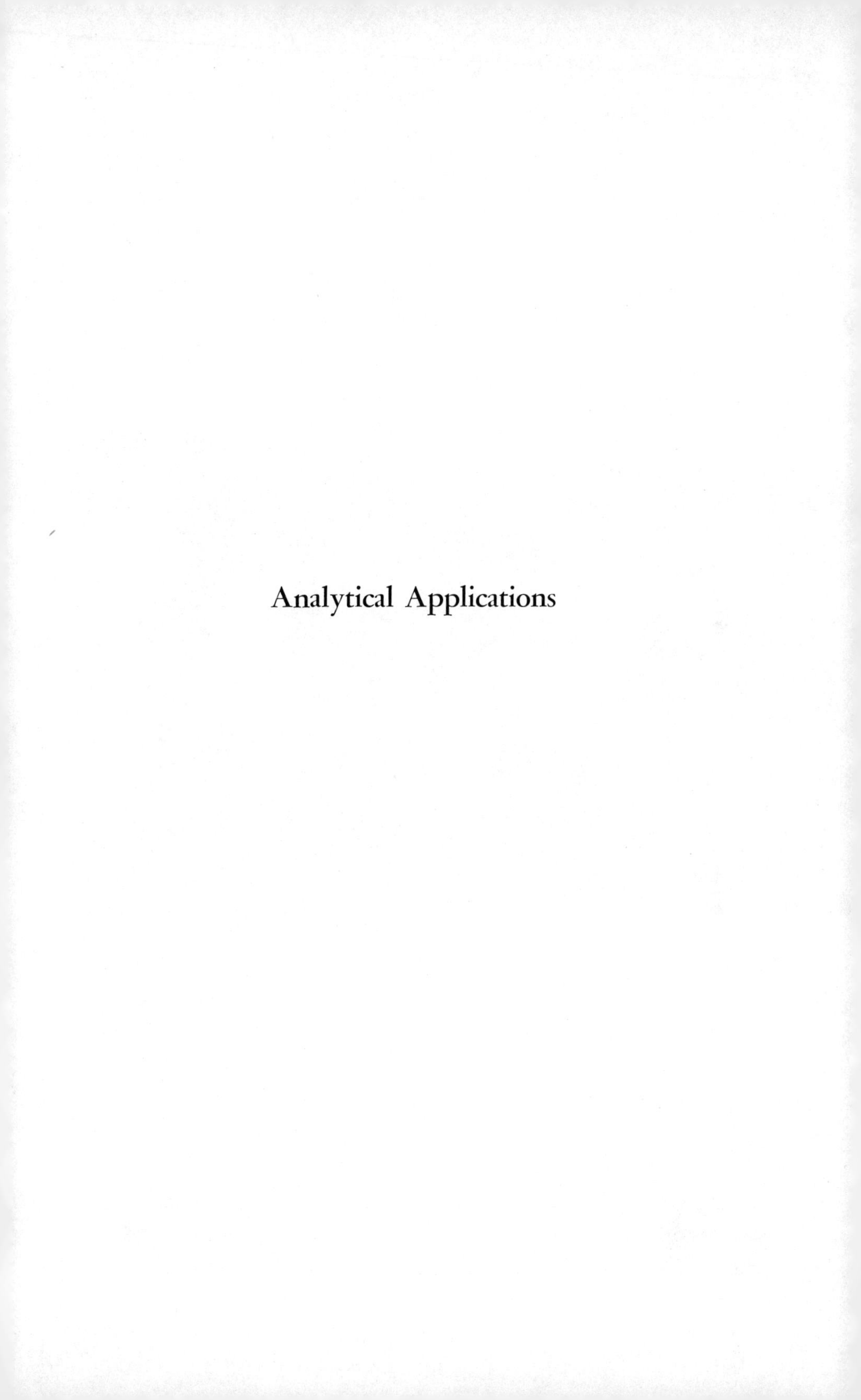

Analytical Applications

Organic Trace Analysis Using Direct Probe Sample Introduction and High Resolution Mass Spectrometry

WILLIAM F. HADDON

Western Regional Research Center, U.S. Department of Agriculture, Albany, CA 94710

Many interesting problems in organic trace analysis are inaccessible to combined gas chromatography-mass spectrometry (GC/MS). For example, a recent treatise on the analysis of organic pollutants in water points out that between 80 and 90 percent of the extractable organic compounds in polluted water fail, even after derivitization, to pass through a gas chromatographic column (1). One approach for introducing less volatile compounds into the mass spectrometer in a highly purified state is the use of a liquid chromatograph coupled directly to the mass spectrometer (2,3,4). However, as an alternative to utilizing chromatographic separation prior to analysis, we can consider the mass spectrometer itself as an analytical instrument which is capable in itself of performing highly efficient separations, based, for example, on mass or energy differences of generated ions or on selective ionization, as well as precise identification and quantitation, based on reference mass spectra obtained separately on known compounds. This use of a mass spectrometer for combined separation-identification functions might be called "mass spectrometer-mass spectrometer" (MS/MS) analysis, with the direct probe serving as the vehicle for sample introduction. A number of laboratories have reported results using this approach, and the methods used to achieve selectivity in complex mixtures have included field ionization (5), chemical ionization (6,7), negative chemical ionization (8,9), collisional activation (10), mass-analyzed ion kinetic energy (11), and high resolution electron ionization (EI) (12-15) techniques. This report describes the use of the high resolution EI method in which particular ions are monitored as a function of temperature as the components of interest volatilize

© 0-8412-0422-5/78/47-070-097$10.00/0

from the direct introduction probe. The term high resolution selected ion monitoring (SIM) is used to describe this type of experiment (16), and the ion abundance profile recorded as a function of temperature is called an exact mass fragmentogram.

In many organic trace analysis problems, the preparation of highly purified samples amenable to low resolution mass spectral identification requires extensive effort. Employing increased mass resolution to achieve higher selectivity can reduce the extent of sample purification significantly in many cases, and recent examples of GC-high resolution SIM for both biofluid analysis (17,18) and environmental monitoring (19) are illustrative.

In many laboratories concerned with analyzing particular components of complex mixtures, the savings in the cost of sample preparation made possible by the availability of high resolution SIM capability may justify the higher initial cost of a suitable high resolution mass spectrometer. Furthermore, small radius (6-inch or less) magnetic sector mass spectrometers having both the fast scan rates required by GC/MS applications and the capability for high resolution (10-20,000) operation with good sensitivity are becoming available commercially. On at least one small radius instrument of this type, image curvature correctors have been employed to enhance the high resolution performance, and further improvements in resolution and sensitivity for these small radius instruments will undoubtedly be forthcoming.

Two examples described below from the fields of clinical and food chemistry illustrate the utility of the high resolution direct probe method for trace analysis. In addition, pesticides in human tissue have been measured at high resolution on a photo-plate equipped mass spectrometer (15), and similar high resolution methods have been employed to measure pesticides in food samples (9) and biofluids (8) using negative chemical ionization, and to quantitate drug metabolites rapidly in blood plasma (7) using chemical ionization mass spectrometry. These papers contain excellent descriptions of the practical problems of preparing samples and quantitating the results.

Experimental

High resolution SIM operation can be achieved on a magnetic sector mass spectrometer by incorporating a switched voltage divider circuit, or by employing computer-driven power supplies.

The over-all electronic stability necessary for quantitative work depends on the precision required and on the mass resolution selected for the analysis. These stability requirements can be derived from the Gaussian curve of Figure 1, which represents a mass spectral peak. The 10 percent valley definition of mass spectrometer resolution corresponds to the ordinate value at a distance 2.44 σ from the center of the peak (5 percent of maximum height for a single peak), where σ is the standard deviation of the curve. The over-all instrument stability, S, in parts per million (ppm) for a particular analysis is

$$S = (\sigma_x/2.44)(10^6/R) \quad (1)$$

where σ_x is the number of standard deviations to the ordinate at the required percent accuracy, and R is the mass resolution (M/ΔM, 10 percent valley definition)

Table I. Mass Spectrometer Stability in Parts per Million Necessary for Quantitative High Resolution Selected Ion Monitoring.[a]

Resolution[b]	Desired Quantitative Accuracy (Percent):			
	99	95	90	80
3,000	16.	45.	61.	89.
5,000	9.8	27.	36.	53.
10,000	4.9	13.	18.	26.
15,000	3.3	9.0	12.	18.
20,000	2.4	6.7	9.2	13.
50,000	.98	2.7	3.6	5.3

[a]Calculated from Eq. 1.

[b]M/ΔM based on 10 percent valley definition.

Table I lists typical calculated values of S. These values show that quantitative precision of 90 percent or greater should be easily achieved on most commercial mass spectrometers which have been designed for accurate mass measurement applications.

When only one or two peaks need to be monitored, the circuitry of an existing peak matcher can be modified to monitor both peaks by switching between peak tops. (7,19) On one widely used commercial high resolution instrument, this is done by disconnecting the peak matcher scan coils (7). Alternatively, repetitive scans over narrow mass ranges can be utilized, but at reduced sensitivity. An advantage of this latter dynamic method of recording data is that the peak profile information is continuously available during the experiment.

An increase beyond two channels can be achieved without a computer system by incorporating a switchable voltage divider circuit (13) (20) (21). It is important for a circuit of this type to employ sample-and-hold amplifiers for each channel to provide continuous output signals suitable for driving an oscillographic or multiple pen recorder.

Computer-Controlled Power Supplies for High Resolution SIM. A dedicated computer can perform both data collection and instrument control functions for high resolution SIM when used in conjunction with a programmable power supply for the mass spectrometer accelerating and electric sector voltages. The computer provides some additional benefits over a hard-wired system: The number of mass channels can be large and can be changed during the analysis; the fraction of time at a particular mass value can be varied according to the expected abundance of the peak; and the system, when suitably programmed, can sweep short mass regions to provide peak profile information which can be used both to ascertain the degree of interference from isobaric ions near the elemental composition of interest, and to make automatic periodic adjustments of the accelerating voltage to compensate for drift in voltages.

Figure 2 shows the programmed power supply circuitry used in our laboratory for high resolution SIM on a CEC 21-110A double-focussing mass spectrometer. Many of the electronic and programming details, including choice of suitable power supply units, are from the Washington University system of Holmes (22) for low resolution SIM. In the USDA system a single programmer (D/A converter) with a 1-volt output drives a pair of power supplies in series, one with a voltage gain of 100 (KEPCO OPS-2000, Kepco, Inc., Flushing, N.Y.) followed by a second with a gain of -1 (KEPCO NTC-2000). These power supplies directly replace the electric sector supply formerly used with the instrument. The 21-110 requires plus and minus 375 volts for nominal 8 kv. operation. We derive a fixed portion of this voltage from the internal reference supply of the OPS-2000 through R_2 and R_3, and a variable portion from the D/A converter (DATEL DAC-169, 16 bit precision, Datel Systems, Inc., Canton, Ma.) through R_1. The

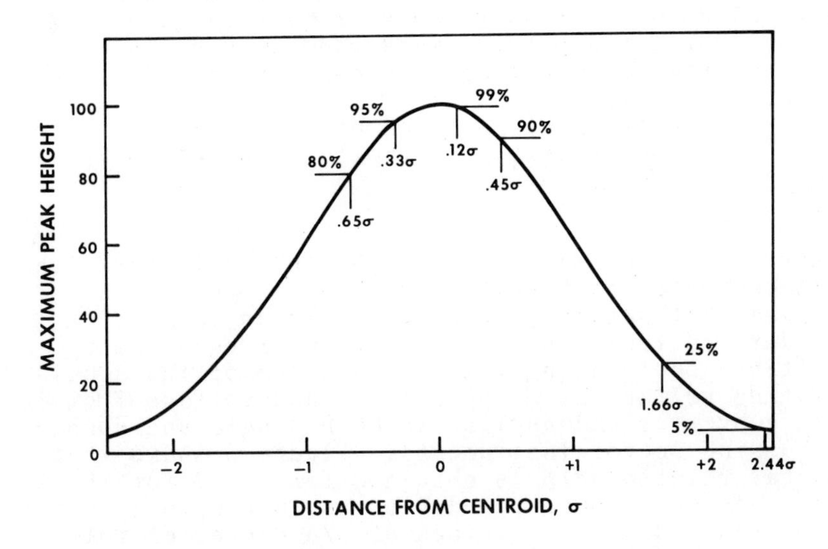

Figure 1. Gaussian model for mass spectral peak

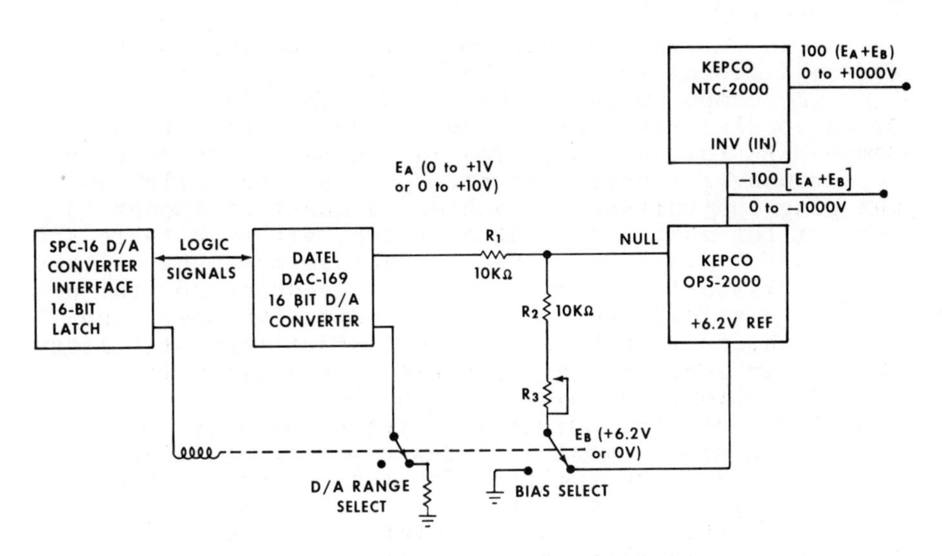

Figure 2. Programmed power supply interface for CEC 21-110 mass spectrometer

setting of R_3 determines the fraction of fixed voltage according to the requirements of mass range and precision. For a ten percent mass range, each digital step of the programmed voltage corresponds to a 1.5 ppm change in mass.

The relationship between mass and voltage output for programmed accelerating voltage operation is

$$\frac{1}{m} = \frac{1}{m_o} + k\ V_{D/A} \qquad (2)$$

where m_o is the mass in focus at the collector with the D/A converter at 0 volts, and $V_{D/A}$ is the voltage, or digital bit setting, required to focus mass m. Eq. 2 is derived from the mass spectrometer equation by writing accelerating voltage as a sum of the accelerating voltage at m_o plus the added voltage from the D/A converter and applies to both single and double focussing sector instruments. Figure 3 shows that a linear relationship is obtained for Eq. 2 for the 21-110 mass spectrometer modified as in Figure 2 over the mass range 293-331. Values of D/A converter voltage which focus particular exact masses can be determined by locating internal reference peaks, typically derived from perfluorokerosene (PFK), and utilizing them to derive the slope and intercept by least squares fit to Eq. 2. Alternatively, the spectrometer can be set to the center of a known peak, which gives m_o directly, and k can be calculated by manually locating one or more additional reference peaks as $V_{D/A}$ is varied.

The computer can be programmed to adjust the applied accelerating voltage during an experiment to compensate for instabilities in the mass spectrometer by sweeping a single reference peak periodically, and changing the voltage at each m/e a constant amount to correct for shifts in peak centroid, although there are small errors in this procedure because mass and voltage are not linearly related (Eq. 2). Feedback control of this type allows the use of less expensive power supplies, which often have excellent short term stability but poorer long term or temperature stability (22).

Microprocessor-based SIM modules are becoming available, and one commercial system suitable for high resolution measurement includes a peak stabilizing capability and the facility for monitoring up to 8 m/e-values (23). As with all SIM circuits limited to peak switching, the one drawback relative to a computer-based system is that peak profile information is not easily obtained. A solution to this problem is to devote three channels of the peak selector to the same peak, one at the center, and two at the 10 percent

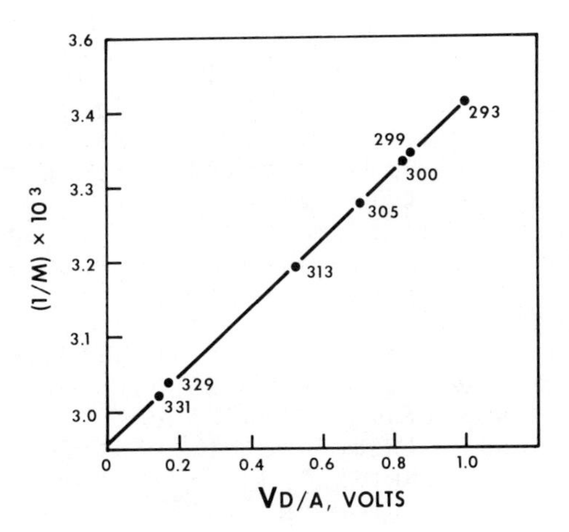

Figure 3. Mass-voltage relationship for programmed accelerating voltage operation of 21-110 mass spectrometer. M = m/e *value focused at collector.* $V_{D/A}$ = *voltage output of D/A converter.*

intensity points on the high and low mass side of the peak. Three exact mass fragmentograms coincident in time (temperature) and having 1:10:1 intensity ratios confirm the absence of interfering isobaric ions during the analysis (24).

Applications

Aflatoxin Analysis.

In our laboratory we have measured small amounts of aflatoxins in foods and feeds, and in the biofluids of experimental animals (25-27). Aflatoxins are metabolites of the mold *Aspergillus Flavus*, and one of the metabolites, aflatoxin B_1, is of particular interest because it is a naturally occurring carcinogen which is sometimes present at up to several parts-per-billion (ppb) in some agricultural products (28-30). The structures and partial EI mass spectra of six aflatoxins are shown in Figure 4, and the mass spectra of about 50 additional mycotoxins are available elsewhere (31). Purification and subsequent transfer of aflatoxins into the mass spectrometer is difficult because the compounds are unstable when highly purified, bond strongly to glass surfaces, and cannot be separated by gas chromatography. These factors restrict the use of low resolution mass spectral analysis for structurally specific verification or quantitation of aflatoxins. The use of high resolution SIM on partially purified aflatoxin containing mixtures introduced *via* the direct probe significantly lowers the limits of detection for mass spectral analysis of these compounds, and has the additional advantage that prior isolation procedures can be simplified (12).

By using high resolution SIM we have detected aflatoxins B_1 and M_1 in experimental milk samples at 19 and 140 ppb, respectively. An analytical extraction scheme for obtaining partially purified aflatoxins from freeze-dried milk and suitable for use in conjunction with high resolution mass spectral analysis is shown in Figure 5. Extract A contains all of the aflatoxins. In extracts B and C, which are more highly purified, aflatoxins B_1 and M_1 are partially separated because of differences in partition coefficients in the extraction solvents (32).

High resolution SIM data for extracts A and C are shown in Figures 6 and 7. These data of intensity *vs.* mass represent scans over 0.3 amu for each appropriate integral m/e at mass resolution of 7,000. The middle scan for extract A (Fig. 6), which was recorded at the temperature of maximum rate of volatilization of the aflatoxins, shows the M_1 molecular ion at m/e 328, the

M-16 ion of M_1 and molecular ion of B_1 (same elemental composition, $C_{17}H_{12}O_6$) at m/e 312, and the M-29 ion of M_1 at m/e 299. The abundance ratio $[M]^{+\cdot}/[M-29]^+$ agrees within experimental error with the ratio of these peaks in the reference low resolution mass spectrum of aflatoxin M_1 shown in Figure 4 and confirms the absence of interference at a resolution of 7000 for this particular extract. Figure 7 gives equivalent data for extract C obtained from the same sample of freeze-dried milk. The additional peak at m/e 314 is from 12 ng aflatoxin B_2, added as an internal standard. The abundance ratios between aflatoxin ions at m/e 312 and 328 in these scans reflect the decreased amount of aflatoxin B_1 for extract C which resulted from solvent partitioning during extraction. Thus for extract A (Fig. 6) the intensity ratio of $C_{17}H_{12}O_6$ to $C_{17}H_{12}O_7$ is 0.32, compared to 0.06 for pure M_1 (see Fig. 4e), but for extract C (Fig. 7) the ratio is 0.092, which indicates less aflatoxin B_1, as expected. The methodology to quantitate the compounds is currently being developed.

A notable feature of the direct probe method for aflatoxin analysis is an enhancement of sensitivity which occurs when partially purified samples of aflatoxin are run on the mass spectrometer, relative to measurements on highly purified samples. Apparently, this is because bonding to the surface of the sample containers is substantially reduced in a complex mixture for these compounds. The data of Figure 8 illustrate this enhancement of sensitivity for aflatoxin B_1 added to an extract of pooled human urines. The curves of Fig. 8 are exact mass fragmentograms, recorded during rapid temperature programming of the direct probe from about 50°C to 250°C, with the mass spectrometer tuned to the molecular ion of B_1, m/e 312.0635, at 7000 resolution. A comparison of the response obtained in curve A of Fig. 8 for 1.3 ng pure B_1 with that of curve D, obtained from 0.03 ng B_1 added to the urine extract, illustrate the 100-fold enhancement of sensitivity for detecting aflatoxins B_1 in a complex mixture. The sensitivity enhancement is somewhat greater for aflatoxin M_1, for which negligible response is obtained for 30 ng of purified compound, compared to a signal to background ratio of better than 100 for 13 ng introduced in a mixture, as shown in Figs. 6 and 7.

Quantitative High Resolution SIM for Analyzing Purines in Blood and Tissue.

Quantitative direct probe measurements at the 2-100 ppm level for five purine components of human blood

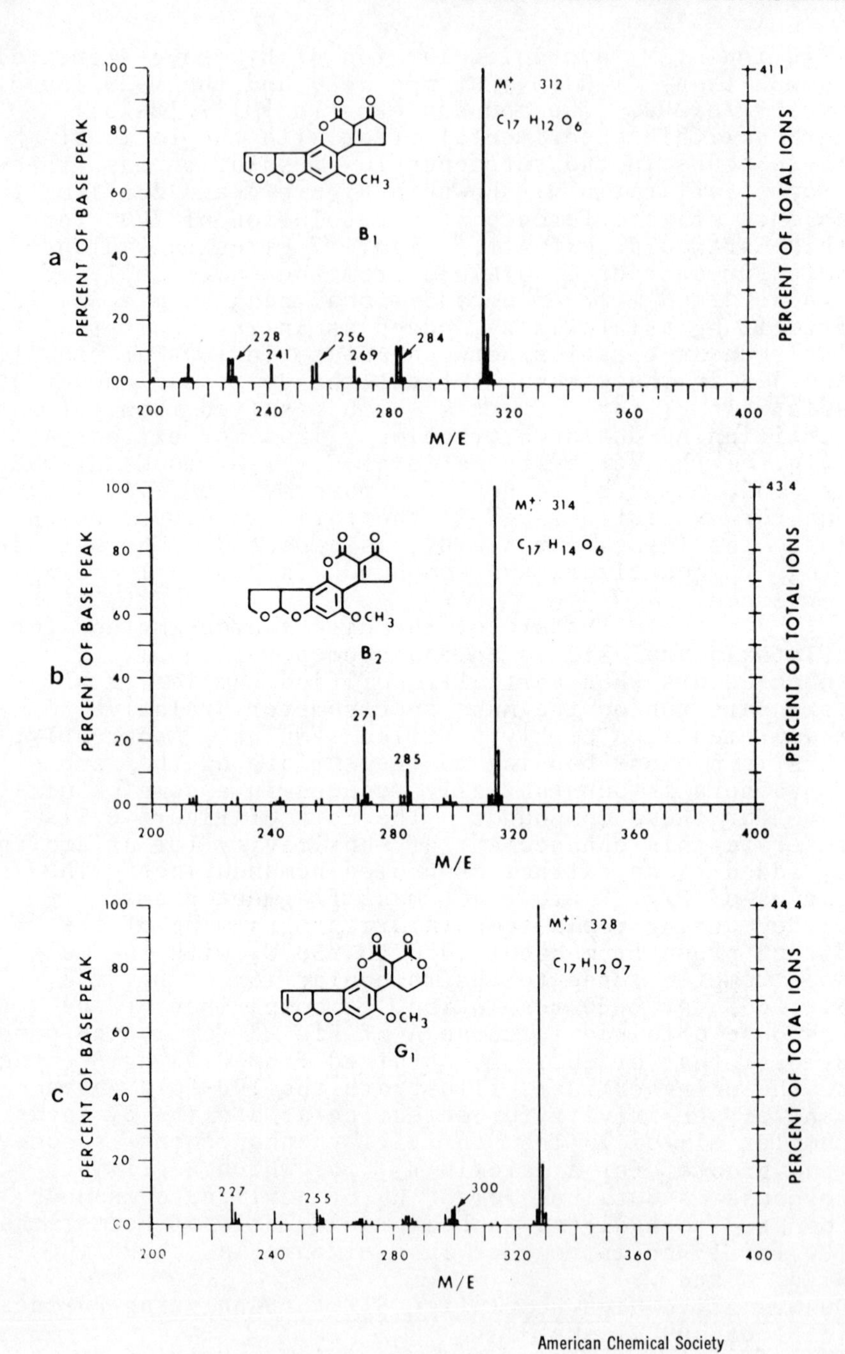

American Chemical Society

Figure 4a, b, c. 70 eV mass spectra of aflatoxins

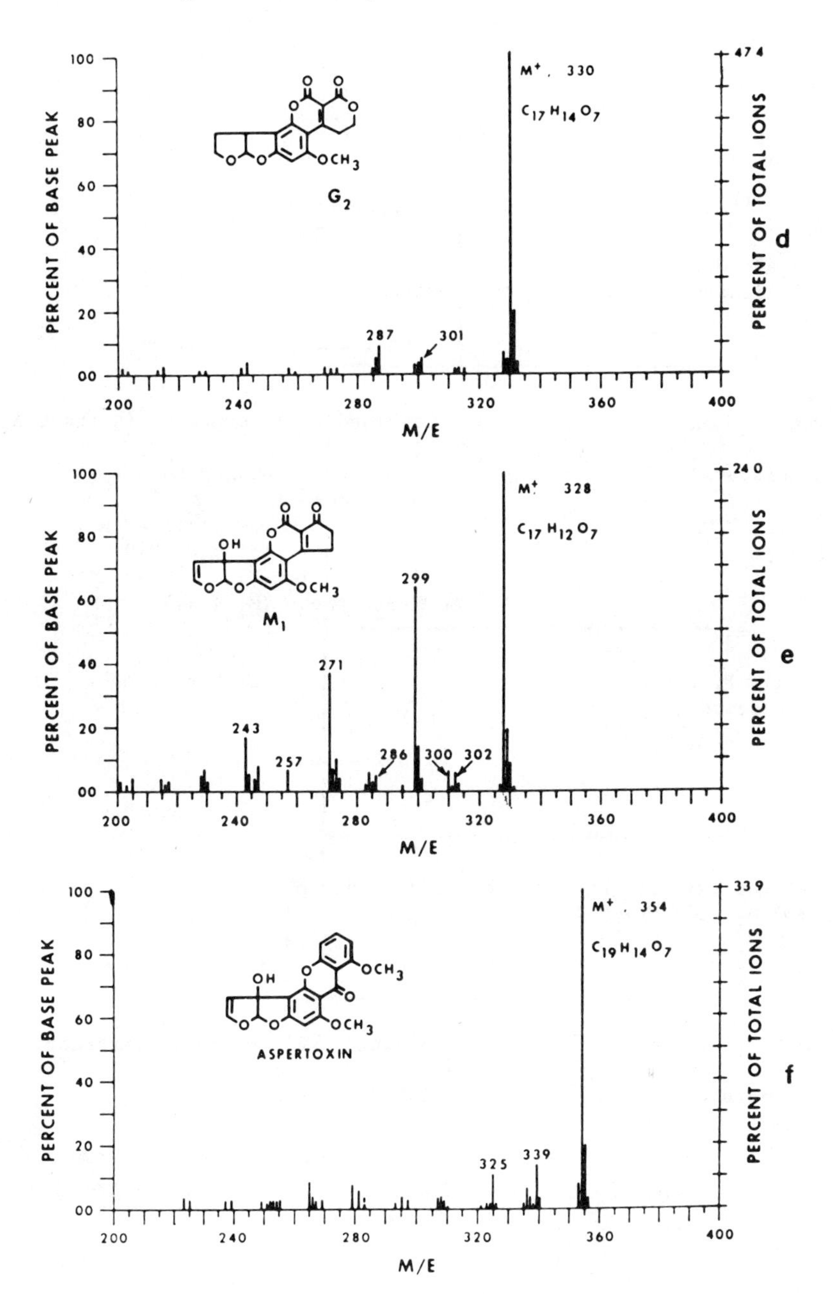

American Chemical Society

Figure 4d, e, f. 70 eV mass spectra of aflatoxins

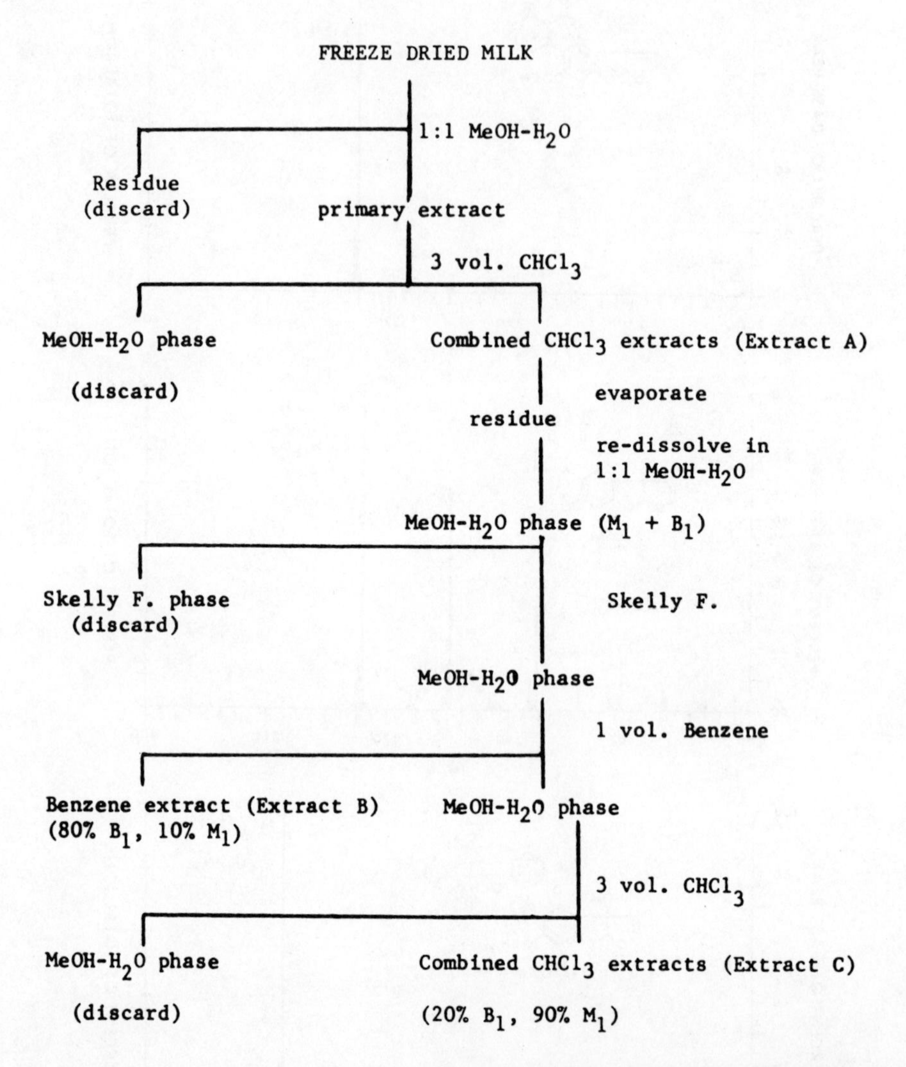

Figure 5. Analytical extraction of aflatoxins B_1 and M_1 from freeze-dried milk

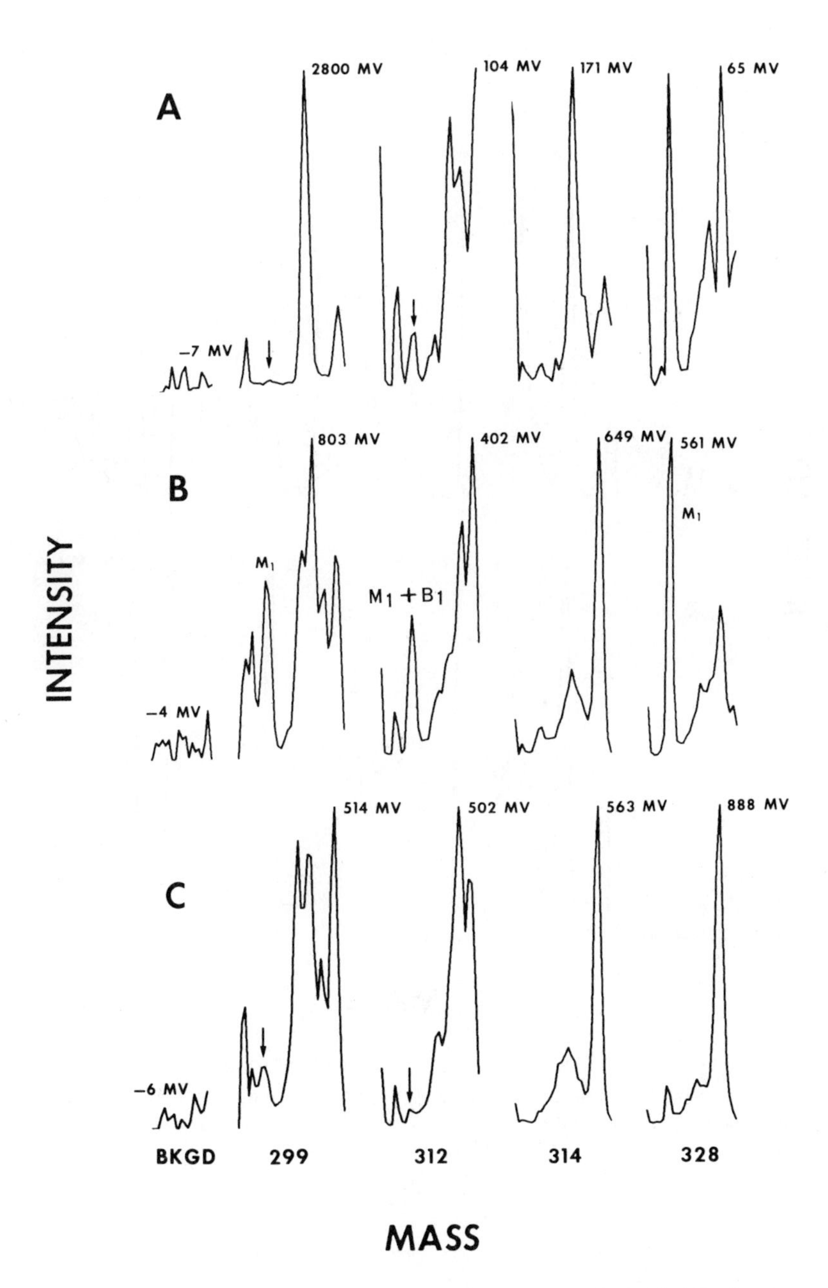

Association of Official Analytical Chemists, Inc.

Figure 6. High resolution SIM scans of intensity in millivolts (mV) vs. mass for 13 ng aflatoxin M_1 and 1.7 ng aflatoxin B_1 in 100μg dried milk extract A. Direct introduction probe temperature increases from A to C.

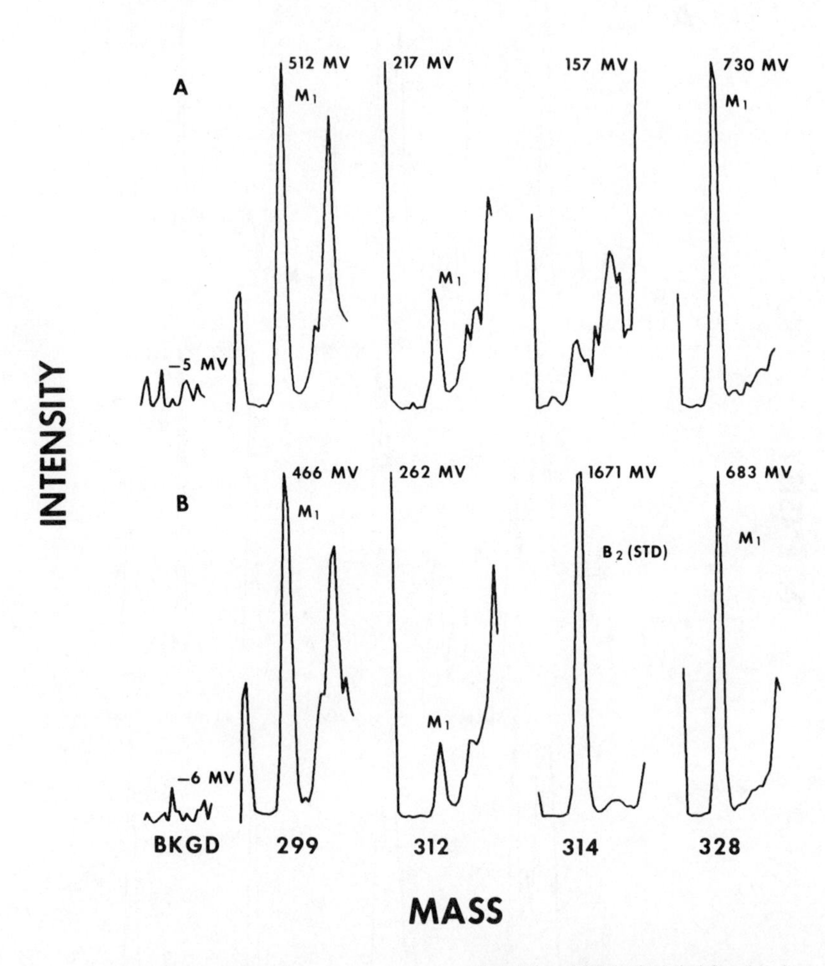

Association of Official Analytical Chemists, Inc.

Figure 7. High resolution SIM scans of intensity in millivolts (mV) vs. mass for extract C. 12.4 ng aflatoxin B_2 (mass 314) added as internal standard. Direct probe temperature increases from A to B.

and tissue have been reported by Snedden and Parker using high resolution SIM (13) (33). The high resolution mass spectral analysis for the purines provided quantitative measurement of hypoxanthine (I), xanthine (II), uric acid (III), allopurinol (IV) and purinol (V) (see Figure 9), which are important in the study and treatment of gout. The use of high resolution mass spectrometry made possible accurate analysis at these levels without pre-purification of the samples. More recently the same group has used high resolution mass spectrometry to quantitate oestrogen and progesterone in human ovarian tissue at 10-50 ppm, again using procedures that require no chemical separation or derivitization prior to mass spectral analysis (14).

To insure that their method is valid, the authors recorded a high resolution spectrum of the mixture to be analyzed at the chosen resolution, in this case 20,000, and compared the abundances of characteristic mass peaks in the experimental sample with those in the high resolution reference spectrum of the pure compound. Figure 10 illustrates the excellent agreement obtained at five of the six masses examined for III in human muscle tissue. Similar experiments with the other purines of Figure 9 yielded a set of masses appropriate for analyzing I-V together.

Integrating the selected ion profiles (exact mass fragmentograms) at five masses as a function of temperature gave a set of relative peak areas. These peak areas were related to the relative amounts of I-V using a set of five linear simultaneous equations and response coefficients developed from runs on purified compounds, according to established procedures of quantitative mass spectrometry (34). The introduction of accurately weighed (mg amounts) samples facilitated the calculation of absolute concentrations.

An evaluation of the sensitivity and measurement error for the analysis reveals that different methods of sample preparation give different quantitative precision. Fig. 11 shows that the best results were obtained using powdered samples of dessicated blood and muscle tissue. Depositing the samples on the direct probe surface (gold) by evaporation of an aqueous solution gave significantly lower precision for a given sample amount, apparently because of interactions between the sample and the surface of the probe for the more polar compounds. When a less polar substance, caffeine, was analyzed, there was no dependence of precision on the method of preparing the sample, as shown in Fig. 11.

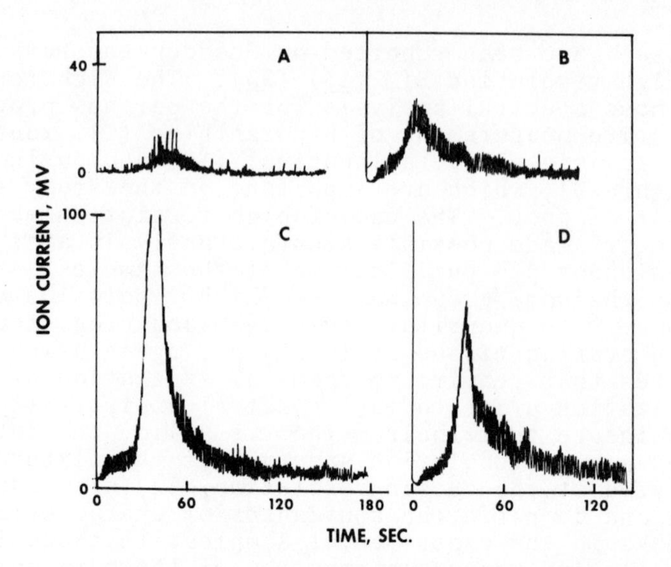

Association of Official Analytical Chemists, Inc.

Figure 8. Exact mass fragmentogram for $C_{17}H_{12}O_6$ (aflatoxin B_1 molecular ion, m/e *312.0635): A, 1.65 ng B_1 from reference solution; B, blank, extract B; C, urine extract plus 0.13 ng B_1; D, urine extract plus 0.030 ng B_1.*

American Chemical Society

Figure 9. Structures of purines analyzed by high resolution SIM: I, hypoxanthine; II, xanthine; III, uric acid; IV, allopurinol; and V, oxipurinol.

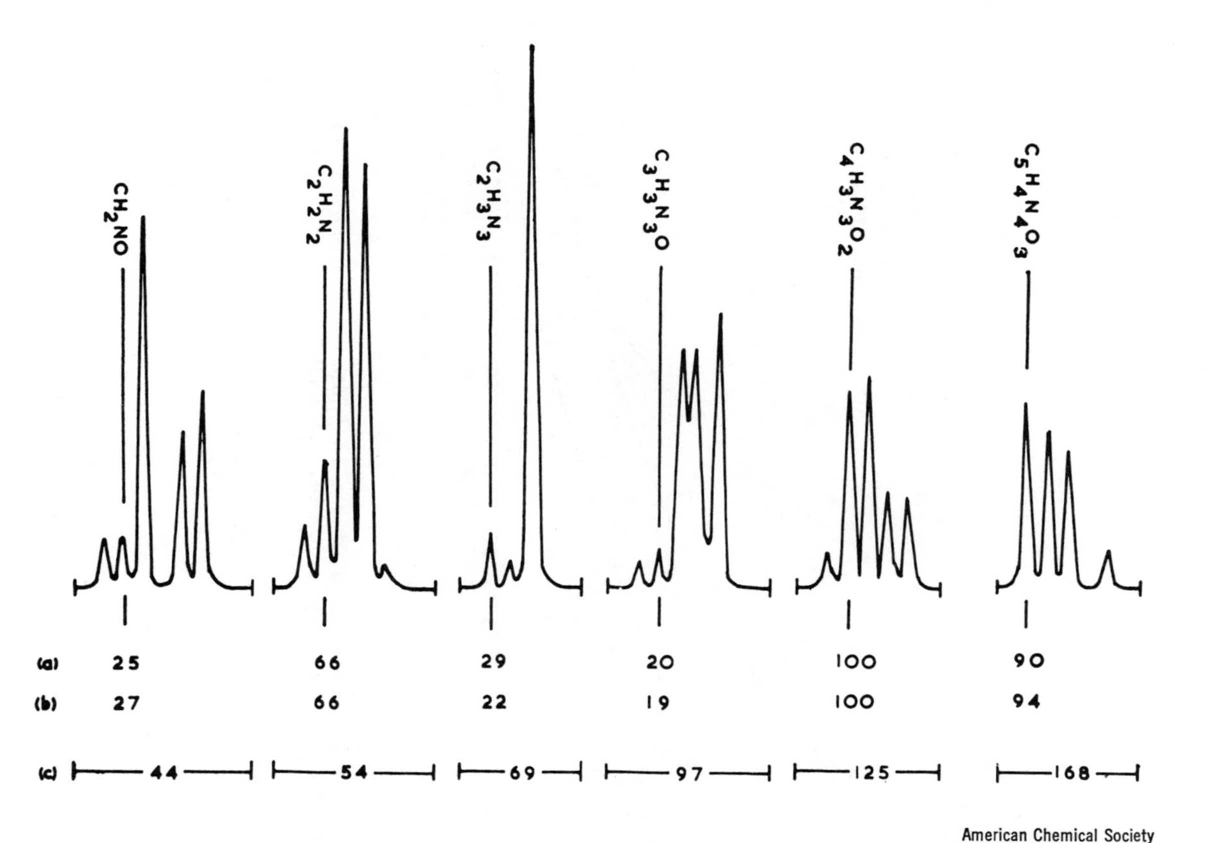

American Chemical Society

Figure 10. Partial high resolution mass spectrum of substances evaporated from human muscle, showing peaks attributable to uric acid. (a) Relative intensities of uric acid peaks from muscle; (b) relative intensities of same peaks from uric acid; (c) nominal mass of multiplet.

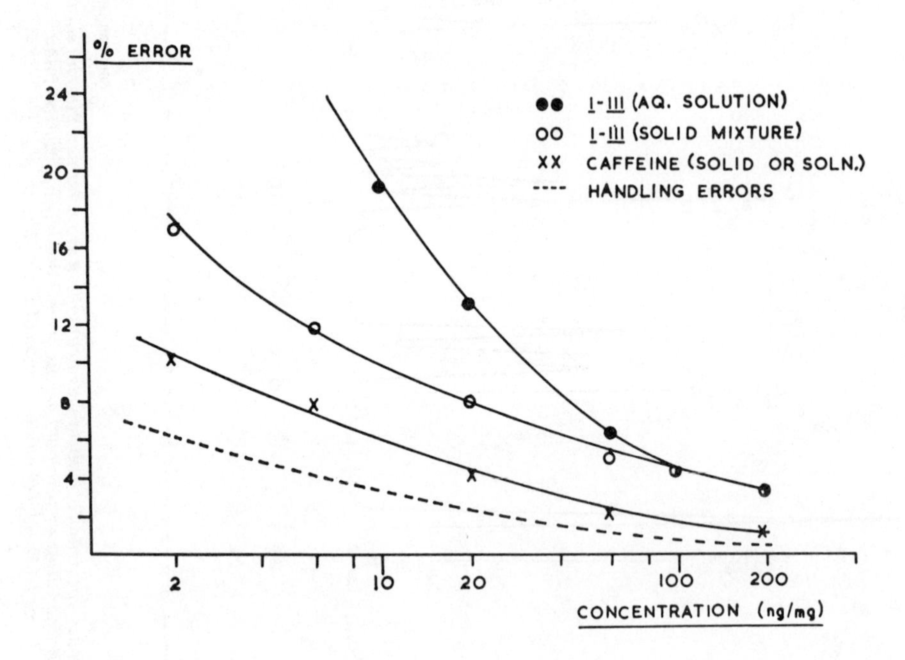

American Chemical Society

Figure 11. Dependence of analytical precision of high resolution SIM results on concentration and diluent for caffeine, hypoxanthine, xanthine, and uric acid in solution and solid mixtures

Discussion

The mass spectrometer is well suited to trace analysis of the type discussed here for two reasons: The process of ion production by the mass spectrometer is generally linear over many orders of magnitude of sample concentration; and the selectivity of the method can be extremely great when either high mass resolution, a selective ionization method, or a combination of these are employed to effect separation between different components of mixtures. The wide dynamic range necessary for quantitation is well accomodated by electron multiplier detection. Photoplate detection will be generally less suitable for trace analysis problems of this nature because of extensive blackening of the photographic plate near intense lines, and non-linearity of response.

The aflatoxins are favorably analyzed by high resolution SIM in complex mixtures in part because of their low ratio of hydrogen to carbon, which leads to an exact mass value below the masses of other substances, primarily lipids, typically present in food extracts. Many other environmental pollutants including halogenated pesticides and many of the common mycotoxins (35) have comparable or greater mass defects and should be amenable to analysis by this method. We have frequently observed peaks below the exact mass positions of the aflatoxins, as shown for the scans at m/e 299 and 312 of Figure 6. These peaks may arise from halogenated pesticide contaminants. High resolution SIM may be a valuable complement to GC/MS and to negative chemical ionization techniques (8,9) for such compounds.

Surface effects. Interactions with the surface of containers used to introduce samples *via* the direct probe can affect detection limits adversely, and an understanding of surface effects appears to be important for fully utilizing the method in trace analysis. In a detailed study of surface interactions for small peptides, Friedman (36-38) describes the rate of volatilization (dn/dt) of a pure compound from a surface by the equation

$$\frac{dn}{dt} = N_o A e^{-E/RT} \qquad (3)$$

where N_o is the number of sample molecules initially on the surface, A is a frequency factor, and the exponential term $e^{-E/RT}$ gives the fraction of molecules at temperature T with sufficient energy to overcome surface bonding. E is the activation energy for removing a molecule from the surface. A high rate of volatilization is realized by increasing N_0, for example by dispersing the sample by evaporation from solution and by reducing E, by using an inert direct probe surface.

The enhanced sensitivity for aflatoxins in mixtures that we observe is consistent with Friedman's model for surface volatilization. Apparently other compounds codeposited from solution along with the aflatoxins either occupy active sites on the sample tubes preferentially, or shield the aflatoxins from them, resulting in more efficient volatilization arising from a reduction in E. The curves of measurement error for the purines (Fig. 11), which indicate lower precision for more polar compounds, are suggestive of surface interactions, but the interpretation is less obvious, because dispersing the compounds over the gold surface of the sample probe actually lowered the precision of the analysis. The presence of high concentrations of inorganic salts probably accounts for this effect for the purines. In Friedman's study of small peptide volatilization, traces of inorganic salts lowered the rate of volatilization substantially at a given temperature (38).

The surface effect problem is an important area for further study. It has been suggested recently that a reduction in surface bonding explains the spectra obtained from "non-volatile" compounds absorbed on activated (surface-prepared) tungsten emitter wires using field desorption mass spectrometry (39,40). Minimizing surface interactions appears promising as a general method to enhance sensitivity for mass spectral analysis of less volatile substances. Of course, it is important to recognize that the chemical instability of a particular substance may be significantly decreased relative to its stability in a solution or solid sample matrix; for example, aflatoxins dispersed on a surface show greatly enhanced reactivity to UV light.

Ion Source Contamination. Contamination of the ion source, which may occur rapidly in some cases, can appreciably reduce the sensitivity of the method. This factor must be weighed with the advantages of using the direct probe when alternative GC/MS procedures are applicable. The rapidity of source contamination varies widely in the few examples reported. Snedden and Parker experienced less than 20 per cent decline in sensitivity in six months of purine analysis (13). Conversely, Dougherty (8) found that sensitivity for polychlorinated biphenyl (PCB) detection was severely reduced after running ten samples of unpurified human urine extract. Minimal solvent extraction and column chromatography prior to direct probe analysis solved this problem, as it has in our laboratory for aflatoxin analysis. Other approaches to reducing ion source contamination may be applicable in some cases. For example, it should be possible to mechanically vent the volatilized sample away from the ionization chamber until the temperature of volatilization for the compounds of interest is reached (see Fig. 8), thereby reducing the quantity of material which enters the ion source.

Chemical Binding Effects. The chemical state of the sample can affect its volatility. A comparison of the quantitative mass spectral measurements for uric acid in blood plasma with the concentration values obtained by enzymatic analysis suggests that the mass spectrometer detects only the uric acid which is bound loosely to blood proteins, and not the uric acid in solution, which may exist primarily as the sodium salt. These data, which must be considered preliminary, suggest that mass spectral analysis, when performed without prior sample treatment, may indicate indirectly the extent of chemical binding for small molecules in some cases.

Conclusions

High resolution selected ion monitoring measurements on samples introduced on the direct probe can provide accurate and highly specific qualitative and quantitative information about compounds present in trace amounts. This mixture analysis method is applicable to some compounds which cannot be separated by gas chromatography because of low volatility or chemical instability. The highest sensitivities are obtained for substances having large mass defects and this embraces a number of classes of compounds, including halogenated pesticides and mycotoxins, which are of particular interest in environmental and food chemistry. Magnetic sector mass spectrometers having sufficient electronic stability for accurate mass measurement can be modified for accurate quantitative high resolution selected ion monitoring measurements.

References

1. Keith, L.H., in Identification and Analysis of Organic Pollutants in Water, L.H. Keith, ed. (Ann Arbor Science Publishers, Ann Arbor) 1976, p. iii.
2. McFadden, W.H., Schwartz, H.L. and Evans, S., J. Chromatogr. (1976) 122, 389.

3. Arpino, P.J., Dawkins, B.J., and McLafferty, F.W., J. Chromatogr. Sci. (1974) 12, 574.
4. Carroll, D.I., Dzidic, I., Stillwell, R.N., Haegele, K.D. and Horning, E.C., Anal. Chem. (1975) 47, 2369.
5. Sphon, J.A., Dreifuss, P.A. and Schulten, H.R., J. Assoc. Official Anal. Chemists, (1977) 60, 73.
6. Hunt, D.F. and Ryan III, J.F., Anal. Chem., (1972) 44, 1306.
7. Weinkam, R.J., Rowland, M. and Meffin, P.J., Biomed. Mass Spectrom. (1977) 4, 42.
8. Dougherty, R.C. and Piotrowska, K., Proc. Nat'l. Acad. Sci. USA, (1976) 73, 1777.
9. Dougherty, R.C. and Piotrowska, K., J. Assoc. Official Anal. Chemists, (1976) 59, 1023.
10. Levson, K. and Schulten, H.R., Biomed. Mass Spectrom. (1976) 3, 137.
11. Soltero-Rigau, E. Kruger, K.L. and Cooks, R.G., Anal. Chem. (1977) 49, 435.
12. Haddon, W.F., Masri, M.S., Randall, V.G., Elsken, R.H., and Meneghelli, B.J., J. Assoc. Official Anal. Chemists (1977) 60, 107.
13. Snedden, W. and Parker, R.B., Anal. Chem. (1971) 43, 1651.
14. Snedden, W. and Parker, R.B., Biomed. Mass Spectrom. (1976) 3, 275.
15. Hutzinger, O., Jamieson, W.D., and Safe, S., Nature (London) (1974) 252, 698.
16. Fenselau, C., Anal. Chem. (1977) 49, 563A.
17. Millington, D.S., Buoy, M.E., Brooks, G., Harper, M.E. and Griffiths, K., Biomed. Mass Spectrom. (1975) 2, 219.
18. Millington, D.S., J. Steroid Biochem. (1975) 6, 239.
19. Evans, K.P., Mathias, A., Mellor, N., Silvester, R. and Williams, A.E., Anal. Chem. (1975) 47, 821.
20. Gallegos, E.J., Anal. Chem. (1975) 47, 1150.
21. Hammar, C.G., and Hessling, R., Anal. Chem. (1971) 43, 298.
22. Holmes, W.F., Holland, W.H., Shore, B.L., Bier, D.M. and Sherman, W.R., Anal. Chem. (1973) 45, 2063.
23. Vacuum Generators, Ltd., Altrincham, Cheshire, England.
24. Millington, D.S., personal communication.
25. Haddon, W.F., Wiley, M. and Waiss, A.C., Anal. Chem. (1971) 43, 268.
26. Masri, M.S., J. Am. Oil Chem. Soc. (1970) 47, 61.
27. Masri, M.S., Haddon, W.F., Lundin, R.E., and Hseih, D.P.H., J. Agric. Food Chem. (1974) 22, 512.

28. Wilson, B.J. and Hayes, A.W. in Toxicants Occurring in Foods, National Academy of Sciences, Washington, D.C., pp. 372-423 (1973).
29. Heildelberger, C., Annu. Rev. Biochem. (1975) 44, 79.
30. Ong, T.M., Mutat. Res. (1975) 32, 35.
31. Aflatoxin Mass Spectral Data Bank, Pohland, A.E. and Sphon, J.A., ed., U.S. Food and Drug Adm., Bureau of Foods, Division of Chem. and Phys., Washington, D.C.
32. Masri, M.S., Page, J.R., and Garcia, V.G., J. Assoc. Official Anal. Chemists (1969) 52, 641.
33. Parker, R.B., Snedden, W., and Watts, R.W.E., Biochem. J. (1969) 115, 103.
34. Kiser, Introduction to Mass Spectrometry and Its Applications, Prentice-Hall, Inc., Englewood Cliffs (1965), p. 216.
35. Kingston, D.G.I., J. Assoc. Official Anal. Chemists (1976) 59, 1016.
36. Beuhler, R.J., Flanigan, E., Greene, L.J. and Friedman, L., Biochem. and Biophys. Res. Commun. (1972) 46, 1082.
37. Beuhler, R.J., Flanigan, E., Greene, L.J. and Friedman, L., Biochem. (1974) 13, 5061.
38. Beuhler, R.J., Flanigan, E., Greene, L.J. and Friedman, L., J. Am. Chem. Soc. (1974) 96, 3990.
39. Soltman, B., Sweeley, C.C. and Holland, J.F., Anal. Chem. (1977) 49, 1164.
40. Hunt, D.F., Shabanowitz, J. and Botz, F.K., Anal. Chem. (1977) 49, 1160.

RECEIVED December 30, 1977

7

Introduction to Gas Chromatography/High Resolution Mass Spectrometry

B. J. KIMBLE

University of California–Berkeley, Berkeley, CA 94720

The combination of a gas chromatograph (GC), a mass spectrometer (MS) and an on-line computer is now well-established as an exceptionally powerful and versatile tool for identifying and quantitating the components present in organic mixtures. As with any successful technique, many improvements have been incorporated in recent years in order to provide the maximum amount of information that can be derived from a GC/MS analysis. Significant developments include the use of:

a) glass capillary columns (1-3), for minimized surface effects and increased chromatographic resolution;

b) direct interfacing of capillary columns to the mass spectrometer ion source (4-7), to reduce sample losses in a molecular separator and to preserve chromatographic resolution by eliminating large-volume connectors;

c) high pressure ionization methods (e.g., chemical ionization, 8-14), to provide complementary information to assist in the interpretation of electron impact spectra;

d) improved computer techniques, for automated data acquisition, data handling (15-17), spectrum recognition (18-23) and interpretation (24-26).

An additional approach for the improved utility of the GC/MS technique would be to determine the recorded mass values in each spectrum with sufficient accuracy to be able to compute the exact elemental composition of each of the ion species contributing to the spectrum (27). Such an achievement would provide the maximum amount of structural information that can be directly derived from a mass spectrum — i.e., a data

© 0-8412-0422-5/78/47-070-120$10.00/0

set comprised of exact elemental compositions and their relative abundances.

The routine availability of elemental composition spectra from a GC/MS analysis would substantially assist the interpreter/analyst in several ways, all of which derive from the increased structural specificity inherent in such data. In the case of mass spectra which are, in fact, interpretable by a trained spectrometrist, exact elemental composition information would eliminate the guesswork involved in relating mass values and mass differences to structural units, and thus, would result in more rapid and reliable interpretations. This is particularly true when the recorded spectrum results from the co-elution of two or more components, and the interpretive process has to include the correct assignment of masses into their appropriate "sub-spectra". Elemental composition information would also facilitate the location and quantitation of specific components in mixtures, since the intensities for a particular elemental composition could be plotted as a function of scan number; such an elemental composition chromatogram would constitute a more specific structural indicator than do nominal mass chromatograms.

Of particular utility, elemental composition spectra would provide direct structural information in the case of "hard-to-interpret" spectra of truly unknown components — that is, when little or no chemical information is available to assist in the interpretation. Also, unexpected compounds can arise in a variety of situations as artifacts or contaminants, or due to inadequate fractionation and/or derivatization procedures, etc.; and of course, "difficult" spectra occur particularly in analyses of mixtures containing several hundred components. With today's increasing need to study more and more complicated samples (particularly environmental mixtures such as sewage effluents, and samples for which minimal separation and fractionation steps are desired), it is this aspect of complex mixture analysis that would particularly benefit from the availability of elemental composition spectra.

To date, the concept of elemental composition assignments in GC/MS analyses has been explored in a variety of ways, and applications of the technique seem likely to increase as suitable commercial instrumentation becomes more widely available. Unfortunately, a somewhat ambiguous terminology has already evolved concerning use of the phrases "accurate mass measurement" and, in particular, "high resolution". In general, these terms are used to contrast with the measurement of nominal masses (*e.g.*, m/e 100 or 101) and instrument operation at nominal mass resolution (*e.g.*, $M/\Delta M <$ 1,000); however, it should be recognized that in practice the so-called accurate mass measurements may not be particularly "accurate", nor the mass resolution particularly "high". For simplicity, the acronyms GC/LRMS and GC/HRMS will be used here to denote GC/MS analyses involving the continuous recording of complete mass spectra at nominal (low) resolution and higher mass resolutions respectively. It should also be noted that the terms "accurate mass measurement" and "high resolution" are not synonymous; this may be illustrated by summarizing the approaches that have been reported in the literature concerning the use of elemental composition information in GC/MS analyses.

Elemental composition monitoring of GC effluents. In this approach, the objective is to utilize the mass spectrometer as a highly structurally specific GC detector for quantitative studies, particularly when the compound of interest is present in a highly complex mixture(28-35). The mass spectrometer is operated at some increased mass resolution (*e.g.*, 10,000) with a single accurate mass (characteristic of the components of interest) focussed continuously on the detector. The increased mass resolution aims to reduce the interferences from other ions having the same nominal mass but of a different elemental composition, thus enhancing compound specificity with respect to co-eluting components from the gas chromatograph, instrument background, column bleed, etc.

Elemental composition assignments from spectra recorded at nominal mass resolution. Here, the masses of the recorded peaks in each spectrum are determined with greater than nominal mass accuracy, and elemental compositions are computed from the mass measurements (36; see also 37, 38). However, as the mass resolution is not increased, problems arise if the nominal mass results from ions of more than one elemental composition — *i.e.*, if the peak is actually an unresolved doublet or multiplet. This disadvantage represents a particular problem in the analysis of complex mixtures when the components of the mixture may not be chromatographically resolved, and more complicated spectra are often produced. The use of a double-beam mass spectrometer for GC/MS analyses (39-41) — in which a mass calibrant and the sample are analysed separately but simultaneously using two separate ion sources, a common mass analyzer and two separate detectors — should improve mass measurement accuracy but does not solve the problem of inadequate resolution of doublet or multiplet masses in the sample spectra. Another technique, which utilizes a quadrupole mass spectrometer to "simultaneously" produce positive and negative ion spectra, may also be developed for accurate mass measurement in GC/MS studies (42).

Elemental composition assignment from spectra recorded at increased mass resolution. In this case, the aim is to continuously record complete mass spectra throughout a chromatographic analysis at a sufficiently high mass resolution to eliminate unresolved peaks and with sufficient mass measurement accuracy to exactly determine all of the elemental compositions present in each spectrum. In principle, this GC/HRMS method can be undertaken using mass spectrometers equipped for either photoplate detection (43-45) or electrical detection (44-51, see also 52). In practice, however, it is the latter method (which parallels nominal mass GC/MS operation) which has the potential for providing the maximum flexibility and speed in terms of computerized operation, repetitive spectrum acquisition, large spectrum capacity, and the generation of total ionization and mass chromatograms.

It is this GC/HRMS approach utilizing magnetic scanning and electrical detection which will be discussed more fully here. It is not intended to describe details of instrument design and set-up procedures, but rather to present for the chemist/user the principal parameters involved, and their requirements, limitations and inter-relationships; it is this knowledge that becomes important in practical terms for the effective and efficient interpretation of GC/HRMS data.

Background

The primary parameters of interest in obtaining elemental composition spectra are mass measurement accuracy (a measure of ability to accurately determine the true mass) and dynamic mass resolution (the degree to which masses can be resolved under magnetic scanning conditions). Investigation of these parameters depends, of course, on the specific mode of spectrometer and computer operation; the discussion here is based on developmental GC/HRMS studies undertaken using a modified AEI MS-902 double-focussing mass spectrometer together with on-line, real-time data acquisition and reduction (47-49, 53).

In outline, a spectrum is produced by exponentially scanning over the mass range from high mass to low mass. The resulting analog signal from the detector amplifier is digitized at equal time intervals and the digital values for each peak detected above a constant preset threshold value are saved for analysis, together with the absolute time of occurrence of the peak with respect to the start of the scan (54). The digital values for each peak define the peak profile, from which the peak intensity is determined from its area (55, the sum of the digital values for that peak) and the peak position in time is determined from the time computed for the peak center-of-gravity or centroid (54).

Mass/time function. In the ideal case, therefore, the mass (M) of any peak is an exponential function of the elapsed time (t) from the start of the scan (Figure 1). That is,

$$M \propto e^{-t/\tau} \tag{1}$$

where τ is the "time constant" of the exponential function, i.e., the time taken to scan between any mass, M_x, and the mass M_x/e (where e=2.718). At the start of the scan, t = 0 and M is equal to the maximum mass value, M_{max}, so that equation (1) becomes

$$M = M_{max}\, e^{-t/\tau} \tag{2}$$

Scanrate. The scanrate in the commonly quoted units of seconds/mass decade (a mass decade covers the range M_x to $M_x/10$) can be derived from this equation by considering the time difference $(t_2 - t_1)$ for two masses M_1 and $M_1/10$:

$$M_1 = M_{max}\, e^{-t_1/\tau} \quad \text{and} \quad M_2 = \frac{M_1}{10} = M_{max}\, e^{-t_2/\tau}$$

These equations combine to give

$$t_2 - t_1 = \tau \log_e 10 = \tau_{10} \tag{3}$$

where τ_{10} is the time to scan one mass decade, i.e., the scanrate in seconds/decade.

Resolution. As normally defined, the mass resolution, R, at 10% valley separation of two equal mass peaks is given by

$$R = M/\Delta M \qquad (4)$$

where M is the nominal mass and ΔM is the accurate mass difference for the two mass peaks (Figure 2). The time difference, Δt, between the maxima of these two peaks can be determined from the rate of change of mass with time (from equation 2) giving

$$dM = \frac{-M_{max}}{\tau} \cdot e^{-t/\tau} \cdot dt$$

which approximates to

$$\Delta M = \frac{-M_{max}}{\tau} \cdot e^{-t/\tau} \cdot \Delta t$$

for small values of ΔM and Δt. Dividing by equation (2) gives

$$\frac{\Delta M}{M} = \frac{-\Delta t}{\tau}$$

and using equation (4), this reduces to

$$\Delta t = \frac{-\tau}{R} \qquad (5)$$

From symmetry considerations, the time difference, Δt, between the peak maxima is equal to the time taken to scan across one of these peaks between the points which are at 5% of the maximum peak height (Figure 2). Thus, if τ and Δt are known, then the resolution can be calculated for each single peak throughout the mass range. Also, equation (5) shows that for constant mass resolution, R, and scanrate τ_{10} (=τ $\log_e 10$), the widths of all singlet peaks measured at 5% of their maximum height should be constant.

Equations 3 and 5, combined to give

$$t = -\tau_{10}/R \log_e 10 \text{ ,} \qquad (6)$$

permit various instrument parameters to be calculated and compared with experimental results. For example, for a constant resolution (10% valley) of 10,000 and a scanrate of 6 seconds/decade, the peakwidth for all masses at their 5% levels is given by

$$\Delta t = \frac{6}{10^4 \times 2.3026} \text{ sec} = 260.6 \ \mu\text{sec}$$

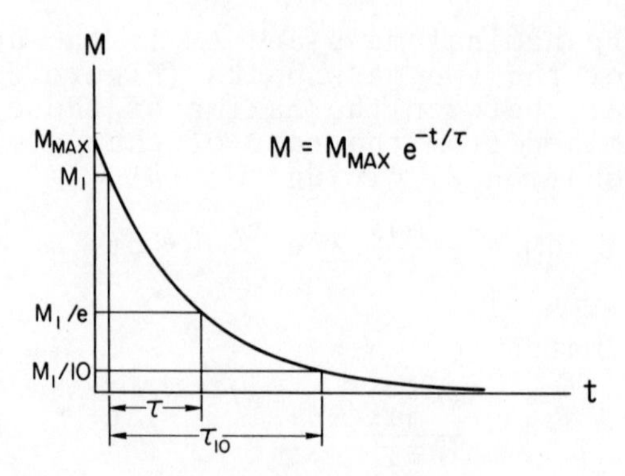

Figure 1. The mass/time function for exponentially scanned mass spectrometers

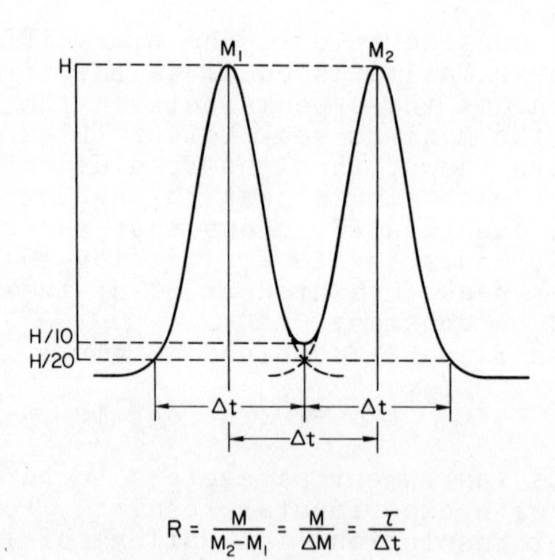

Figure 2. Mass resolution (10% valley definition)

For a digitization rate of 50 kHz, the data points are separated in time units of 1/50,000 sec or 20 μsec, so that in this example, the widths of all singlet peaks should be defined by 260/20 = 13 datapoints between their 5% levels.

Mass measurement accuracy. It is the ability to measure masses in a spectrum with high accuracy that is the basis for determining the correct elemental compositions. As perfectly exact mass measurements cannot be achieved in practice, the approach used to generate a high resolution spectrum is to compute, for each measured mass, a list of possible elemental compositions (56) whose accurate masses fall within a small mass "window" around each measured mass. In order to restrict the elemental composition listing to somewhat manageable proportions, the user is required to select the types and maximum numbers of each element that he considers relevant on chemical grounds. He is also required to input the "width" of the mass window to be used for the calculations — *i.e.*, the measured mass ± δppm, or a mass window of 2δ.

The value of δ must be chosen carefully so that the correct elemental composition for a particular measured mass is included in the list of possibilities. If the chosen value of δ is too small, then the correct composition may not be listed; however, if the chosen value of δ is significantly larger than necessary, then the resulting list of compositions rapidly becomes too unwieldy to evaluate. The selection of the appropriate value of δ for any particular set of instrument operating parameters depends directly on a knowledge of the mass measurement accuracy determined experimentally under those conditions.

The attainment of high mass measurement accuracy is dependent upon many factors (57-60) such as adequate definition of actual peak shapes, the absence of abnormal peakshapes, definition of the actual mass/time function, software algorithms for mass calculation, and mass resolution. The dependency on mass resolution relates to the presence of unresolved peak doublets or multiplets (Figure 3); for these peakshapes, the position of the center-of-gravity of the overall peak profile is inaccurate, resulting in inaccurate mass measurements for the ion species actually present. In general, improved mass measurements can be obtained by averaging the measurements from several spectra (61, 62). This process reduces the effect of the random errors associated with accurate mass determinations, and, thus, a smaller mass window can be used in the generation of possible elemental compositions for averaged mass measurements. Statistically, the mass measurement accuracy would be expected to be improved by a factor of $\sqrt{n}$ where n is the number of measurements averaged for a particular accurate mass.

Practical Considerations and the Evaluation of Instrument Performance

GC and fast-scanning HRMS. In practical terms, the ability to determine accurate mass measurements for spectra recorded from chromatographic effluents centers on the performance of the fast-scanning high resolution mass spectrometer and its data system (57, 63-67). As for GC/LRMS, the gas chromatograph and GC/MS interface may be of various designs provided that:

a) the increased pressure in the ion source due to the carrier gas permits operation of the mass spectrometer;
b) each chromatographic peak is of sufficient duration so that spectra will be recorded for each component;
c) a sufficient amount of each component is introduced *via* the GC inlet system into the ion source so that recognizable spectra can be recorded.

As these design considerations have been the subject of many publications (*e.g.*, 68), only the most important aspects for GC/HRMS will be discussed further here.

The main parameters for GC/HRMS operation are summarized in Table 1; many of these parameters, of course, also pertain to GC/LRMS analyses, but they become somewhat more critical for the concomitant measurement of accurate masses under fast-scanning high resolution conditions. The table also indicates the primary optimum requirements for each parameter. For several of these, there are conflicting requirements due to the inter-relationships that exist between the different parameters, and the actual values utilized for analyses will necessarily be a set of values that achieve the optimum compromise.

Scan cycle time. The overall cycle time for recording a single spectrum is a function of the scanrate, the mass range, and the time required to reset the magnetic field between the minimum and maximum mass values.

As described above, the theoretical scanrate, τ_{10}, is a constant and is directly proportional to the time constant of the exponential function, τ. In practice,

Table 1. Primary Considerations for GC/HRMS

Parameter	Requirement
mass measurement accuracy	maximize to limit number of possible elemental compositions generated
dynamic mass resolution	maximize to reduce number of unresolved peaks, hence increase mass measurement accuracy minimize to increase sensitivity
sensitivity	maximize to reduce sample quantity requirements
scan time (function of mass range and scanrate)	minimize scan time to preserve chromatographic resolution in total ionization and mass chromatograms maximize mass range for useful information maximize scanrate (sec/decade) for increased sensitivity and mass measurement accuracy
magnet reset time	minimize to reduce "wasted" time and preserve chromatographic resolution in total ionization and mass chromatograms
GC peak elution time	minimize for higher chromatographic resolution maximize with respect to scan cycle time (*i.e.*, scan time + magnet reset time) to ensure sufficient time for recording mass spectra

the mass/time function is not perfect over the whole mass range; that is, τ is not constant but varies with mass as illustrated in Figure 4. The use of an internal mass standard such as perfluorokerosine (PFK), which contributes peaks of known accurate mass throughout the mass range, provides a means for determining the actual mass/time function over small sections of the mass range. Thus, a value of τ can be assigned to each of the calibration masses, and from each of these values the measured scanrate, τ_{10}, in seconds/decade can be calculated from equation 3. The variation observed for a nominal scanrate of 6 seconds/decade follows the profile for τ shown in Figure 4, and ranges between about 6.5 seconds/decade at m/e 580 and 5.5 seconds/decade at m/e 80; the actual overall scantime between m/e 800 and m/e 65 is about 7.2 seconds (48, 49). It should be noted that the upper section of the mass range (~ m/e 800-650) is not easily usable for accurate mass assignments due to the low intensity of the PFK calibration masses in this region, and the rapid rate of change of τ in this region precludes accurate extrapolation of τ above the highest reliable calibration mass.

The time taken to return to the initial maximum mass value for the next scan (the magnet reset time) should, of course, be minimized as far as possible to reduce "wasted" time and minimize the overall scan cycle time. As reported previously (48), magnet circuit modifications undertaken on a MS-902 mass spectrometer have permitted this time to be reduced to 2.4 seconds whilst maintaining mass calibration up to ~ m/e 650 for the subsequent scan. With these instrument parameters, therefore, the overall scan cycle time (*i.e.*, scantime + magnet reset time) is just under 10 seconds.

GC peak elution time. The value for the overall scan cycle time is important for the adjustment of the gas chromatographic operating parameters (*e.g.*, flowrate, column temperature and programming rate). In order to ensure that mass spectra can be recorded for any GC peak, the time for elution of a peak has to be adjusted so that it is at least twice the scan cycle time. As illustrated in Figure 5, the minimum factor of two derives from the unknown time relationship between the start of a scan and the commencement of elution of a GC peak. It is necessary, therefore, in this case, to ensure that all chromatographic peaks elute into the mass spectrometer ion source over a time period of at

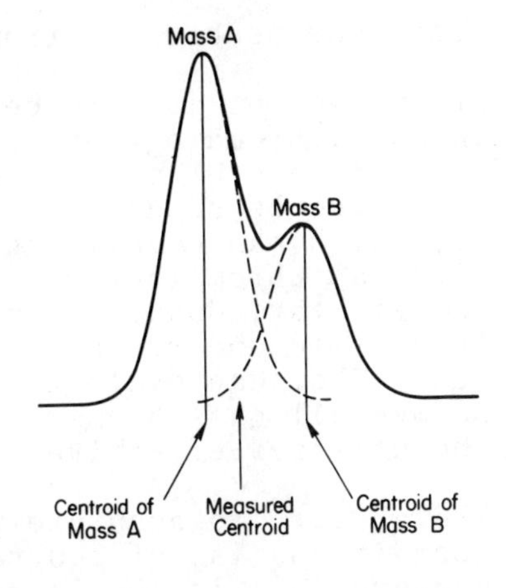

Figure 3. Effect of inadequate mass resolution on mass measurement accuracy

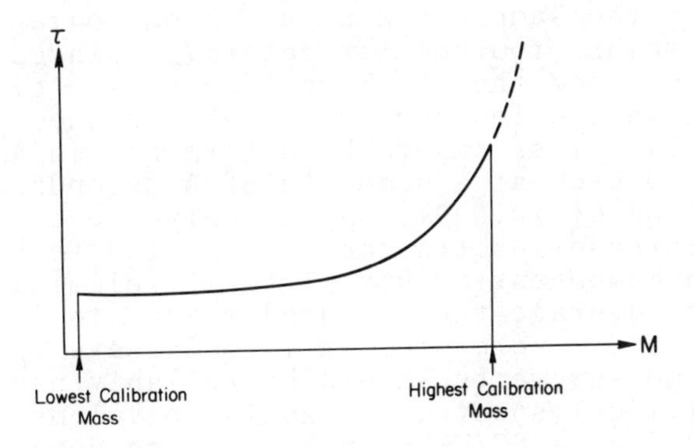

Figure 4. The variation of τ as a function of mass

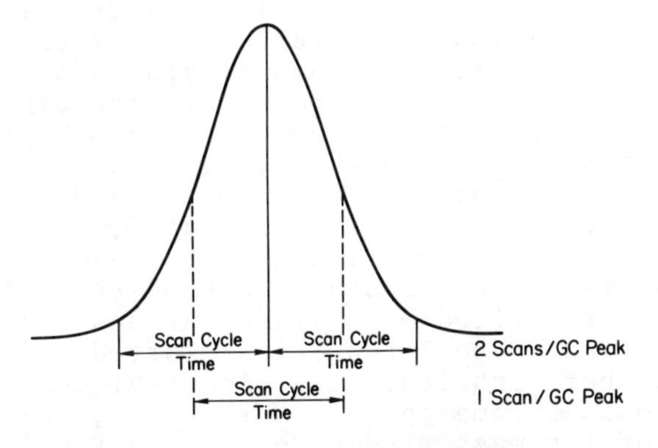

Figure 5. The extreme relationships between scan cycle time and GC peak elution time; scan cycle time equals one-half of the GC peak elution time

least 20 seconds so that at least one usable spectrum can be recorded.

The use of a GC peak elution time of just twice the scan cycle time results in mass chromatograms which show a variety of profiles from the same GC peak (see Figure 6). Similar features may also be seen with total ionization chromatograms, which can complicate the matching of such output with gas chromatograms recorded by a flame ionization detector; quantitative comparisons are also more difficult to undertake with such variations in peak heights. The scan cycle time, therefore, directly limits the overall gas chromatographic performance that can be utilized effectively.

Mass measurement accuracy. The ability to accurately measure masses under GC/HRMS conditions is, of course, the object of the exercise. However, it is the introduction of the sample *via* a gas chromatograph that presents major problems for determining accurate masses. Because of the short residence time of each component in the ion source, a well-defined peak profile for each mass has to be determined in a very short time (~ 260 μsec at a scanrate of 6 seconds/decade and a resolution of 10,000). Also, only one or two reasonably intense spectra can be expected to be recorded from each component. This virtually eliminates the ability to average the measured masses to improve accuracy, and, therefore, it is necessary that high accuracy measurements should be *reliably* produced for *each* individual spectrum. The large volume of data generated from a GC/HRMS analysis also demands a high degree of reliability as detailed investigations into anomalous features are extremely time-consuming.

In addition to the generation of elemental composition listings for measured accurate masses, a detailed knowledge of attainable mass measurement accuracy is also important for GC/HRMS studies in order to generate accurate mass chromatograms (or elemental composition chromatograms) to locate the elution times of specific components or compound classes for further detailed study (47). In this case, the user selects the elemental composition of interest and the corresponding accurate mass is calculated. An appropriate mass window (δ) is specified and, for each spectrum, the intensity of the accurate mass which is located within this specified mass window around the calculated mass, is saved. The intensities plotted as a function of scan number constitute the elemental composition or accurate mass chromatogram. Here again, selection of the size of the mass window depends on knowledge of the mass measurement accuracy; the choice is critical to ensure that the correct masses are located but that "extraneous" masses are excluded as far as possible. Mention should also be made of the generation of nominal mass spectra and chromatograms from accurate mass data. Here the intensities of all accurate masses that fall within the mass range between (X.0000-0.4000) amu and (X.0000+0.6000) amu are summed to give the intensity of nominal mass X. The mass measurement accuracy becomes important for the accurate masses that may lie close to the limits of this range, and consideration should be given to the conventions used for defining the nominal mass of an ion (69).

In view of all these basic considerations, it is essential that a detailed evaluation of accurate mass measurement capability should be undertaken for a given set of operating parameters to ensure reliable single-scan spectra and accurate mass chromatograms.

The practical determination of mass measurement accuracy is undertaken for a particular set of instrument operating parameters by recording a number of spectra of a selected compound whose principal masses are uniquely known. Because of the continuous variation in sample flowrate which results from samples introduced *via* a gas chromatograph, the basic evaluation is made under GC/HRMS operating conditions (*i.e.*, with normal carrier gas flowrate, column temperature, *etc.*), but with introduction of the evaluation compound *via* a batch inlet system at constant sample flowrate. Actual GC/HRMS evaluation can then be checked by introducing a known compound *via* the gas chromatographic inlet and verifying that the resulting mass measurements are consistent with the basic evaluation for similar masses of similar intensities.

The compound selected for evaluation of mass measurement accuracy should produce ion masses that are singlet peaks well-separated from other masses at the operating mass resolution. Ideally, they should also adequately cover the mass range of interest, and should be of approximately the same relative intensities to permit evaluation of mass measurement accuracy as a function of both mass and intensity. In practice, a compromise has to be made with respect to these latter requirements, and a variety of compounds for this purpose have been reported in the literature (*e.g.*, perchlorobutadiene, 61, and heptacosafluorotributylamine, 48,49).

Evaluation of the measurements obtained for the selected masses from several scans can be undertaken by a variety of related methods. One approach (61) described the production of a histogram which plots the percentage of the total number of mass measurements as a function of their mass measurement errors (the difference between the measured mass and the true mass). This method, however, suffers from the drawback that only a limited number of scans were evaluated, and the resulting histogram was generated from composite error measurements from ions of differing masses and also differing relative intensities.

As GC/HRMS analyses necessarily involve the production of large numbers of spectra and require a corresponding computer capability, a more detailed statistical evaluation can be undertaken if a large number (e.g., > 100) of spectra are recorded and evaluated automatically. In this approach (70), for each selected accurate mass, M_t, to be evaluated, the measured mass is located in each spectrum within an estimated mass window around the true mass, M_t. For each value of M_t, several parameters are calculated from all of the recorded spectra:

a) The number of spectra (N) for which the measured mass was found within the mass window. (The estimated mass window should be chosen so that it represents the absolute maximum mass measurement error than can be tolerated for spectral interpretation — e.g., ±20 ppm. Thus, for all values of M_t, N should correspond to the number of spectra recorded; i.e., all of the measurements will be within 20 ppm of their true value).

b) The mean mass error, E. (This may be positive or negative with respect to M_t).

c) The standard deviation of the mass measurement errors (σ ppm).

d) The standard deviation of the mean mass error ($\sigma_m = \sigma/\sqrt{N}$).

e) Mean peak intensity.

f) Standard deviation of the intensity measurements.

(For additional evaluation of instrument performance, mean and standard deviation values are also calculated for a variety of other parameters associated with each selected mass; these include τ, $W_{5\%}$, W_{th}, R, as explained below).

These values permit a more detailed evaluation of the relative effects of the systematic and random errors associated with mass measurement accuracy. In the absence of systematic errors, the mean error E should approach zero (i.e., the mean measured mass should equal the true mass), and there is a 99.78%

chance that the mean mass will be within $3\sigma_m$ of the true value. Also, for a normal distribution, 99.78% of the measurements for each mass should fall within 3σ of the mean measured value, i.e., virtually all of the measurements should fall within a spread of 6σ ppm. A suitable choice for determining the mass window (2δ) for elemental composition listings and chromatograms would result, therefore, from using a δ value equal to the sum of the maximum value of 3σ for the masses considered, and the maximum absolute value of E for the same masses (see Figure 7).

The use of the 3σ values (corresponding to 99.78% of the measurements for a normal distribution) rather than 2σ values (95% of the measurements) becomes important for GC/HRMS analyses because the large number of spectra (several hundred, each possibly containing several hundred accurate masses), that are generally recorded necessitates a high degree of reliability as well as accuracy. In general, the number of elemental compositions computed for a typical measured mass (M) becomes unmanageable for a mass window larger than about M ± 15 ppm, particularly for more complex molecules that may contain many types of atoms (e.g., C, ^{13}C, H, N, O, Si, F). Large mass measurement errors are also problematic for the generation of elemental composition chromatograms since the mass measurement distribution (spread) may overlap with the distribution of measured masses for another elemental composition present in some of the spectra. Thus, intensities of extraneous masses may be included in the final output even though the "correct" mass window has been selected, and this would result in one or more artifact peaks in the elemental composition chromatogram. In practical terms for spectral interpretation, it is also important to know the minimum absolute mass intensity value for which the desired mass measurement accuracy can be assured; data output containing lower intensity values can then be evaluated with appropriate caution.

Finally, it is important to remember that the consideration of mass measurement accuracy relates only to well-defined normal peak shapes. Any significant distortion of such peak shapes, such as unresolved doublets, should be recognized and the effect on mass measurement accuracy taken into account.

Sensitivity. In GC/HRMS, an evaluation of overall system sensitivity must consider not only the detection of specified masses from a given quantity of a component injected into the gas chromatograph (as for GC/LRMS), but also the mass measurement accuracy that can be

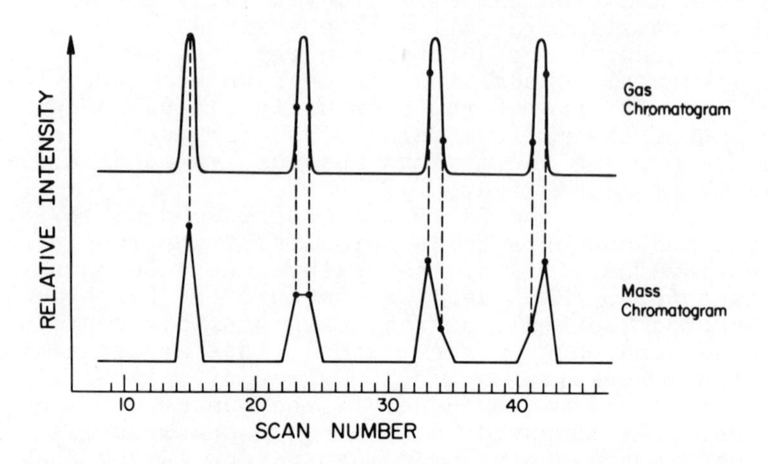

Figure 6. Representation of an accurate mass chromatogram for equal GC peak shapes; scan cycle time equals one-half of the GC peak elution time

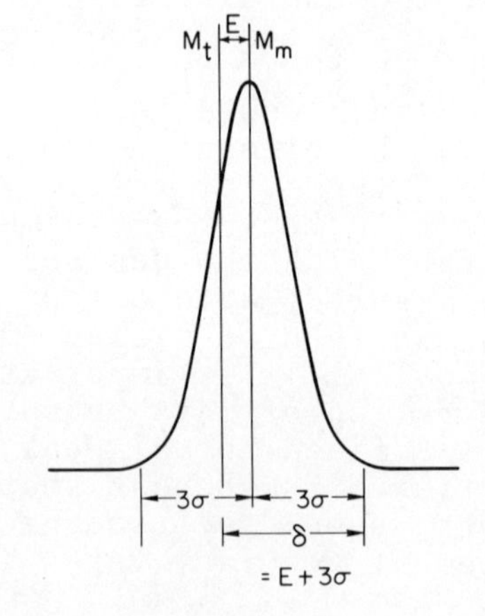

Figure 7. The evaluation of mass measurement accuracy; M_t = *true mass,* M_m = *mean measured mass,* E = *mean error,* σ = *standard deviation of the measured masses.*

obtained for these masses. This more stringent criterion results in somewhat less sensitivity compared to detecting the presence of a particular mass, since the peak profiles must be sufficiently well-defined that the peak center-of-gravity or centroid can be accurately and reliably determined.

Several discussions have appeared in the literature concerning the theoretical aspects of sensitivity and mass measurement accuracy (57, 71). Basically, the limiting factor concerns the statistics associated with the ion arrival times at the detector, and this determines how well the resulting digitized peakshape is defined; the frequency of digitization or the number of datapoints/mass peak is also a consideration here (72). It has been predicted (71) that for a mass resolution of 10,000, mass measurement accuracies within 10 ppm (=3σ, or 99.78% confidence level) should be achieved with 30 ions/peak, and within 17 ppm (=3σ) for 10 ions/peak. Estimates made from actual mass measurements (73) indicate that 10-15 ions/peak are required to give a mass measurement accuracy of 30 ppm (=3σ). The relationship between the absolute peak intensity, I (the sum of the digital values for a peak profile), and the number of ions, N, comprising that peak may be determined (61) from the equation

$$\frac{I}{N} = G\left(\frac{Ref}{v}\right)$$

where G is the gain of the multiplier — the primary variable under normal operating conditions. (The other parameters are: R = input resistance of the multiplier amplifier in ohms, e = charge on an electron in coulombs, f = frequency of digitization in sec^{-1}, v = voltage increment represented by one binary "bit" on the A/D converter in volts).

As with GC/LRMS, it is important that the overall sensitivity is determined for a selected component introduced through the gas chromatograph, particularly to take into account any losses that may occur when using a molecular separator interface. It is also necessary for GC/HRMS that one or more ion masses (*e.g.*, basepeak and/or molecular ion, *etc.*) are selected in order to specify the mass measurement accuracy that can be obtained from an injection of a given amount of the compound. In addition, the limited number of high resolution spectra that can be recorded for any particular GC peak under GC/HRMS conditions also has to be considered. In the limiting case, when the scan cycle time (*e.g.*, ~10 seconds) is about one-half of the GC peak elution time, the maximum recorded intensity for any mass may be expected to vary by a factor of about two (see Figure 6), depending on the time of starting a scan in relation to the time when the GC peak begins to elute —— a relationship which is unknown in practice. For an injection of N ng of the selected component, the maximum sample flowrate at the top of the GC peak corresponds to about N/10 ng/sec for a GC peak elution time of 20 seconds. Measured mass intensities, therefore, will be proportional to sample flowrates between about N/20 and N/10 ng/sec. If the variation of mass measurement accuracy with respect to a range of measured intensities has been determined under the same conditions, then an assessment can be made for the mass measurement accuracy expected for selected masses recorded from an injection of N ng under a given set of experimental conditions (*i.e.*, scanrate, mass resolution, multiplier gain, *etc.*).

Dynamic mass resolution. A knowledge of the actual mass resolution of each recorded peak is an important measure of the overall performance of a fast-scanning high resolution mass spectrometer, and directly relates to the ability to accurately measure masses. Also, it is frequently necessary to be able to determine whether or not certain mass doublets would be expected to be resolved under the instrument conditions used.

As discussed in the section above on mass resolution, the mass resolution (10% valley definition) of any peak can be determined from its width at 5% maximum peak height ($W_{5\%}$) and τ (equation 6), where both $W_{5\%}$ and τ are measured in the same units (*e.g.*, seconds or number of data points). The actual variation of τ throughout the mass range (see Figure 4) means that the resolution calculation for a particular peak should incorporate the appropriate value of τ for the mass peak considered; this value is interpolated for each peak during the mass assignment process based on the values calculated for the surrounding calibration masses.

The determination of the actual peak width $W_{5\%}$ for any peak is somewhat more involved as it cannot be directly measured from the set of digital values that constitute a peak profile. For ideal (Gaussian) peak-shapes, however, it is possible to calculate $W_{5\%}$ in terms of the peak area, A (the sum of the digital values for that peak profile), and the maximum peak height, H (the maximum digital value for the peak profile). From the general equation for a Gaussian function, the width at 5% of the maximum peak height (*i.e.*, at H/20) can be shown to be given by

$$W_{5\%} = 2\sigma\sqrt{2\log_e 20} = 4.8955\ \sigma \qquad (7)$$

where σ is the standard deviation of the function. The area A under this curve is given by

$$A = H\sigma\sqrt{2\pi}$$

where H is the maximum peak height, so that, eliminating σ gives

$$W_{5\%} = \frac{A}{H}\sqrt{\frac{4}{\pi}\log_e 20} = 1.953\ \frac{A}{H} \qquad (8)$$

(*cf.* $W_{5\%}$ = 1.90 A/H from consideration of a triangular peakshape).

Therefore, the resolution $R_{5\%}$ (*i.e.*, 10% valley definition) is given by

$$R_{5\%} = \frac{\tau}{\Delta t} = \frac{\tau}{W_{5\%}} = \frac{H}{A} \cdot \frac{\tau}{1.953} \qquad (9)$$

(where τ is measured in number of data points).

Preliminary evaluation of this calculation using actual data (70) indicates that for well-defined normal peak shapes, the values calculated for $W_{5\%}$ are approximately constant, as would be expected, and also correspond to values measured directly from plots of individual peak profiles.

In practice, the utility of the calculation of $R_{5\%}$ is limited to the evaluation of peak profiles. This is because each peak is recorded as a series of digital values that exceed a constant preset threshold which is chosen to eliminate "noise" but to maximize the overall dynamic range. If the digital values for two close mass peaks do not cut the threshold value at the valley (Figure 8a), then the profile will be recognized as the profile of a single peak and the resulting calculations will be in error. In order that the profile be recognised as two distinct peaks, the maxima for equal-sized peaks must be separated by W_{th} — the width of each peak measured at the *threshold* (Figure 8b). This threshold peakwidth is easily determined from the number of datapoints recorded for each peak, and this number can be used to calculate a "threshold resolution", R_{th}, for each mass peak from the equation

$$R_{th} = \frac{\tau}{W_{th}} \qquad (10)$$

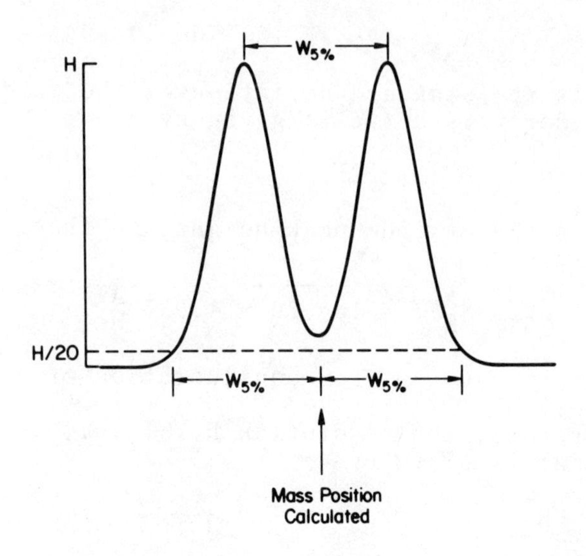

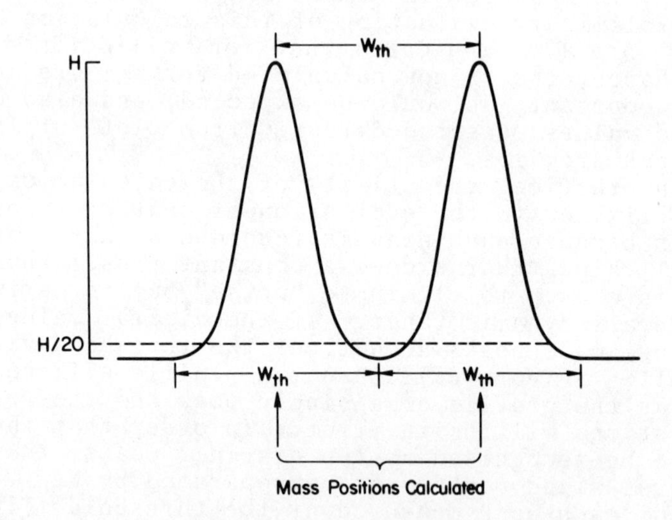

Figure 8. (a) Peak profile for doublet separated with resolution $R_{5\%} = \tau/W_{5\%}$ *(10% valley definition); (b) Peak profiles for peaks resolved at threshold,* $R_{th} = \tau/W_{th}$.

In terms of the interpretation of actual data, the threshold or "effective" resolution will more accurately reflect the overall instrumental ability to resolve closely positioned mass peaks and thus to subsequently determine their accurate masses and elemental compositions.

The decrease in effective resolution due to the use of the threshold peakwidth for the calculation rather than $W_{5\%}$, can be illustrated by assuming that the threshold peakwidth for a Gaussian peakshape is approximated by

$$W_{th} \simeq 6\sigma \qquad (11)$$

where 6σ is the base width which defines 99.78% of the peak area. Therefore, from equations (7) and (11), W_{th} is given by

$$W_{th} = \frac{6}{4.8955} W_{5\%} = 1.23\ W_{5\%} \qquad (12)$$

Calculated values for $R_{5\%}$, W_{th} and R_{th} are shown in Table 2 for the case of a nominal scan rate of 6 seconds/decade and a nominal resolution ($R_{5\%}$) of 10,000; corresponding values are also given for the range of τ encountered for this nominal scan rate. These calculations demonstrate a significant decrease in effective resolution for $W_{th} = 6\sigma$, a decrease of 18.4% compared to the resolution at 5% maximum peak height, and a decrease of up to 25% compared to the supposed resolution of 10,000. Further decreases in effective resolution are to be expected for large peaks since a peakwidth of 6σ for a Gaussian peakshape corresponds to the width at 1.11% of the maximum peak height; for large peaks, the actual threshold will be significantly below 1.11% of maximum peak height, and the threshold peakwidth will thus be larger than 6σ.

Table 2. Comparison of resolution values calculated from peakwidth at 5% of maximum peakheight (10% valley definition) and peakwidth at 1.11% of maximum peakheight.

Scanrate, τ_{10}, sec/decade	τ, sec.	$W_{5\%}$ $=4.8955\sigma$, μsec	W_{th} $=6\sigma$, μsec	$R_{5\%}$	R_{th}
5.5	2.389	260.6	319.4	9,167	7,479
6.0	2.606	260.6	319.4	10,000	8,159
6.5	2.823	260.6	319.4	10,833	8,838

In addition to the actual determination of the effective resolution for each recorded peak, these calculations can be utilized to identify "abnormal" peak-shapes that can occur when operating a high resolution mass spectrometer under GC/HRMS conditions (60, 74). A variety of such peak shapes can be rapidly identified by comparison of the calculated peakwidth at 5% height ($W_{5\%}$) and the measured threshold peakwidth (W_{th}); abnormal peakshapes include: small "noisy" peaks, flat-topped ("saturated") peaks, tailing peaks, and unresolved doublets or multiplets. All of these peak-shapes will adversely affect the accurate determination of the peak center-of-gravity and hence the measurement of accurate mass.

The effect of abnormal peak shapes on the quality of actual data output is illustrated in Figure 9. The plot shows the accurate mass chromatogram for m/e 119.0861 ($\equiv C_9H_{11}$) from a GC/HRMS analysis of the neutral fraction of an extract of petroleum refinery wastewater (75). The complexity of this fraction is evidenced by the unresolved "hump" which is also present in the flame ionization gas chromatogram and many of the other mass chromatograms generated from the GC/HRMS data. Notable in the chromatogram shown are the two abrupt zero intensity values located at scan numbers 96 and 143, indicating that no accurate masses were located in these scans within the mass range m/e 119.0861 ± 20ppm, i.e., corresponding to the elemental composition C_9H_{11}. Detailed investigation of these two scans revealed that C_9H_{11} ions were indeed present in the spectra and that the absolute intensities were consistent with the overall profile observed in the accurate mass chromatogram. Plots of the appropriate peak profiles, also shown in Figure 9, illustrate the abnormal peak profiles that were actually recorded; the vertical lines denote the positions of the calculated centers-of-gravity for the two profiles. In both cases the calculated center-of-gravity was shifted sufficiently so that the corresponding accurate mass was outside the mass window used to generate the chromatogram. Both of these peak profiles would be easily recognized as being abnormal by comparison of their area/maximum height ratios (giving the calculated $W_{5\%}$ values) and the measured threshold peakwidths, W_{th}. In both cases, the availability of software algorithms for detecting profile minima and rethresholding would effectively overcome the errors introduced by these unresolved peak profiles.

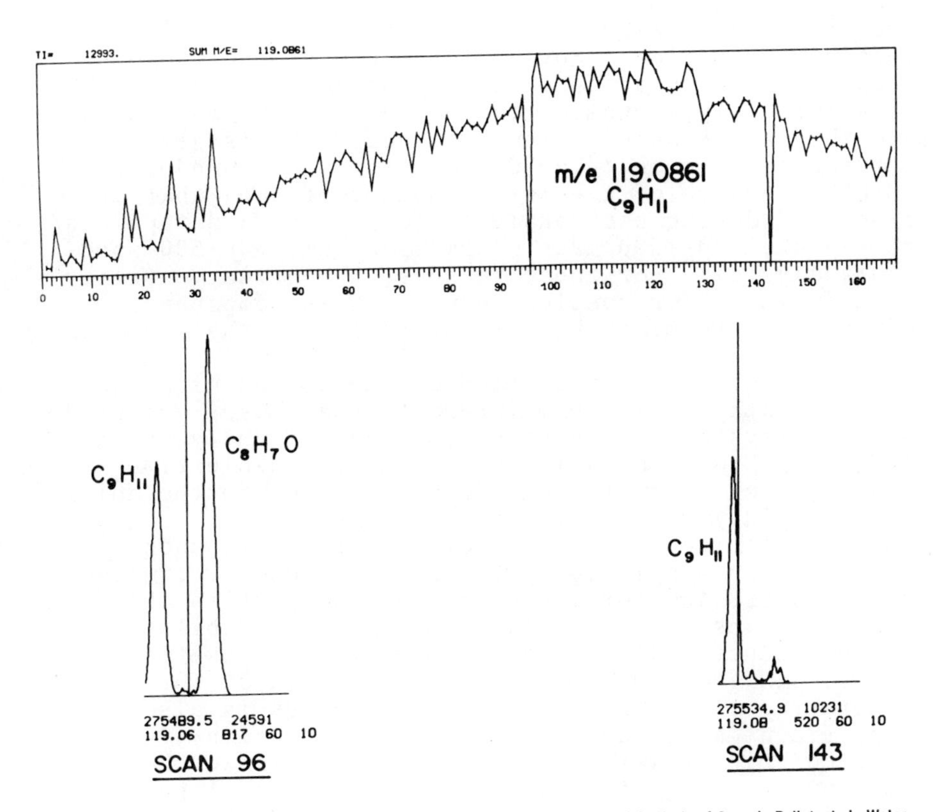

Figure 9. Accurate mass chromatogram for m/e *119.0861 (≡C_9H_{11}). Peak profiles plotted from scans 96 and 143.*

Conclusions

It has been the intention of this chapter to summarize for the chemist/user the primary aspects of GC/HRMS which need to be considered when undertaking evaluation and interpretation of such data. The degree of data evaluation necessary in high resolution mass spectrometry is significantly greater than for low resolution studies. Any evaluation must include mass measurement accuracy and its dependency on mass and peak intensity, dynamic resolution for each mass peak, and the many possible elemental compositions generated for each mass peak. The difficulty is compounded in GC/HRMS because of the large amounts of data that are generated during each analysis (*e.g.*, 15-25 datapoints/mass peak, 500-1500 mass peaks/spectrum, 300-600 spectra/analysis). For these reasons, the primary utility of GC/HRMS is for complex samples whose components cannot be determined by interpretation of their nominal mass spectra.

Even with the production of guaranteed high-quality data, a flexible interactive and user-oriented software system is essential in order to fully utilize the available information within a reasonable time-frame. Such software considerations include the ability to rapidly review elemental composition listings that are limited by user-specified criteria, such as listings for single masses, mass differences, limited mass ranges, and for masses having intensities above a specified absolute or relative value. The ability to include functional groups of elemental compositions (*e.g.*, C_6H_5, CO_2CH_3, $OSiC_3H_9$, *etc.*) as input for generating elemental composition listings is also a significant assistance in data interpretation (*76*). In particular, software algorithms which utilize chemical concepts to limit the number of compositional possibilities generated for each mass (*26*, *77*-*79*) are necessary to speed-up the interpretation process and assist in the generation of a self-consistent set of elemental compositions for each high resolution spectrum of interest. The use of elemental compositions generated for accurate mass differences is also useful in this context, since the elemental compositional possibilities for a relatively small mass difference are substantially fewer than for the accurate masses themselves. As for GC/LRMS, the automatic identification of the spectra of interest would also significantly speed-up the overall data interpretation process. Algorithms such as that described by Biller and Biemann (*15*) should be implementable for reasonably intense, accurately measured masses; however, more complex algorithms (17) would require the ability to record a substantial number of spectra throughout the time of elution of a gas chromatographic peak.

In essence, therefore, the routine utility of the GC/HRMS technique demands an extensive software capability for the rapid evaluation of instrument performance, for the production of reliable high-quality data, and for the efficient handling of the detailed information contained in the elemental composition spectra recorded of gas chromatographic effluents. In particular, the routine use of such software for instrument evaluation and performance monitoring is vital if the potential of the GC/HRMS technique is to be exploited in a meaningful way.

REFERENCES

1. Novotny, M., and Zlatkis, A., Chromatographic Reviews, (1971), 14, 1.
2. Grob, K., and Grob, K., Jr., J. Chromatog., (1974), 94, 53, and references therein.
3. Grob, K., and Grob, G., p. 75 in "Identification and Analysis of Organic Pollutants in Water", ed. L.H. Keith; Ann Arbor Science Publishers, Ann Arbor, Michigan, 1976.
4. Henderson, W., and Steel, G., Anal. Chem., (1972), 44, 2302.
5. Grob, K., and Jaeggi, H., Anal. Chem., (1973), 45, 1788.
6. Leferink, J. G., and Leclercq, P. A., J. Chromatog., (1974), 91, 385.
7. Henneberg, D., Henrichs, U., and Schomburg, G., Chromatographia, (1975), 8, 449.
8. Schoengold, D. M., and Munson, B., Anal. Chem., (1970), 42, 1811.
9. Arsenault, G. P., Dolhun, J. J., and Biemann, K., Anal. Chem., (1971), 43, 1720.
10. Arsenault, G. P., p. 817 in "Biochemical Applications of Mass Spectrometry", ed. G. R. Waller; Wiley-Interscience, 1972.
11. Blum, W., and Richter, W. J., Tetrahedron Letters, (1973), 835.
12. Fentiman, A. F., Foltz, R. L., and Kinzer, G. W., Anal. Chem., (1973), 45, 580.
13. Jelus, B., Munson, B., and Fenselau, C., Anal. Chem., (1974), 46, 729.
14. Ryhage, R., Anal. Chem. (1976), 48, 1829.
15. Biller, J. E., and Biemann, K., Anal. Letters, (1974), 7, 515.

16. Nau, H., and Biemann, K., Anal. Chem., (1974), 46, 426.
17. Dromey, R. G., Stefik, M. J., Rindfleisch, T. C., and Duffield, A. M., Anal. Chem., (1976), 48, 1368.
18. Mathews, R. J., and Morrison, J. D., Aust. J. Chem., (1974), 27, 2167, and references therein.
19. Sweeley, C. C., Young, N. D., Holland, J. F., and Gates, S. C., J. Chromatog., (1974), 99, 507.
20. Abramson, F. P., Anal. Chem., (1975), 47, 45.
21. Mellon, F. A., p. 117 in "Mass Spectrometry, Volume 3", ed. R. A. W. Johnstone; Specialist Periodical Reports, The Chemical Society, London, 1975.
22. Pesyna, G. M., Venkataraghavan, R., Dayringer, H. E., and McLafferty, F. W., Anal. Chem., (1976), 48, 1362.
23. Smith, D. H., Yeager, W. J., Anderson, P. J., Fitch, W. L., Rindfleisch, T. C., and Achenbach, M., Anal. Chem., (1977), 49, 1623.
24. Lederberg, J., p. 193 in "Biochemical Applications of Mass Spectrometry", ed. G. R. Waller; Wiley-Interscience, 1972.
25. Kwok, K.-S., Venkataraghavan, R., and McLafferty, F. W., J. Am. Chem. Soc., (1973), 95, 4185.
26. Dromey, R. G., Buchanan, B. G., Smith, D. H., Lederberg, J., and Djerassi, C., J. Org. Chem., (1975), 40, 770.
27. Beynon, J. H., in Adv. Mass Spectrom., Volume 1, ed. J. D. Waldron; Pergamon Press, NY, 1959.
28. Millington, D. S., J. Steroid Biochem., (1975), 6, 239.
29. Millington, D. S., Buoy, M. E., Brooks, G., Harper, M. E., and Griffiths, K., Biomed. Mass Spectrom., (1975), 2, 219.
30. Telling, G. M., Bruce, T. A., and Althorpe, J., J. Agr. Food Chem., (1971), 19, 937.
31. Gough, T. A., and Webb, K. S., J. Chromatog., (1972), 64, 201, and (1973), 79, 57.
32. Crathorne, B., Edwards, M. W., Jones, N. R., Walters, C. L., and Woolford, G., J. Chromatog., (1975), 115, 213.
33. Sen, N. P., Miles, W. F., Seaman, S., and Lawrence, J. F., J. Chromatog., (1976), 128, 169.
34. Gallegos, E. J., Anal. Chem., (1975), 47, 1150.
35. Evans, K. P., Mathias, A., Mellor, N., Silvester, R., and Williams, A. E., Anal. Chem., (1975), 47, 821.

36. Prater, T. J., Harvey, T. M., and Schuetzle, D., Twenty-Fourth Annual Conference on Mass Spectrometry and Allied Topics, San Diego, 1976, p. 625.
37. Hammar, C.-G., and Hessling, R., Anal. Chem., (1971), 43, 299.
38. Haddon, W. F., and Lukens, H. C., Twenty-Second Annual Conference on Mass Spectrometry and Allied Topics, Philadelphia, 1974, p. 436.
39. Barber, M., Chapman, J. R., Green, B. N., Merren, T. O., and Riddoch, A., Eighteenth Annual Conference on Mass Spectrometry and Allied Topics, San Francisco, 1970, p. B299.
40. Chapman, J. R., Merren, T. O., and Limming, H. J., Nineteenth Annual Conference on Mass Spectrometry and Allied Topics, Atlanta, 1971, p. 218.
41. Wolstenholme, W. A., and Elliott, R. M., American Laboratory, (1975), Nov., 68.
42. Hunt, D. F., Stafford, G. C., Shabanowitz, J., and Crow, F. W., Anal. Chem., (1977), 49, 1884; see also chapter by D. F. Hunt, this volume.
43. Watson, J. D., and Biemann, K., Anal. Chem., (1965), 37, 844.
44. Habfast, K., Maurer, K. H., and Hoffmann, G., Twentieth Annual Conference on Mass Spectrometry and Allied Topics, Dallas, 1972, p. 414.
45. Habfast, K., Maurer, K. H., and Hoffmann, G., p. 408 in "Techniques of Combined Gas Chromatography/Mass Spectrometry," ed. W. H. McFadden; John Wiley and Sons, N.Y., 1973.
46. McMurray, W. J., Green, B. N., and Lipsky, S. R., Anal. Chem., (1966), 38, 1194.
47. Kimble, B. J., Cox, R. E., McPherron, R. V., Olsen, R. W., Roitman, E., Walls, F. C., and Burlingame, A. L., J. Chromatog. Sci., (1974), 12, 647.
48. Kimble, B. J., Walls, F. C., Olsen, R. W., and Burlingame, A. L., Twenty-Third Annual Conference on Mass Spectrometry and Allied Topics, Houston, 1975, p. 503.
49. Kimble, B. J., McPherron, R. V., Olsen, R. W., Walls, F. C., and Burlingame, A. L., Adv. Mass Spectrom., Volume 7, in press; Proceedings of the Seventh International Mass Spectrometry Conference, Florence, Italy, August 1976.
50. Rapp, U., Schröder, U., Meier, S., and Elmenhorst, H., Chromatographia, (1975), 8, 474.

51. Dobberstein, P., Zune, A., and Wolstenholme, W. A., Twenty-Fourth Annual Conference on Mass Spectrometry and Allied Topics, San Diego, 1976, p. 234.
52. Chapman, J. R., Evans, S., and Holmes, J. M., Sixteenth Annual Conference on Mass Spectrometry and Allied Topics, Pittsburgh, 1968, p. 313.
53. Burlingame, A. L., Olsen, R. W., and McPherron, R. V., p. 1053 in Adv. Mass Spectrom., Volume 6, ed. A. R. West; Applied Science, 1974.
54. Smith, D. H., Olsen, R. W., Walls, F. C., and Burlingame, A. L., Anal. Chem., (1971), 43, 1796.
55. Burlingame, A. L., Smith, D. H., and Olsen, R. W., Anal. Chem., (1968), 40, 13.
56. Burlingame, A. L., p. 15 in Adv. Mass Spectrom., Volume 4, ed. E. Kendrick; The Institute of Petroleum, London, 1968.
57. McMurray, M. J., p. 43 in "Mass Spectrometry: Techniques and Applications", ed. G. W. Milne; Wiley-Interscience, 1971.
58. Klimowski, R. J., Venkataraghavan, R., and McLafferty, F. W., Org. Mass Spectrom., (1970), 4, 17.
59. Boettger, H. G., and Kelly, A. M., Seventeenth Annual Conference on Mass Spectrometry and Allied Topics, Dallas, 1969, p. 337.
60. Banner, A. E., Thirteenth Annual Conference on Mass Spectrometry and Allied Topics, St. Louis, 1965, p. 193.
61. Burlingame, A. L., Smith, D. H., Merren, T. O., and Olsen, R. W., p. 17 in "Computers in Analytical Chemistry", ed. C. H. Orr and J. A. Norris; Progress in Analytical Chemistry, Vol. 4, Plenum Press, N.Y., 1969.
62. Venkataraghavan, R., Klimowski, R. J., and McLafferty, F. W., Accounts Chem. Res., (1970), 3, 158.
63. Merritt, C., Issenberg, P., Bazinet, M. L., Green, B. N., Merren, T. O., and Murray, J. G., Anal. Chem., (1965), 37, 1037.
64. Habfast, K., p. 3 in Adv. Mass Spectrom., Volume 4, ed. E. Kendrick; The Institute of Petroleum, London, 1968.
65. McMurray, M. J., Lipsky, S. R., and Green, B. N., p. 77 in Adv. Mass Spectrom., Volume 4, ed. E. Kendrick; The Institute of Petroleum, London, 1968.
66. Bowen, H. C., Clayton, E., Shields, D. J., and Stanier, H. M., p. 257 in Adv. Mass Spectrom., Volume 4, ed. E. Kendrick; The Institute of Petroleum, London, 1968.

67. Hilmer, R. M., and Taylor, J. W., Anal. Chem., (1973), 45, 1031.
68. McFadden, W. H., "Techniques of Combined Gas Chromatography/Mass Spectrometry"; Wiley-Interscience, N.Y., 1973.
69. Fales, H. M., Heller, S. R., Milne, G. W. A., and Sun, T., Biomed. Mass Spectrom., (1974), 1, 295.
70. Kimble, B. J., and McPherron, R. V., unpublished work.
71. Campbell, A. J., and Halliday, J. S., Thirteenth Annual Conference on Mass Spectrometry and Allied Topics, St. Louis, 1965, p. 200.
72. Halliday, J. S., p. 239 in Adv. Mass Spectrom., Volume 4, ed. E. Kendrick; The Institute of Petroleum, London, 1968.
73. Green, B. N., Merren, T. O., and Murray, J. G., Thirteenth Annual Conference on Mass Spectrometry and Allied Topics, St. Louis, 1965, p. 204.
74. Barber, M., Green, B. N., McMurray, W. J., and Lipsky, S. R., Sixteenth Annual Conference on Mass Spectrometry and Allied Topics, Pittsburgh, 1968, p. 91.
75. Burlingame, A. L., Kimble, B. J., Scott, E. S., Wilson, D. M., and Stasch, M. J., p. 587 in "Identification and Analysis of Organic Pollutants in Water", ed. L. H. Keith; Ann Arbor Science Publishers, Ann Arbor, Michigan, 1976.
76. Kundred, A., Spencer, R. B., and Budde, W. L., Anal. Chem., (1971), 43, 1086.
77. Biemann, K., and McMurray, W., Tetrahedron Letters, (1965), 647.
78. Hilmer, R. M., and Taylor, J. W., Anal. Chem., (1974), 46, 1038.
79. Venkataraghavan, R., McLafferty, F. W., and Van Lear, G. E., Org. Mass Spectrom., (1969), 2, 1.

RECEIVED December 30, 1977

8

Analytical Applications of Postive and Negative Ion Chemical Ionization Mass Spectrometry

DONALD F. HUNT and SATINDER K. SETHI

Department of Chemistry, University of Virginia, Charlottesville, VA 22901

As early as 1916 Dempster (1) observed an ion at m/e=3, which was correctly identified as H_3^+. By 1925 it was well established that this ion was produced by a secondary process resulting from collision between ion (H_2^+) and neutral species (H_2) in the mass spectrometer ion source. Studies of such ion molecule collisions were largely neglected until 1952, when interest was revived by the observation of the ion, CH_5^+ formed by the reaction, (2)

$$CH_4^{+\cdot} + CH_4 \longrightarrow CH_5^+ + CH_3^{\cdot}$$

The birth of Chemical Ionization Mass Spectrometry (CIMS) took place when Field (3) and Munson (3a) realized that an ion such as CH_5^+ could ionize sample molecules by transferring a proton to them in the gas-phase. Such an ionization process is totally different from ionization of a molecule by removal of an electron, as is done in most other mass spectrometric methods. Here it is a chemical reaction between the primary ion (reagent ion) and the sample molecule which is responsible for ionization of the sample. It is possible to control both the energetics of sample ion formation as well as the type of structural information obtained in the resulting mass spectrum. Different CI reagents undergo different ion-molecule reactions with the same sample molecule, and each ion-molecule reaction affords different structural information about the sample in question.

Ion molecule reactions have been developed to identify different organic functional groups and to differentiate, primary, secondary and tertiary alcohols (4), 1°, 2°. and 3° amines (5), cyclic alkanes from

© 0-8412-0422-5/78/47-070-150$10.00/0

olefins (4), sulphur containing aromatics from non-sulphur containing aromatics, (6) and even in some cases the oxidation state of the heteroatom in poly-aromatic hydrocarbons (7). Recently there have been attempts to distinguish stereoisomers in the gas-phase using optically active reagent gases (8).

In addition to an efficient technique for the production of a wide variety of positive ions, CI is also an excellent method for generating negatively charged sample ions. Due to the large concentration of thermal or near thermal energy electrons, produced during ionization of the CI reagent gas, the resonance electron capture mechanism operates efficiently to produce large concentrations of negative sample ions under CI conditions. A description of our initial research effort in negative ion CI will be presented later in the chapter. Particularly noteworthy is the development of methodology which facilitates simultaneous detection of both positive and negative ions on a quadrupole mass spectrometer (6).

In addition to the above work we have also recently developed methodology for obtaining CI mass spectra of nonvolatile salts and thermally labile molecules, under CI conditions using quadrupole instruments with field desorption emitters as solid probes but in the absence of an externally applied field (9).

We have also demonstrated that accurate mass measurements (<10 ppm) can be made using GC-MS conditions on quadrupole spectrometers operating in the pulsed positive negative ion configuration (10).

Comparison of EI and CI Methods:

In order to fully appreciate the limitations of EI method and the potential of CIMS both in analytical chemistry and in the study of fundamental processes in gas phase, a comparison of EI and CIMS is given below.

Under EI conditions sample molecules are placed in the ion source under high vacuum (10^{-5} to 10^{-6} torr) and are ionized by impact of an energetic (>50eV) electron beam. Since a 50eV electron travels with a velocity of 4.2×10^{8} cm/sec, it transverses a molecular diameter in *ca.* 2.4×10^{-16} sec. Ionization of the sample molecule occurs on this time scale. Since the fastest molecular vibration, a C-H stretching vibration, has a period of about 10^{-14} sec, all atoms can be considered to be effectively at rest during this period (11). EI ionization therefore involves electronic excitation by Frank-Condon type of process. During this ionization process, the ion produced acquires energy in the range of 1-8eV, and, thus, frequently undergoes extensive fragmentation. Since a high vacuum is employed under EI conditions, ion-molecule collisions are effectively precluded. The internal energy of the ions, therefore, remains in non-equilibrium distribution from the instant of ionization. Formation of fragment ions from the excited parent ion, is explained by the Quasi-equilibrium theory (QET) (12) which assumes that initial excitation energy is randomized throughout the molecule at a rate which is fast relative to the rate of bond dissociation. QET predicts that fragment ions will be formed by a series of competing consecutive, unimolecular decomposition reactions. It is important to realize here that the EI process stands in contrast to the usual kinetic situation encountered both in solution and under CI conditions. In these situations, molecules are continually energized and deenergized by collisions, and a Maxwell-Boltzman type distribution of energies is either approached or realized.

In CIMS, a set of reagent ions are first generated by bombarding a suitable reagent gas at pressures between 0.5 and 1 torr, with high energy electrons (100 to 500eV). Sample molecules are introduced in the usual manner but at a conc. below 0.1% that of the reagent gas. Under these conditions only the reagent gas is ionized by EI and sample molecules are ionized only by ion-molecule reactions.

In general, gas-phase ion molecule reactions are appreciably faster than reactions between neutral species. Reactions proceeding at or near diffusion controlled rates are not uncommon under chemical ionization conditions. An explanation for these large cross-sections for reaction can be found in the treatment by Langevin (13). Long range attractive forces, which result from polarization of the neutral molecule by the approaching ion, are produced. Depending upon the proximity and the relative velocity of the two species, these attractive forces may cause the distance of closest approach to be sufficiently small for a reaction to occur. Furthermore, if the ion approaches the target molecule to within a certain range of distances, the trajectory takes on a spiral or orbiting nature around the molecule. The orbiting behavior of the two species increases the duration of the interaction, *i.e.* the lifetime of the ion-molecule complex, permits the ion and molecule to perturb each others' electronic structure, and to sample several possible activated complexes. Ionization of sample does not occur by a Frank-Condon process, since the lifetime of the ion-molecule complex can be long compared with vibrational time periods.

Under CI conditions the amount of energy imparted to the sample ion is dependent in part on the exothermicity of the ion-molecule reaction employed. In CI with methane and iso-butane, the most commonly used reagent gases, the ionization occurs by either proton transfer to, or hydride abstraction from, the sample. Since the exothermicity of gas-phase proton transfer and hydride abstraction reactions is usually low (0-3eV), the resulting even-electron ions are relatively stable towards further fragmentation. Those ions that do fragment generally do so by pathways different from those available to the odd electron species generated initially under EI conditions. Accordingly, the structural information obtained from EI and CI spectra of the same sample is usually complementary. The opportunity for sample ion to undergo stabilizing collisions with neutral reagent gas molecules under CI conditions also contributes to the reduced fragmentation observed in the CI mode. A lower limit for the number of collisions experienced by an ion in the CI source can be estimated, by using the expression

$$Z_c = K \cdot N$$

where Z_c is the number of collisions, K is the rate constant for the reaction and N is the number density of gas molecules. Typical values for $K \simeq 10^{-9} cm^3 \cdot molecule^{-1} \cdot sec^{-1}$ and $N = 2 \times 10^{16}$ (150°C, 1 torr) yield a value of 2×10^7 collisions/sec or ≃1 collision every 10^{-7} sec. (3b)

The proton affinity (P.A.) of a molecule is defined as the heat liberated on protonation. The higher the P.A. of a reagent molecule, more stable is its protonated form (reagent ion). The exothermicity of proton transfer will, of course, depend both on the acidity of the reagent molecule and the basicity of the sample molecule. For a given sample, the proton affinity values given below,

$H_2 + H^+ \rightarrow H_3^+$ $\Delta H = -101$ Kcal/mole; P.A.(H_3^+) = 101

$CH_4 + H^+ \rightarrow CH_5^+$ $\Delta H = -127$ Kcal/mole; P.A.(CH_5^+) = 127

$NH_3 + H^+ \rightarrow NH_4^+$ $\Delta H = -207$ Kcal/mole; P.A.(NH_4^+)=207

show that reaction with H_3^+ will produce protonated

sample ion, $[M + H]^+$, having 26 Kcal/mole more energy than those generated by proton transfer from CH_5^+. Due to very high proton affinity of ammonia, the NH_4^+ will only transfer a proton to molecules which are more basic than ammonia. Accordingly, the NH_4^+ ion finds utility as a reagent for selectively ionizing basic components in a mixture of organic compounds.

In many cases extensive fragmentation of the sample is desirable in order to obtain as much structural information as possible. Fragmentation under EI is due to high internal energy of the molecular ion although the free radical character of $M^{+\cdot}$ lowers activation energy for many otherwise inaccessible decomposition pathways. Fortunately EI-type spectra can be obtained under CI conditions by using powerful one electron oxidizing agents like $N_2^{+\cdot}$. The nitrogen radical cation formed by electron impact on N_2 gas at 1 torr

$$N_2^* + e(80eV) \rightarrow N_2^{+\cdot} + N_2^* + e \quad (1)$$

$$AB + N_2^{+\cdot} \rightarrow AB^{+\cdot} + N_2 \quad \Delta H \simeq -(2\text{-}8)eV \quad (2)$$

$$AB + N_2^* \rightarrow AB^{+\cdot} + N_2 + e \quad \Delta H \simeq -(0\text{-}3)eV \quad (3)$$

can transfer 2-8eV of energy to the sample molecule (AB) during the ionization step (Eq-2). Extensive fragmentation of the resulting $M^{+\cdot}$ ion results and a spectrum identical to that produced by EI methodology is obtained. Metastable (N_2^*) can also ionize the sample as shown in (Eq-3).

Selective Reagent Gases for Positive Ion CIMS

Argon-Water: When an argon-water mixture is employed as the CI reagent gas, the spectra obtained exhibit features characteristic of both conventional EI and Brönsted acid CI spectra (14). Use of this reagent gas mixture is particularly valuable when an ion characteristic of the sample molecular weight and abundant fragment ions characteristic of molecular structure are both required to solve the analytical problem at hand.

Electron bombardment of Ar/H_2O (20/1) at 1 torr produces ions at m/e 40(Ar^+), 80(Ar_2^+), and 19(H_3O^+) as well as a population of metastable argon neutrals (Ar^*). Proton transfer from H_3O^+ to a sample molecule is usually only slightly exothermic and seldom results in extensive fragmentation of the resulting M+1 ion. In contrast electron transfer from sample to Ar^+ is highly exothermic (4-6eV) and produces from the sample an energy rich radical cation which suffers fragmentation to produce a EI-type spectrum. The ability to record both EI- and CI-type spectra in a single scan is particularly useful when the maximum structural information possible is desired and the quantity of sample available for analysis is only sufficient for a single experiment. A mixture of nitrogen and water affords spectra identical to those obtained with argon and water as the CI reagent. For the purpose of comparison, conventional EI and CI(Ar-H_2O) spectra of di-n-pentylamine are shown in Figure 1. Reaction of the amine with H_3O^+ affords a single ion $[M+1]^+$. In contrast the EI spectrum displays a relatively weak molecular ion.

Deuterium oxide: When D_2O is employed as the CI reagent, all active hydrogens attached to N,S, or O atoms in an organic sample undergo exchange during the lifetime of the sample in the ion source of the mass spectrometer. Aromatic hydrogens have also been shown to undergo exchange (15). If the mol. wt. of the sample is already known from previous CI(CH_4) spectra, the number of active hydrogens in the molecule can be counted by inspection of the mol. wt. region of the CI (D_2O) spectrum. Differentiation of 1°, 2°, and 3° amines is easily accomplished in this manner (5). In the CI (D_2O) spectrum of 6-ketoestradiol (Figure 2), the M+1 peak observed in the water CI spectra at m/e=287 is shifted to m/e=290. This latter ion corresponds to d_2-ketoestradiol+D^+ and results from exchange of the two active hydrogen atoms in the diol followed by deuteration.

Ammonia

Electron bombardment of ammonia generates NH_4^+ along with $(NH_3)_2H^+$ and $(NH_3)_3H^+$. These ions function as weak Brönsted acids and will only protonate strongly basic substances like amides (16), amines (17), and some α,β-unsaturated ketones (18). The resulting sample ions seldom undergo fragmentation because of the low exothermicity associated with proton transfer reactions (Figure 3a). Aldehydes, ketones, esters, and acids, which are not sufficiently basic to accept a proton from NH_4^+, show ions in CI(NH_3) spectra resulting from the electrophilic attachment of NH_4^+ to the molecule (Figure 3b) (19).

Another interesting aspect of CI (NH_3) research is the finding that ammonia can be employed as a reagent gas for the direct analysis of organics in water. The

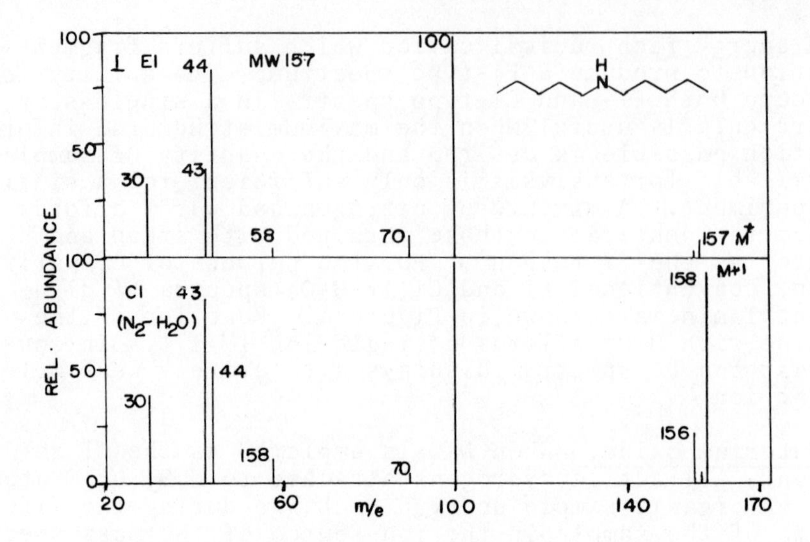

Figure 1. EI and CI (N_2 + H_2O) mass spectra of di-n-pentylamine. The intensity of reagent ions is 50 to 100 times greater than as shown.

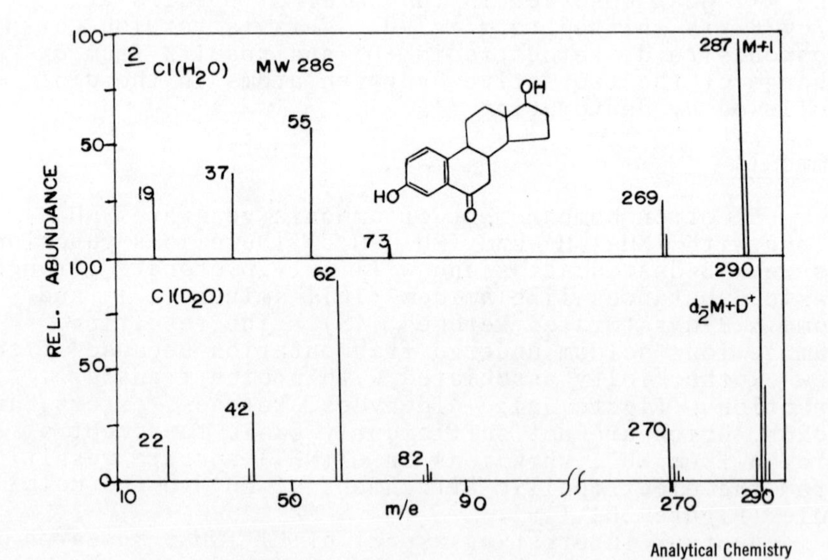

Figure 2. CI (H_2O) and CI (D_2O) mass spectra of 6-ketoestradiol. The intensity of reagent ions is 50 to 100 times greater than as shown.

ammonium ion is not sufficiently acidic to protonate water. Accordingly, organics in water can be selectively ionized when ammonia is employed as the CI reagent gas.

The NH_4^+ ion can also be used as stereochemical probe of organic structures. Simple alcohols are not ionized under CI (NH_3) conditions. In contrast, diols in which the two hydroxyl groups can simultaneously form intramolecular hydrogen bonds to the NH_4^+ ion are ionized (20). Differentiation of trans diaxial diols from the diequatorial or axial-equatorial isomers is easily accomplished by CI (NH_3) mass spectrometry (7) (Figure 3c).

Like ammonia, methylamine, is also a useful reagent gas. Aldehydes and ketones react with $CH_3NH_3^+$ in the CI source to form protonated Schiff bases (21). The reaction is quite sensitive to the steric environment of the carbonyl group (7).

Nitric Oxide: Nitric oxide is one of the most versatile reagent gases for positive ion CIMS. Electron bombardment of nitric oxide affords NO^+ which functions as an electrophile, hydride abstractor, and one electron oxidizing agent toward organic samples.

Depending on the type of organic functional groups present, any or all of the above reactions may be observed. We find that nitric oxide CI spectra are particularly useful for identifying organic functional groups in sample molecules, for differentiating olefins from cycloalkanes, and for fingerprinting hydrocarbon mixtures. Of particular interest is the finding that CI (NO) spectra can be employed to differentiate primary, secondary, and tertiary alcohols. Nitric oxide CI spectra of tertiary alcohols contain only $(M-17)^+$ ions formed by abstraction of the hydroxyl group to form nitrous acid. Spectra of secondary alcohols exhibit three ions; $(M-1)^+$, which corresponds to a protonated ketone; $(M-17)^+$; and $(M-2+30)^+$. The latter ion is generated by the oxidation of the alcohol followed by addition of NO^+ to the resulting ketone. Spectra of primary alcohols also exhibit ions corresponding to $(M-1)^+$ and $(M-2+30)^+$. In addition, however, an ion, $(M-1)^+$, unique for primary alcohols is observed. This ion is produced by hydride abstraction from C_1 of the aldehyde formed on oxidation of the primary alcohol. Figures 4a, b, c show CI(NO) spectra of three isomers of pentanol.

Scheme: NO^+ as a Selective CI Reagent.

TERTIARY ALCOHOLS:

$$(R)_3C\text{-}OH \xrightarrow{NO^+} R_3C^+ \quad (M\text{-}17)^+$$

SECONDARY ALCOHOLS:

$$(R)_2CH\text{-}OH \xrightarrow{NO^+} (R)_2\overset{+}{C}\text{-}OH \;(M\text{-}1)^+ \rightarrow (R)_2C{=}O \xrightarrow{NO^+} (R)_2CO{\cdot}NO^+ \;(M\text{-}2\text{+}30)^+$$

$$\rightarrow (R)_2CH^+ \quad (M\text{-}17)^+$$

PRIMARY ALCOHOLS:

$$RCH_2\text{-}OH \xrightarrow{NO^+} RC^+H\text{-}OH \;(M\text{-}1)^+ \longrightarrow RCH{=}O \xrightarrow{NO^+} RCHO{\cdots}NO^+ \;(M\text{-}2\text{+}30)^+$$

$$\rightarrow RC^+{\equiv}O \quad (M\text{-}3)^+$$

Differentiation of olefins and cycloalkanes having the same M. W. is also easily accomplished using nitric oxide as reagent. Spectra of cycloalkanes exhibit only an $[M\text{-}1]^+$ ions whereas those of olefins contain both $[M\text{-}1]^+$ and $[M\text{+}30]^+$ ions (19). The latter species results from electrophilic addition of NO^+ to the double bond (Figure 4d, e). CI(NO) spectra of hydrocarbons closely resemble those obtained under field ionization conditions. Over 80% of the ion current in CI(NO) spectra of most hydrocarbons is carried by the $[M\text{-}1]^+$ ion. This situation stands in sharp contrast to that obtained under either EI or CI(CH_4) conditions, where extensive fragmentation of hydrocarbon molecule is observed.

In addition to the above results, it is possible to use CI(NO) spectra to identify many functional groups in organic molecules. Spectra of acids, aldehydes and ketones show M+30, M-17; M+30, M-1; and M+30 ions respectively. One drawback to the use of nitric oxide as a CI reagent is that it is a strong oxidizing agent and therefore rapidly destroys hot metal filaments used to produce the beam of ionizing electrons. To overcome this problem we have developed a Townsend electric discharge (filamentless) source for producing a beam of ionizing electrons or ions (6).

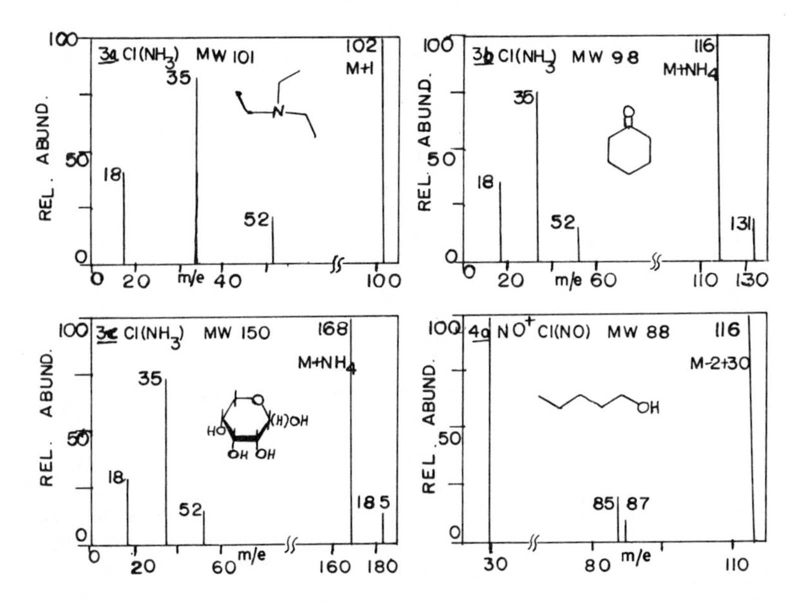

Figure 3. CI (NH_3) mass spectra of: (a) triethylamine, (b) cyclohexanone, (c) D-(−) ribose. The intensity of reagent ions is 50 to 100 times greater than as shown.

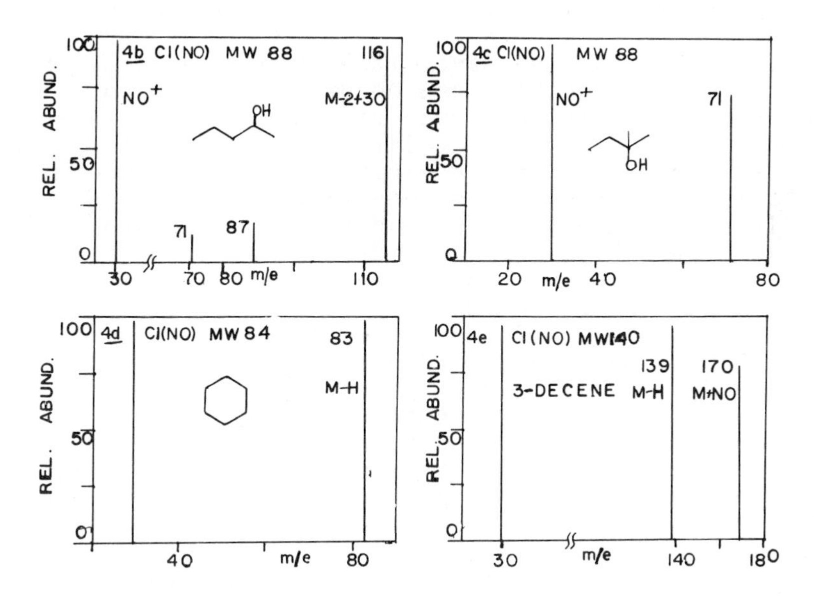

Figure 4. CI (NO) mass spectra of: (a) 1-pentanol, (b) 2-pentanol, (c) 2-methyl-2-butanol, (d) cyclohexane, (e) 3-decene. The intensity of reagent ions is 50 to 100 times greater than as shown.

NEGATIVE ION CHEMICAL IONIZATION MASS SPECTROMETRY (NICIMS)

Negative ions can be formed in the gas phase by the following three mechanisms, depending upon the energy involved (22).

$AB + e(\tilde{}0eV) \longrightarrow AB^{\cdot -}$		Resonance electron capture
$AB + e(0\text{-}15eV) \longrightarrow A^{-} + B^{\cdot}$	$B^{-} + A^{\cdot}$	Dissociative electron capture
$AB + e(>10eV) \longrightarrow A^{-} + B^{+} + e$	$A^{+} + B^{-} + e$	Ion pair production

With the exception of a small population of low energy secondary electrons produced under EI conditions during positive sample ion formation, most of the electrons available under EI conditions possess energy in excess of 10eV. Accordingly, most negative sample ions are produced by either ion-pair formation or by dissociative electron capture mechanisms, and most of the sample ion current is carried by low mass fragments, species like $O^{\cdot -}$, HO^{-}, Cl^{-} and CN^{-}, etc. Ions of this type provide little structural information about the sample molecule in question. In contrast to the above situation, Wurman and Sauer (23) showed that the thermalization of electrons can occur in a fraction of micro-second in the presence of gases like methane and iso-butane. The resulting large population of thermal electrons makes resonance electron capture the dominant mechanism for formation of negative ions under CI conditions. Once formed the negatively charged sample ions suffer up to several hundred stabilizing collisions with neutral reagent gas molecules before they exit the ionization chamber.

Unlike the results obtained by high energy electron impact, spectra recorded under CI conditions exhibit abundant molecular anions ($M^{\cdot -}$) for many types of molecules. Further, those molecules that fragment under negative ion CI conditions generally do so by elimination of small moieties from the parent anion. Since the structural features which stabilize a negative charge on an organic molecule are not usually the same as those that stabilize a positive charge, electron capture negative ion CI spectra tend to provide structural information complementary to that available in the positive ion mode.

Perhaps the most exciting feature of negative ion CIMS is the finding that the sensitivity associated with ion formation by electron capture in the CI source can be 100-1000 times greater than that available by any positive ion methodology. This result suggests that negative ion CIMS will soon become the method of choice for the quantitation of many organics in complex mixtures by GCMS. Key to the success of the negative ion technique is the development of chemical derivatization procedures which facilitate introduction of groups into the sample under analysis that enhance both formation of molecular anions, $M^{\bar{\cdot}}$, by electron capture and stabilization of the resulting $M^{\bar{\cdot}}$ toward undesirable fragmentation. Preliminary studies indicate that pentafluorobenzaldehyde and pentafluorobenzoyl chloride are excellent reagents for this purpose. Reaction of these two reagents with primary amines and phenols facilitates detection of these classes of compounds at the femtogram (10^{-15} g) level by negative ion GC-CIMS methodology (Table I).

TABLE I. Detection Limits for Derivatives of Primary Amines and Phenols

Compound	GCMS Detection Limit	Signal/Noise
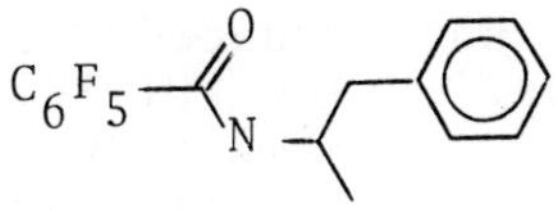 Pentafluorobenzoyl amphetamine	10×10^{-15} g	4/1
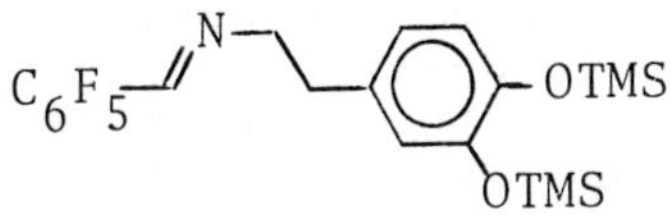 Pentafluorobenzylidene dopamine-bis-trimethyl silyl ether	25×10^{-15} g	4/1
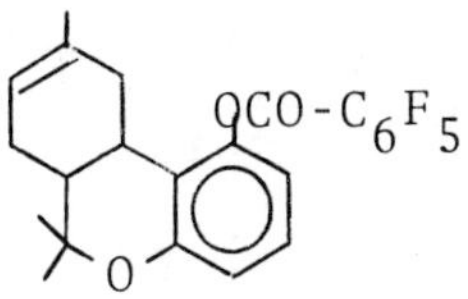 $\Delta^{1,6}$-Tetrahydrocannabinol pentafluorobenzoate	20×10^{-15} g	4/1

Pulsed Positive and Negative Ion CI (PPINICI):

Simultaneous recording of positive and negative ion CI mass spectra on Finnigan Model 3200 and Model 3300 quadrupole mass spectrometers is accomplished by pulsing the polarity of the ion source potential (±1-10V) and focusing lens potential (±10-20V) at a rate of 10 kHz as illustrated in Figure 5. Under these conditions, packets of positive and negative ions are ejected from the ion-source in rapid succession and enter the quadruple mass filter. Unlike the magnetic instruments, ions of identical m/e, but different polarity, traverse the quadrupole field with equal facility and exit the rods at the same point. Detection of ions is accomplished simultaneously by two continuous diode multipliers operating with first dynode potentials of opposite polarity. The result is that positive and negative ions are recorded simultaneously as deflections in opposite direction on a conventional light beam oscillograph.

Electron Capture - EI Type Spectra:

As noted earlier when N_2 or argon is used as reagent gas, the positive ion CI spectra are essentially identical to that obtained under EI conditions. In the negative ion mode, sample ions are formed by electron capture. Negative ions are not produced from nitrogen or argon under CI conditions. Using PPINICI technique with N_2 as a reagent gas we can simultaneously detect and record EI type spectra on the positive ion trace and ions produced by resonance electron capture on the negative ion trace. Since the structural features that stabilize positive and negative fragment ions are not usually the same, the above methodology permits one to simultaneously record spectra which contain complementary structural information. An example of this case (Figure 6a) is the EI-EC spectrum of amytal.

Electron Capture - Brönsted Acid Type Spectra:

If methane or isobutane is used as reagent gas for PPINICIMS, Brönsted acid type CI spectra are produced on the positive ion trace and electron capture spectra are generated on the negative ion trace. Negative ions derived from methane or isobutane are not observed in this mode of operation.

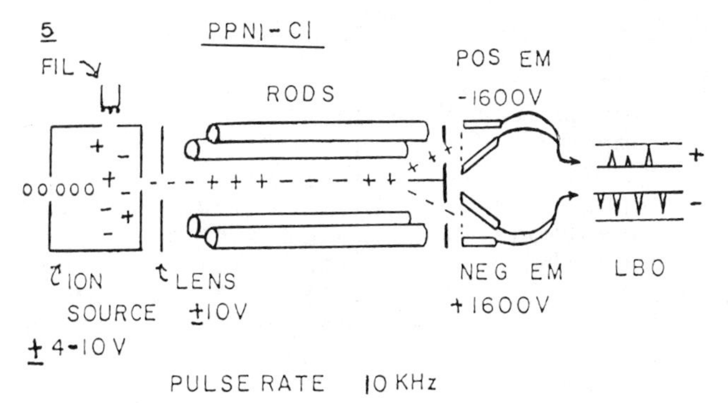

Analytical Chemistry

Figure 5. Pulsed positive negative ion chemical ionization (PPNICI) mass spectrometer: FIL–filament, EM–electron multiplier, and LBO–light beam oscillograph

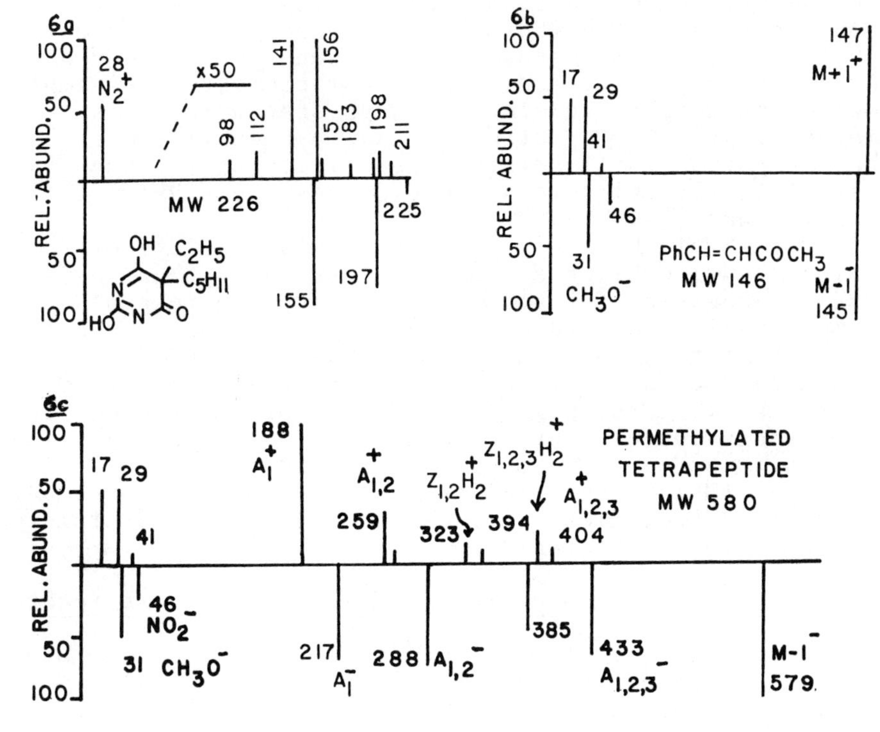

Analytical Chemistry

Figure 6. Pulsed positive negative ion CI mass spectra: (a) electron capture-impact type spectrum of amytal; (b) Brönsted acid–Brönsted base spectrum of trans-4-phenyl-*3-buten-2-one; (c) Brönsted acid–Brönsted base electron capture spectrum of the N-acetyl-per-methylated-methyl ester of Met-Gly-Met-Met. The intensity of reagent ions is 50 to 100 times greater than as shown.*

Brönsted Acid - Brönsted Base Type Spectra:

Simultaneous production of Brönsted acid and Brönsted base spectra is accomplished by using a reagent gas composed of methane and methyl nitrite. The quantity of methyl nitrite employed is insufficient to alter the population of methane reagent ions. Accordingly, a conventional methane CI spectrum of the sample is obtained on the positive ion trace. In the negative ion mode methyl nitrite is efficiently converted to CH_3O^- by dissociative electron capture reaction shown below.

$$CH_3ONO \xrightarrow{e^-} CH_3O^- + NO$$

Methoxide then functions as a strong Brönsted base and abstracts protons from organic molecules to produce $(M-1)^-$ ions. In most cases the excess energy liberated in the proton abstraction reaction remains in the new bond that is formed (CH_3OH). Consequently, the sample $(M-1)^-$ ion seldom possesses enough energy to undergo extensive fragmentation. Ions characteristic of sample molecular weight are almost always seen using this methodology. The spectra of trans-4-phenyl-3-buten-2-one (Figure 6b) illustrates the utility of methane-methyl nitrite mixture as reagent-gas under PPINICI conditions.

It should be mentioned that a wide variety of anionic nucleophiles and bases, less basic than CH_3O^-, can be generated for study as negative ion reagents by simply adding a third component to the methane-methyl-nitrite mixture employed above. When the third compound reaches a relative concentration of about 5% of the total mixture, all of the CH_3O^- formed by electron capture is consumed by ion molecule reactions involving proton transfer to CH_3O^- from the third component. In this way $(M-1)^-$ ions from cyclopentadiene, acetone, mercaptans, nitriles, etc. can be generated and employed as negative ion reagents.

Electron Capture - Brönsted Acid - Brönsted Base Spectra:

If a mixture of methane-methyl nitrite is employed as the CI reagent gas and the quantity of methyl nitrite is insufficient to consume the available population of thermal electrons, the reactant species generated in the ion source consist of CH_5^+, CH_3O^- and thermal electrons. Using the PPINICI technique the reagent combination produces simultaneously Brönsted acid CI spectra on the positive ion trace and a mixture of Brönsted base and electron capture CI spectra on the negative ion trace.

This combination of reagents is particularly useful when the problem at hand requires the determination of both sample molecular weight and detailed structural information. Sequencing of polypeptides by MS is a good example of such a problem. For a derivatized polypeptide, the PPINICI [CH_4-CH_3ONO (trace)] spectrum shows an $(M-1)^-$ ion derived from reaction of the sample with CH_3O^-, the positive ion Brönsted acid spectrum provides both the N- and the C-terminal sequence ions, and the electron capture negative ion spectrum shows ions which occur at m/e values 29 units (-N·CH_3-permethylation) higher than the N-terminal acyl sequence ion on the positive ion trace. Thus, N-terminal sequence ions can be easily recognized by the appearance of doublets separated by 29 mass units on positive and negative ion traces. Shown in Figure 6c is the PPINICI (CH_4-CH_3ONO) spectrum of N-acetyl-permethylated-methyl ester of Met-Gly-Met-Met.

Oxygen as a PPINICI Reagent:

When oxygen at 1 torr containing 10% hydrogen is employed as the CI reagent gas, O_2^+, $O_2^{\bar{\cdot}}$ and a population of thermal electrons function as the reactants in the positive and negative ion modes, respectively.

$$O_2 \xrightarrow{e^-} O_2^+ + e^-$$

$$O_2 \xrightarrow{e^-} O_2^{\bar{\cdot}} \quad O^{\bar{\cdot}}$$

$$O^{\bar{\cdot}} + H_2 \longrightarrow H_2O + e^-$$

We find that this reagent gas mixture is ideally suited for the analysis of tetrachlorodibenzodioxin (TCDD), polyaromatic hydrocarbons, and alcohols (6). Positive ion CI (O_2) spectra of aromatic molecules usually consist of a single ion corresponding to M^+. This ion is formed by electron transfer from the sample to $O_2^{+\cdot}$. Positive ion CI(O_2) spectra are, therefore, analogous to low voltage EI spectra except that under CI conditions there is no loss in sample sensitivity. Under low voltage EI conditions sample sensitivity may drop by 1 or 2 orders of magnitude.

In the negative ion mode, polyaromatic hydrocarbons either react with $O_2^{\bar{\cdot}}$ or capture an electron and then suffer reaction with a diradical oxygen molecule. Depending on the structure of the molecule the resulting ions may correspond to $M^{\bar{\cdot}}$, $(M-1)^-$, $(M+14)^-$,

$(M+15)^-$, $(M+31)^-$, or $(M+32)^-$. PPINICI with oxygen as the reagent gas is particularly suited for differentiation of isomeric polyaromatics such as the $C_{24}H_{12}$ pair, benzo(ghi)perylene and indeno(1,2,3-cd)pyrene (Figure 7), and molecules such as RC_3 and RSH_4 which have the same molecular weight but different elemental compositions. In the case of the $C_{24}H_{12}$ pair, both isomers exhibit a single ion corresponding to M^+ in the positive ion mode. On the negative ion trace benzoperylene shows ions corresponding to $M^{\cdot -}$ and $(M+15)^-$ in a 1/1 ratio. The latter species is probably a phenolic anion resulting from the reaction of $M^{\cdot -}$ with oxygen followed by the elimination of $OH^{\cdot}$. The negative ion spectrum of indenopyrene also exhibits the same two ions, $M^{\cdot -}$ and $(M+15)^-$, but in a ratio of 10/1. Further, the ratio of the total positive to total negative sample ion current is 0.5 for the benzoperylene and 22 for the indenopyrene. The presence of a five-membered ring in the indenopyrene facilitates formation of a stable negative ion and easy identification of this compound even in the presence of the other $C_{24}H_{12}$ isomers.

Although RSH_4 and RC_3 compounds have the same mol. wt. and exhibit a single ion at the same m/e value on the positive ion trace, these two types of molecules are easily differentiated in the negative ion mode. Sulfur containing molecules undergo attachment of oxygen to $M^{\cdot -}$ and form $(M+32)^{\cdot -}$ ions whereas polyaromatics containing only carbon and hydrogen form $M^{\cdot -}$ and $(M+15)^-$ ions (phenolic anions) under $CI(O_2)$ conditions. If both compound types are present in the same sample mixture, ions from each appear as an unresolved doublet at M^+ on the positive ion trace and as a doublet separated by 17 mass units (M+15 and M+32) on the negative ion trace.

$NICI(O_2)$ is also of value as a technique for analyzing alcohols. Molecules containing alcohol groups form a hydrogen bond to $O_2^{\cdot -}$ to produce $(M+O_2)^{\cdot -}$ ion and also react with $O^{\cdot -}$ to give $(M-17)^-$ ions.

Analysis of Nonvolatile Compounds:

Most mass spectrometric techniques require the sample molecules to be in the gaseous state prior to ionization and thus are severely limited in their application to the analyses of salts and thermally labile compounds. Thermal energy is the most common force used to break the intermolecular bonds and surface-molecule bonds and to facilitate introduction of sample molecules into the gaseous state. Input of energy into

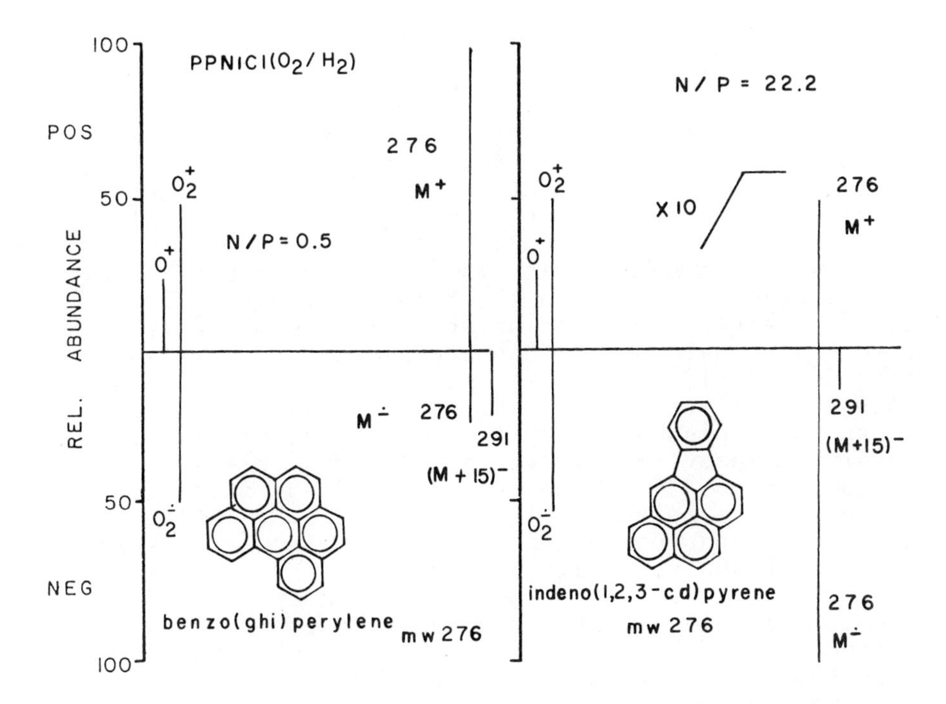

Figure 7. PPNICI (O_2/H_2) mass spectra of two isomers of $C_{22}H_{12}$: (a) benzo(ghi) perylene; (b) indeno(1,2,3-cd-)pyrene. The intensity of reagent ions is 50 to 100 times greater than as shown.

vibrational levels of the molecule can either result in dissociation of intra-molecular, inter-molecular, or surface-molecule bonds. For most salts and thermally labile molecules, the former process becomes dominant, and extensive thermal degradation of the sample occurs during the vapourization step.

A variety of techniques have been developed to overcome this problem. Derivatization of polar group in the molecule to eliminate intermolecular hydrogen bonding, is perhaps the most common and successful method. Unfortunately, derivatization adds an unwanted extra step in sample analysis. Attempts to reduce surface-molecular interactions by volatilizing the sample from relatively inert material such as Teflon also have shown some promise (24).

Two techniques, which increase the rate of sample vaporization relative to the rate of pyrolysis by placing large amounts of energy into the molecule on a time scale that is fast compared with the vibrational time period (10^{-12} to 10^{-13} sec), have also shown considerable potential. The more promising of these techniques, plasma desorption mass spectrometry (25), involves impact of high energy (100 MeV) fission products from ^{252}Cf onto the back side of metal foil coated with a molecular layer of sample. Simultaneous desorption and ionization of sample molecules results. Many polar and very high molecular weight compounds (e.g., Vitamin B_{12}) afford ions characteristic of sample molecular weight when analyzed by this new methodology. The second high energy technique employs a pulsed laser for both desorption and ionization (26). This method has not been applied to the analysis of thermally labile molecules but has shown promise for analysis of salts.

The mass spectrometeric technique which is now most commonly used for the analysis of non-volatile samples is field desorption (27). This technique was introduced by Beckey in 1969, and employs a very strong electrostatic field (1-4V/A°) to ionize molecules absorbed on specially prepared surface, and a combination of this field and thermal energy to desorb the ionized sample molecules. The activated surface consists of dense, highly branched carbon microneedles (30 μm long) grown on 10 μm tungsten wire.

According to Muller (28), the strong field lowers the barrier for electron tunnelling from molecule to the wire surface and increases the rate of ionic desorption by lowering the required heat-of-desorption.

The rate constant K for ionic desorption is given as

$$K = \nu \cdot \exp(-Q/kt)$$

where ν is the vibrational frequency (10^{13} sec^{-1}), and Q is the heat of ionic desorption, reduced from the thermodynamic value Q°

$$Q° = Ha + IP - \phi$$

by a Schottky term, $3.8\eta^{3/2}F^{1/2}$

$$Q = Q° - 3.8\eta^{3/2}F^{1/2}$$

where Ha = Heat of desorption of neutral molecule

IP = Ionization potential

ϕ = Work function of the surface

F = Applied field in V/A°

η = Charge of the evaporating ion

The effect of the field can be more clearly seen by inspecting the changes in the energy levels of the surface-molecule interaction on application of the field. Figure 8 shows such energy levels when IP-ϕ is a large positive value. This is usually the situation for organic molecules adsorbed on the surface. For these molecules the IP is between 9-12 ev and ϕ varies between 4-6 ev. Figure 8b shows that in the presence of a strong field the ionic curve "crosses" the atomic curve at Xc. (In fact the curves repel rather than cross because there are no symmetry or spin differences which permit degeneracy in the two states.) On thermal excitation the desorption from the atomic ground state will result either in the emission of an ion by adiabatic transition, i.e., field ionization at or beyond Xc, or by the emission of a neutral by non-adiabatic transition along the atomic curve. At very high fields (Figure 8c), the intersection point, Xc, comes so close to the surface that Q is greatly reduced. The resulting separation of the two states becomes so large that desorption of the molecule can only occur in ionic form. Also note that the probability of ionic desorption increases as the polarizibility (α) of the molecule increases, due to the lowering of energy of ionic desorption by $1/2\ \alpha F^2$. This latter term corresponds to the polarization energy of the molecule.

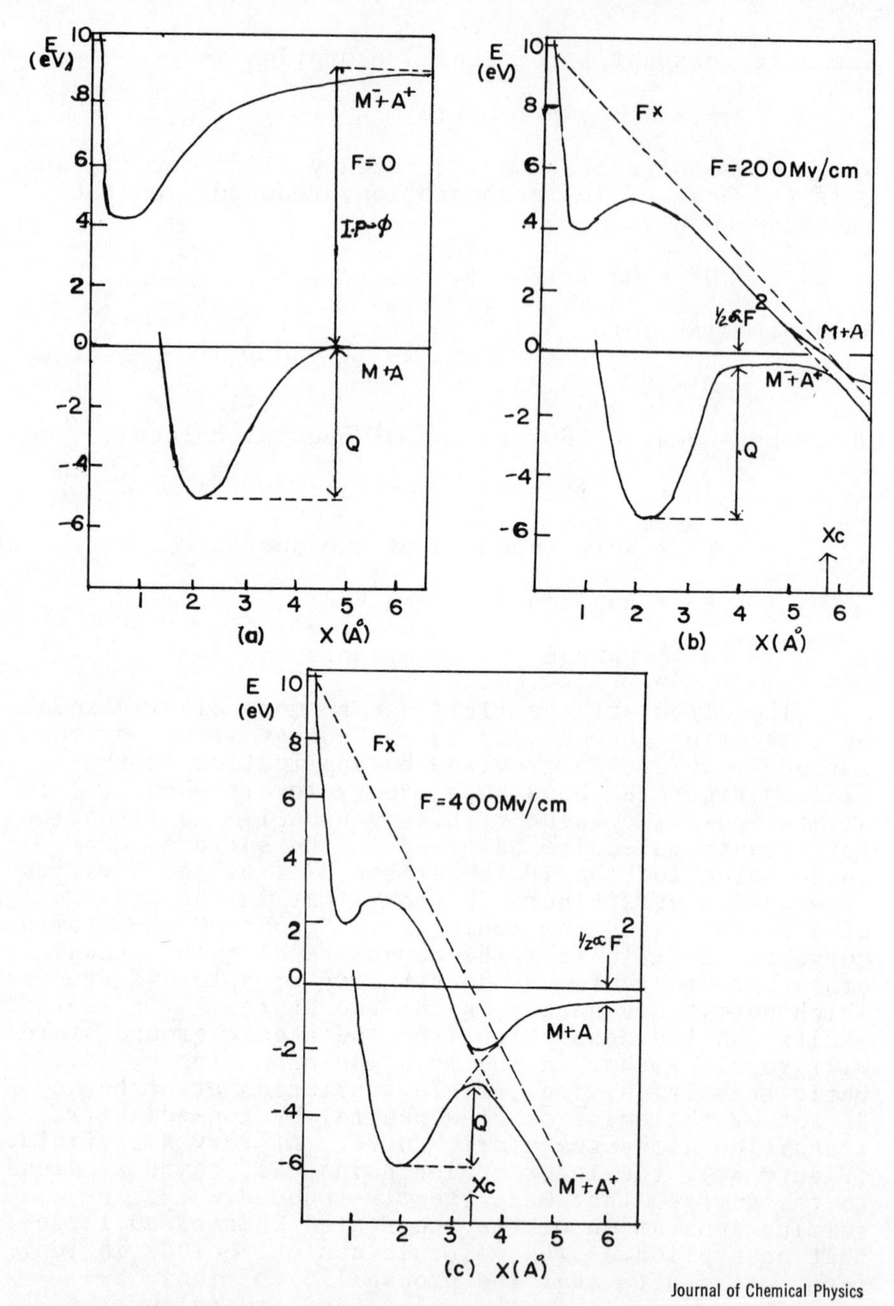

Journal of Chemical Physics

Figure 8. Potential energy diagrams for the field desorption process in the situation where: (a) the ionization potential (IP) is large when compared with the work function (ϕ); (b) in a field of intermediate strength (200 MV/cm), field ionization can occur beyond critical distance Xc; *(c) in the presence of high field (400 MV/cm). Desorption of covalently bound molecules requires a reduced activation energy* Q*, followed by field ionization beyond* Xc.

Despite the above treatment, there does not exist a comprehensive theory which can account for all the results obtained in FD experiments with large organic molecules. Neither is it possible to pinpoint the magnitude of field generated near the surface of the emitter in commercial FD instruments. At a given applied potential the strongest electrical fields are generated at surfaces having the smallest radii (i.e., sharpest points). Therefore, field desorption is thought to occur from the tip of the carbon dendrite on the emitter, and molecules are thought to migrate to the dendrite tip with the aid of thermal energy, through some type of fluid structure formed on the surface.

Recently, Holland et al. (29) observed that a normal field desorption mass spectrum can be obtained, at the usual emitter temperature, without the application of high (10KV) external voltage (the desorption field). This work is reviewed in another chapter in this volume. Since the ion accelerating voltage (3KV) was left on in their experiments, a field on the order of 10^3V/cm is still present in the vicinity of the emitter. From their experiments, Holland et al., concluded that the sample transport and desorption is independent of the applied voltage over the range 7KV to 12KV. To explain their results, Holland et al. postulated an ionization model in which the field, if present, merely acts as a vehicle to remove ions once they are formed by chemical reactions in a semi-fluid layer of sample molecules on the surface of the emitters. In our laboratory, we have recently employed FD emitters as solid probes under CI conditions and agree with Holland that, for many molecules, the presence of a high external field is unnecessary. Our experiments are conducted on a Finnigan quadrupole mass spectrometer without application of an external field to the emitter. It is important to note, however, that our experiments were conducted under CI conditions while Holland et al. carried out their experiments under EI conditions.

All of our studies have been performed on Finnigan model 3200 or 3300 quadrupole instruments equipped with CI ion sources and an INCOS Model 2300 data system. Methane at 0.5 torr pressure was used as the reagent gas, with the ion source temperature between 100-250°C. Normal field desorption emitters were used. Sample preparation involved placing a drop of solution containing the sample in a suitable solvent on the surface of the emitter using a 10 μℓ syringe. The sample was introduced to the ionization chamber by removing the repeller assembly from the CI ion source and by pressing the FD emitters in the hole thus vacated. In this configuration the emitter wire is situated directly on line with the electron entrance hole, 3 mm back of the ion-exit slit. The spectra were generated by simply heating the emitter wire rapidly and scanning the mass spectrum at 4 sec/decade.

Many salts, *e.g.*, sodium and potassium benzoates, creatine and arginine hydrochlorides, choline chloride, and thermally labile compounds like guanosine, cyclic-adenosine monophosphate (C-AMP), sugars, dioxathan (a pesticide) and arginine-containing undervitized peptides, all afford spectra containing ions which facilitate assignment of sample molecular weight as well as fragment ions characteristic of molecular structure. Some typical spectra are shown in Figure 9. None of these compounds gives an ion characteristic of molecular weight under conventional EI, CI, or FI (field ionization) conditions.

At least three mechanisms for the observed desorption and ionization of samples on emitter surfaces under CI conditions deserve consideration.

Mechanism I. *Thermal desorption of sample followed by chemical ionization of the gaseous neutral molecule*.

As the emitter current (temperature) is increased, a point is reached where the crystal lattice of the sample breaks down and migration of sample molecules to the tips of the carbonaceous dendrites occurs in the resulting semi-fluid state. Desorption from the emitter surface occurs at a lower temperature than that required for desorption from conventional solids probes because the surface-sample bonding, and sample-sample interactions are minimized at the tips of the carbonaceous dendrites.

Mechanism II. *Field desorption and ionization mediated by CI reagent ions*

This mechanism is analagous to that operating under conventional field desorption conditions except the required strong field is provided by the ions in the CI reagent gas rather than by an external 10KV potential difference applied to the emitter and draw-out plate. According to Mechanism II, sample molecules migrate to the tips of the emitter dendrites at some critical temperature. There they experience a field generated by nearby ions in the reagent gas plasma. This field

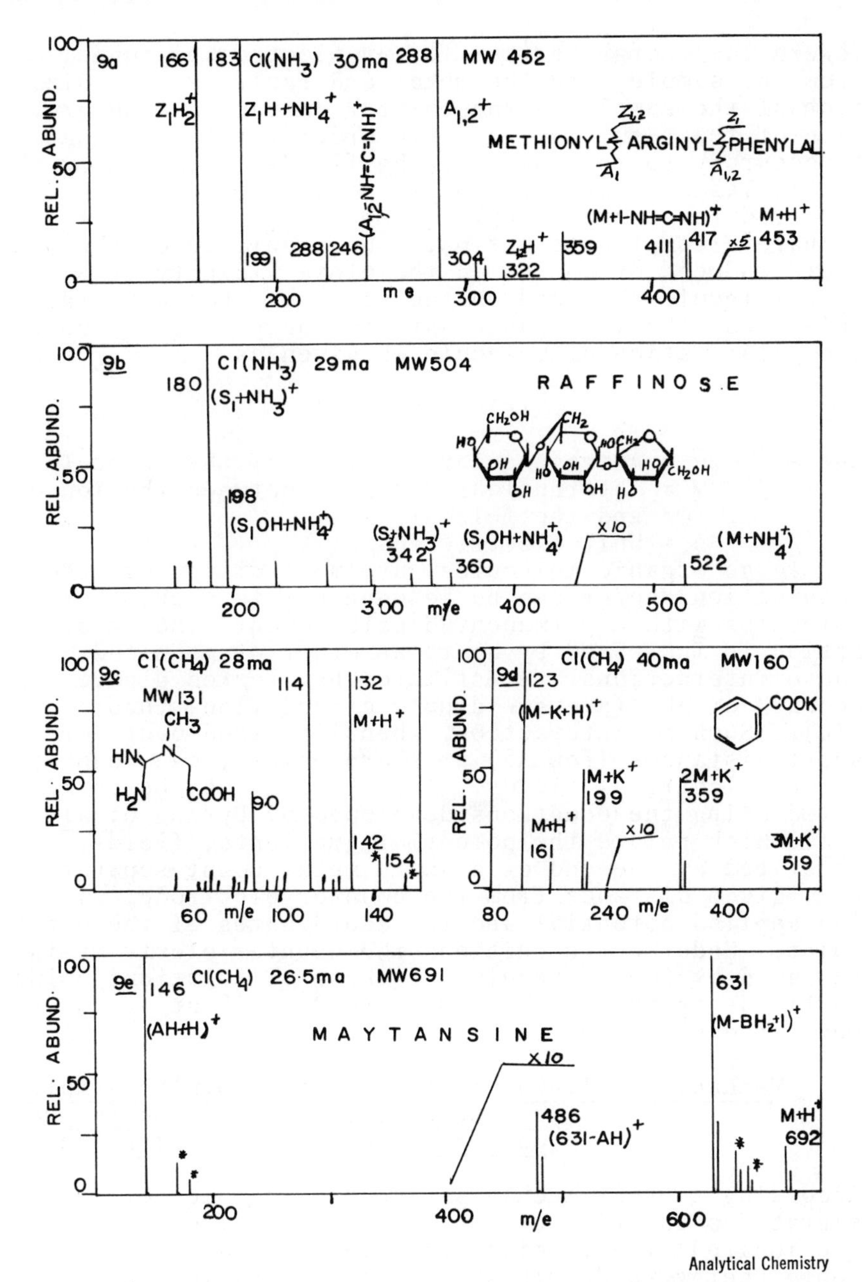

Figure 9. Activated-emitter, solid-probe, NH_3 (0.5 torr) or CH_4 (0.5 torr) CI spectra of: (a) methionyl-arginyl-phenylalanine, (b) raffinose, (c) creatin, (d) potassium benzoate, (e) maytansine—a large macrocyclic natural product with antitumor properties. Ions designated by an asterisk contain a methane reagent ion, $C_2H_5^+$ or $C_3H_5^+$, plus the sample molecule or a neutral fragment derived from the sample molecule.

lowers the energy barrier for an electron to tunnel from the sample into the metal and facilitates ionization of the sample on the emitter surface and desorption of the resulting ion. In order for the ionization to occur by this mechanism, the fields generated by the ionic plasma would have to be comparable to that required for conventional FD experiments (ca. IV/A°). We have attempted to estimate the magnitude of the field induced by an ion in the close vicinity (4-5A°) of a molecule absorbed on the surface. For nonpolar molecules, the long range interaction potential due to the polarization of molecule is given as

$$V(r) = -\bar{\alpha}e^2/2\gamma^4$$

where $\bar{\alpha}$ = Average polarizibility of the molecule
γ = Internuclear distance between the ion and the molecule
e = Unit electric charge

For large organic molecules, at small distances, the interaction energy can be between 0.1 to 1 eV. For molecules with a permanent dipole moment, the interaction is increased by up to an order of magnitude. These interactions are estimated by "locked dipole moment" or ADO (average dipole orientation) theories (30). Such an interaction, when impressed over very short distances (few A°) at sharp points, can generate strong electrical fields. These fields can be estimated using the equations developed by Eyring *et al.* (31) which relate the potential gradients, fields, generated at the end of a sharp metal point separated by a given distance from the counter electrode, with the applied potential and the coordinates of the metal point. Under our conditions the counter electrode is replaced by ions. Results of these calculations point to fields on the order of 0.05 to 0.5V/A° at an internuclear distance of 4A°.

Mechanism III. Chemical ionization of the sample on the emitter surface and thermal desorption of the resulting ion.

According to this mechanism, neutral sample molecules migrate to the tips of the emitter dendrites where they are ionized by a chemical reaction with a CI reagent ion. Energy to desorb the resulting sample ion is provided by the temperature of the emitter, the exothermicity of the ion-molecule reaction or possibly by fields generated by the ion plasma at the emitter surface.

Work is currently in progress in our laboratory to determine which of the above mechanisms is responsible for the observed results.

Accurate Mass Measurement:

Determination of molecular composition by accurate mass measurements is usually accomplished with a double focusing magnetic sector mass spectrometer operating at high resolution (>10,000) in part to resolve sample ions and reference ions produced simultaneously. The accurate mass of the sample ion is then calculated by extrapolation from the measured mass of a nearby reference ion. Drawbacks of the above methodology include the high cost of the necessary instrumentation, low sensitivity which always accompanies operation at high resolution, sample ion suppression due to the large quantities of reference compound required to produce abundant reference ions at high mass, and the difficulty in using the method for GC-MS analysis due to the slow scan speeds usually required to maintain adequate ion current at high resolution.

To overcome this problem, Aspinal et al. (32), in 1975, developed a technique for obtaining accurate mass measurements at low resolution using a double beam magnetic sector mass spectrometer. In this method two positive ion beams, one containing ions from the internal standard and one beam containing ions from the sample are produced in two separate ion sources, passed through the same magnetic field simultaneously, and collected separately at two electron miltipliers. Accurate mass measurements (<10 ppm) under GC conditions have been demonstrated using this methodology, but the cost of the necessary instrumentation is still quite high.

In an effort to make elemental composition data available to users of quadrupole mass spectrometers, we have recently examined the possibility of obtaining accurate mass measurement data using PPNICI methodology. Results to date indicate that accurate mass measurements (<10 ppm) can be achieved while operating at unit resolution on a quadrupole spectrometer using scan speeds compatible with routine GC-MS conditions (5 sec scan cycle time). The experimental technique involves recording spectra of a reference substance (PFK) and sample simultaneously in the negative ion mode and positive ion mode, respectively. This is possible because the PFK negative ion current is 600 times larger than the corresponding positive ion current. Thus, by recording spectra of sample plus a trace quantity of PFK, it is possible to obtain spectra where reference ions only appear on the negative ion trace and only sample ions appear on the positive ion trace. Accurate mass measurement of positively charged sample molecules is then accomplished by using an INCOS Model 2300 data system to scan the spectrum, to acquire data simultaneously from both positive and negative ion multipliers, to determine peak centroids, and to calculate the exact mass of the ions on the positive ion trace based on their positions in time relative to PFK ions in the negative ion trace.

Multiple spectra of sample + PFK (trace) are recorded by repetitively scanning the quadrupole instrument over a mass range 65-650 using a scan cycle time of 5 sec. When cocaine was analyzed by the above procedure and data from any five consecutive scans were averaged, measured values for the $(M+H)^+$ ion at m/e = 304.155 were found to be within 10 ppm (3 mmu) of theoretical value. It is important to note that no reference compound suppression is observed in this method, since only a trace quantity of PFK is used.

Conclusion

In this chapter, we have discussed methods of selective ionization using both positively and negatively charged reagent gases. The combination of both techniques in a single mass spectrometer has been described and some applications presented. A particularly useful application is the capability to obtain exact mass data using a quadrupole mass spectrometer. In addition, a new technique for obtaining mass spectra of nonvolatile samples has been outlined.

Literature Cited.

1. Dempster, A. J., Phil. Mag., (1916), 31, 438.
2. (a.) Tal'yoze, V. L. and Lyubimova, A. K., Dokl. Akad. Nauk SSSR, (1952), 86, 909. (b.) Field, F. H.; Franklin, J. L.; and Lampe, F. W., J. Am. Chem. Soc., (1957), 79, 2419.
3. Field, F. H., Accounts Chem. Res., (1968), 1, 42. Some recent reviews are: (a) Munson, M.S.B., Anal, Chem., (1977), 49, 772A. (b) Field, F. H., "Ion-Molecule Reactions", J. L. Franklin, Ed., Plenum Press, N. Y. (1972).
4. Hunt, D. F. and Ryan, J. F., J.C.S. Chem. Comm., (1972), 620.
5. Hunt, D. F.; McEwen, C. N.; and Upham, R. A., Anal. Chem., (1972), 44, 1292.

6. Hunt, D. F.; Stafford, G. C.; Crow, F. W.; and Russell, J. W., Anal. Chem., (1976), 48, 2098.
7. Hunt, D. F., unpublished results.
8. Fales, H. M. and Wright, G. J., Abstracts, Twenty Fifth Annual Conference on Mass Spectrometry and Allied Topics, Washington, D.C., May 1977, No. W-11.
9. Hunt, D. F.; Shabanowitz, J.; and Botz, F. K., Anal. Chem., (1977), 49, 1160.
10. Hunt, D. F.; Stafford, G. C.; Shabanowitz, J.; and Crow, F. C., Anal. Chem., in press.
11. Cooks, R. G.; Beynon, J. H.; Caprioli, R. M.; and Lester, G. R., "Metastable Ions", Elsevier Company, Amsterdam, 1973.
12. Rosenstock, H. M.; Wallenstein, M. B.; Wahrhaftig, A. L.; and Eyring, E., Proc. Natl. Acad. Sci. U.S., (1952), 38, 667.
13. Knewstubb, P. F., "Mass Spectrometry and Ion-Molecule Reactions", Cambridge University Press, 1969.
14. Hunt, D. F., and Ryan, J. F. III, Anal. Chem. (1972), 44, 1306.
15. Freiser, B. S.; Woodin, R. L.; and Beauchamp, J. L., J. Amer. Chem. Soc., (1975), 97, 6893.
16. Dzidic, I., J. Amer. Chem. Soc., (1972), 94, 8333.
17. Wilson, M. S.; Dzidic, I.; and McCloskey, J. A., Biochim. Biophys. Acta, (1971), 240, 623.
18. Dzidic, I. and McCloskey, J. A., Org. Mass Spectrum, (1972), 6, 939.
19. Hunt, D. F., Adv. Mass Spectrometry, (1974), 1, 517.
20. Hogg, A. M. and Nagabhushan, T. L., Abstracts, Twentieth Annual Conference on Mass Spectrometry and Allied Topics, Dallas, Texas, June 1972, No. P2.
21. Beggs, D., ibid, No. N4.
22. For recent reviews see: (a) Dillard, J. G., Chem. Rev., (1973), 73, 589. (b) K. Jennings, Mass Spectrometry, Vol. 4, Specialist Periodical Reports, The Chemical Society, Burlington House, London, 1977.
23. Warman, J. M. and Sauer, M. C., Jr., J. Chem. Phy., (1970), 52, 6428.
24. Beuhler, R. J.; Flanigan, E.; Greene, L. J.; and Friedman, L., Biochem., (1974), 13, 5060.
25. Macfarlane, R. D. and Torgerson, D. F., Science, (1976), 191, 920.
26. Mumma, R. O. and Vastola, F. J., Org. Mass Spectrom., (1972), 6, 1373.

27. Beckey, H. D. and Schulten, H. R., Angew. Chem. Internat. Edit., (1975), 14, 403.
28. Muller, W. W., Phys. Rev., (1956), 102, 618.
29. Holland, J. F.; Soltman, B.; and Sweeley, C. C., Biomed. Mass Spectrom., (1976), 3, 340.
30. Bowers, M. T. and Su, Timothy, "Interactions Between Ions and Molecules", P. Ausloos, Ed., Plenum Press, N. Y. (1974).
31. Eyring, C. F.; Mackeown, S. S.; and Millikan, R. A., Phys. Rev., (1928), 31, 900.
32. Aspinal, M. L.; Compson, K. R.; Dowman, A. A.; Elliott, R. M.; and Haselby, D., Abstracts, Twenty Third Annual Conference on Mass Spectrometry and Allied Topics, Houston, Texas, May 1975, No. C-2.

RECEIVED December 30, 1977

Investigations of Selective Reagent Ions in Chemical Ionization Mass Spectrometry

K. R. JENNINGS

Department of Molecular Sciences, University of Warwick, Coventry, United Kingdom

The great majority of reagent ions used in chemical ionization mass spectrometry (1) are either charge exchange reagents such as Ar^+ and N_2^+ or ions which react as gas phase acids or bases such as H_3^+ and CH_5^+ or H^- and CH_3O^-. For each group of reagent ions, the amount of fragmentation depends primarily on the exothermicity of the electron or proton transfer reaction. There are, however, other types of reagent ions which undergo different reactions with specific classes of compounds and which, therefore, may give more detailed information about the presence or location of particular functional groups. within the molecule. This paper gives a brief account of work recently carried out in our laboratories in which the reactions of the vinyl methyl ether molecular ion, $CH_3OCH{=}CH_2^{+\cdot}$ and the atomic oxygen negative ion, $O^{-\cdot}$ with a number of substrates were investigated.

Reactions of the Vinyl Methyl Ether Molecular Ion.

Previous work using ion cyclotron resonance (ICR) spectrometry on the reactions of olefinic molecular ions with various fluoro-olefins (2,3) showed that in each system, a new olefinic ion was formed. Deuterium-labelling experiments suggested that these reactions occur by the formation of a substituted cyclobutane ion as an intermediate which fragments with the retention of the structural identity of the substituted methylene groups. A similar reaction was observed (4) in the vinyl methyl ether system in which the two secondary ions of m/e = 88 and 84 correspond to the elimination of ethylene and methanol, respectively. In each system, therefore, an olefinic ion attacks a neutral olefin molecule to yield an ion of even mass by the elimination of an olefin, and in the

© 0-8412-0422-5/78/47-070-179$05.00/0

vinyl methyl ether system a further ion of even mass is formed by the loss of methanol from the reaction complex. The reactions of ions of this type with other unsaturated species were therefore investigated with a view to using this type of reaction to locate the position of double bonds in molecules. No useful results were obtained using $CH_2CF_2^{+\cdot}$ and $C_2F_4^{+\cdot}$ as reagent ions since they reacted mainly by charge transfer. But in a number of systems, reactions of the $CH_3OCHCH_2^{+\cdot}$ ion provided information which enables one to locate the position of a double bond (5). In studies in the high pressure source of the double focussing MS50 mass spectrometer, it was found that a 9:1 mixture of CO_2 and CH_3OCHCH_2 gave a good yield of $CH_3OCHCH_2^{+\cdot}$ and substantially reduced the reaction of this ion with its parent molecule.

The reactions of interest are of the general type

$$CH_3OCH{=}CH_2^{+\cdot} + X\text{-}CH{=}CH\text{-}Y \rightarrow \begin{array}{c} X \quad H \quad H \quad Y \\ | \quad\quad | \\ H \quad H \quad H \quad OCH_3 \end{array}$$

$$\rightarrow XCH{=}CH_2 + YCH{=}CHOCH_3^{+\cdot}$$

in which, in general, the simpler of the two possible olefins is eliminated as a neutral and a substituent containing a heteroatom is retained along with the methoxyl group as part of the olefinic ion. Hence, for example, all terminal olefins react by the formation of a complex from which ethylene is eliminated.

The expected even-mass secondary ions were observed in the mass spectra given by the reactions of $CH_3OCHCH_2^{+\cdot}$ ions with a number of simple hydrocarbon olefins. In each case, a further even-mass ion was observed due to the loss of CH_3OH from the reaction complex, together with odd mass ions which arise from the loss of atoms or radicals from the complex. These reactions readily allow one to distinguish between isomeric species such as 1-butene and 2-butene. In Figure I the very different spectra given by 1-octene and trans-4-octene are illustrated. Both $CH_2{=}CHCH_2CH{=}CH_2$ and $C_6H_5CH_2CH{=}CH_2$ reacted as terminal mono-olefins and gave a characteristic peak due to the loss of C_2H_4 from the reaction complex. Similar reactions were also observed with allyl formate and with $CH_2{=}CH\text{-}CH_2COOH$. When oleic acid was run as an "unknown", the position of its double bond was correctly located by means of the reaction

$$CH_3(CH_2)_7CH{=}CH(CH_2)_7COOH \quad + \quad CH_3OCH{=}CH_2^{+\cdot} \rightarrow$$

$$\begin{array}{l} CH_3(CH_2)_7\underset{|}{C}H\text{-}\underset{|}{C}H\text{-}(CH_2)_7COOH^{+\cdot} \\ \qquad\qquad CH_2\,CH\text{-}OCH_3 \end{array} \rightarrow$$

$$CH_3(CH_2)_7CH{=}CH_2 \quad + \quad CH_3OCH{=}CH\text{-}(CH_2)_7COOH^{+\cdot}$$

$$\underline{m}/\underline{e} = 200$$

The importance of the reaction in which methanol is eliminated increases as the size of the molecule increases and is the only reaction yielding an even mass ion with unsaturated cyclic compounds and with non-conjugated dienes such as 1,4 and 2,6-octadienes. Unsaturated nitriles are simply protonated. The conjugated species 1,3-butadiene undergoes a Diels-Alder reaction in which methanol is eliminated.

It is planned to extend this study to more complicated unsaturated systems. In view of the quite general nature of the reaction, it is hoped that other olefinic ions will be identified as useful reagent ions. It should then be possible to locate the positions of carbon-carbon double bonds in a wide variety of compounds without prior derivatization and in favorable cases, as in the 1-fluoropropenes (6), it may be possible to distinguish cis- and trans-isomers.

Reactions of the Atomic Oxygen Negative Ion.

In recent years negative chemical ionization mass spectrometry (7) has become increasingly important. In certain applications, the type of spectrum produced and the sensitivity of the method make it superior to positive chemical ionization, as will be described elsewhere in this volume. Most negative reagent ions react primarily as strong gas phase bases but the radical anion, $O^{-\cdot}$, has been found to have some unique reactions which yield interesting structural information in a number of cases.

Most of the early studies (8,9) were carried out at low pressures in a ICR spectrometer. N_2O was found to be a convenient source of $O^{-\cdot}$ ions although the fact that they are formed with 0.38 eV of translational energy makes N_2O less satisfactory as a source of $O^{-\cdot}$ ions for determination of rate constants. In the combined EI/CI source of the MS50, further reactions involving N_2O may occur, but these were suppressed by using a 9:1 mixture of N_2 and N_2O at a total pressure of 0.5 Torr. This gave a high yield of $O^{\cdot -}$ ions with only a trace of NO^- ions. The Daly detector of the MS50 was replaced by an electron multiplier for negative ion work, but modifications currently being carried out to the Daly detector will enable it to operate in the negative ion mode.

Early work (8) on the reaction of $O^{\cdot -}$ with CH_2CD_2 had shown that only $CH_2C^{\cdot -}$ and $CD_2C^{\cdot -}$ ions are formed, indicating that the reaction occurs by H_2^+ abstraction from a single carbon atom. This was later shown (9) to be a reaction characteristic of many 1-olefins although H^+ abstraction from alkyl groups predominate unless the terminal $=CH_2$ group is activated by the presence of an electron withdrawing group in the molecule, e.g., F or CF_3. Other classes of compounds were found to undergo the H_2^+ abstraction reaction. In the case of aliphatic nitriles, the fact that the reaction occurs with n-C_3H_7CN but not with $(CH_3)_2CHCN$ provides strong evidence that H_2^+ abstraction takes place at the α-C atom.

In an extensive study of the reactions of $O^{\cdot -}$ ions with carbonyl compounds (10), the major reactions were found to be of the general type

$$M + O^{\cdot -} \rightarrow (M-1)^- + \cdot OH \quad (1)$$

$$\rightarrow (M-2)^{\cdot -} + H_2O \quad (2)$$

$$\rightarrow (M+O-R)^- + R\cdot \quad (3)$$

$$\rightarrow (M-2-R)^- + R\cdot + H_2O \quad (4)$$

$$\rightarrow (M-1)\cdot + OH^- \quad (5)$$

Further reaction of OH^- leads to the production of $(M-1)^-$ so that the intensity ratio $(M-1)^-/(M-2)^{\cdot -}$ rises as the pressure rises. In the high pressure source, the three isomeric pentanones give quite different spectra as shown in Figure 2, each of which can be rationalized in terms of the above reactions. The varying yields of $(M-2)^{\cdot -}$ ions are attributed to the relative ease of abstraction of H_2^+ from the carbon atoms adjacent to the carbonyl group since deuterium-labelling experiments indicated that H_2^+ abstraction occurred only from these positions. In the case of $(CH_3)_2CHCOCH(CH_3)_2$, no $(M-2)^{\cdot -}$ ions are formed, consistent with this interpretation. Hence in compounds of this type, the reactions of $O^{\cdot -}$ ions can be used to detect the presence of activated CH_2 groups.

With aromatic compounds (11), reactions analogous to (1)-(3) occur but a further reaction characteristic of aromatic compounds is:

$$M + O^{\overline{\cdot}} \rightarrow (M\text{-}H\text{+}O)^{-} + H\cdot \qquad (6)$$

In certain cases, low intensity peaks were observed due to the occurrence of the reaction

$$M + O^{\overline{\cdot}} \rightarrow (M\text{-}H\text{+}O\text{-}HX)^{-} + H\cdot + HX \qquad (7)$$

Benzene gives a very simple spectrum consisting of approximately equal intensity peaks at $(M-2)^{\overline{\cdot}}$ and $(M-H+O)^{-}$ only. With 1,3,5-benzene-d_3, the major low mass peak is $(M-3)^{\overline{\cdot}}$ suggesting 1,2 elimination of H_2 although 1,4 elimination cannot be ruled out. The two higher mass peaks were of very similar intensities, indicating a negligible H/D isotope effect. Pyridine yields a spectrum closely resembling that of benzene, but there is in addition a low intensity $(M-1)^{-}$ peak. Since the hydrogen atoms are no longer equivalent, it is of interest to investigate positional selectivity in the displacement reaction. In the case of 4-deuteropyridine, the peak arising from deuterium displacement is only slightly less intense than that due to hydrogen displacement although statistically, an intensity ratio of 1:4 would be predicted. This indicates a marked preference for the displacement of the hydrogen atom in the 4-position. The results obtained with 1,3,5-benzene-d_3 show that this cannot be attributed to an isotope effect.

However, it was with the alkyl aromatic compounds that the most interesting spectra were obtained. Observations on spectra given by deutero-toluenes showed that the two major peaks arise from the abstraction of H^+ from the methyl group to give the $(M-1)^{-}$ ion and the displacement of a ring hydrogen atom to form the $(M-H+O)^{-}$ ions. The three isomeric xylenes give the spectra shown in Figure 3 and it is seen that the *meta*-isomer gives a spectrum quite different from those of the other two isomers. The prominence of the $(M-2)^{\overline{\cdot}}$ ion appears to arise from the presence of the two methyl groups *meta* to each other since when one of the methyl groups is fully deuterated, the $(M-3)^{\overline{\cdot}}$ is formed by loss of HOD, indicating that one hydrogen atom is removed from each methyl group in forming the $(M-2)^{\overline{\cdot}}$ ion. This is also consistent with the observation that the $(M-2)^{\overline{\cdot}}$ ion is the base peak in the spectrum given by mesitylene.

In the spectra of all the alkyl benzenes discussed above, the ratio $I(M-H+O)^{-}/I(M-1)^{-}$ lies in the range 0.07-0.19. However, in the case of *t*-butylbenzene,

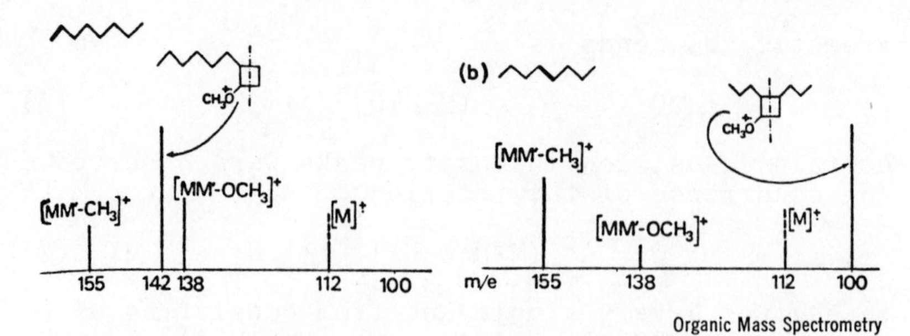

Figure 1. Spectra given by 1-octene and trans-*4-octene on reaction with $CH_3OCHCH_2{\cdot}^+$*

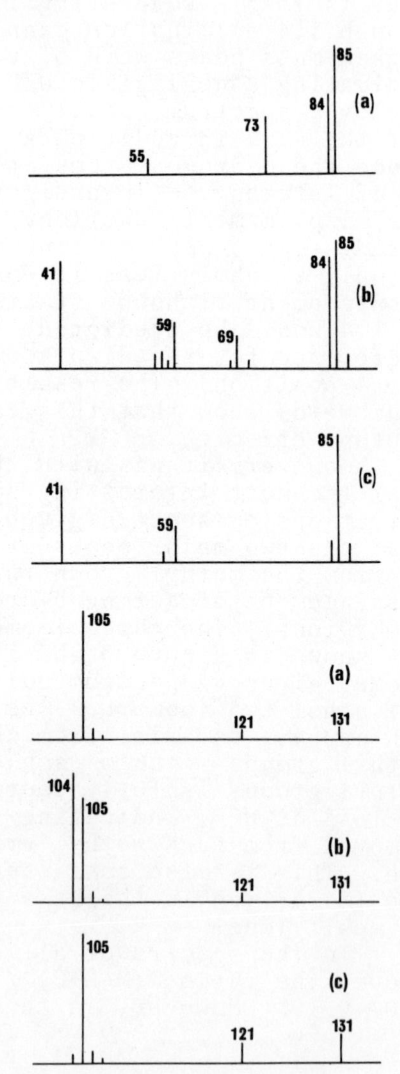

Figure 2. $O{\cdot}^-$ NCI spectra of three isomeric pentanones, (a) C_2H_5CO-C_2H_5, (b) CH_3CO-n-C_3H_7, (c) CH_3-$COCH(CH_3)_2$

Figure 3. $O{\cdot}^-$ NCI spectra of the three isomeric xylenes, (a) o-*xylene, (b)* m-*xylene, (c)* p-*xylene*

this ratio rises to about 5, suggesting that the H^+ abstraction to form the $(M-1)^-$ ion is largely from the carbon atom α to the aromatic ring.

The ability to distinguish positional isomers illustrated by the spectra of the xylenes is also illustrated by the spectra given by the three methylpyridines. The relative intensities of the major peaks are given in the Table I. It is seen that the base peak of the 2-methylpyridine is the $(M-H+O)^-$ ion whereas when the 4-position is blocked, as in 4-methylpyridine, this is a peak of low intensity, consistent with the observations made on 4-deuteropyridine. This was not always the case, however, and the spectra given by the three fluorotoluenes on reaction with $O^{\cdot-}$ are very similar.

Table I. Relative Intensities of the Major Peaks in the O^- NCI Spectra of Methylpyridines

Ion m/e	2-Methylpyridine	3-Methylpyridine	4-Methylpyridine
$(M-H+O)^-$ 108	100	77	26
$(M-3H+O)^-$ 106	13	40	15
$(M-CH_3+O)^-$ 94	13	13	9
$(M-H)^-$ 92	68	100	100
$(M-2H)^{\cdot-}$ 91	95	89	57
$(M-3H)^-$ 90	10	36	9

Conclusion.

This brief account of the reactions of $CH_3OCHCH_2^{+\cdot}$ and $O^{-\cdot}$ indicates the possibilities of the use of ion-molecule reactions in probing structural features of certain classes of compounds. Undoubtedly, other ions will be found which will undergo different types of reactions and one can look forward to the establishing of a series of selective reagent ions which will allow one greatly to extend the type of structural information obtainable.

Experimental.

The work was carried out in part using a Varian V5900 ion cyclotron resonance (ICR) spectrometer for preliminary work at low pressure and in part in an MS50 double focussing mass spectrometer (AEI Scientific Apparatus Ltd) equipped with a dual purpose electron-impact/chemical ionization (EI/CI) source. The source pressure was measured by means of an MKS "Baratron" (Model 77H-10) the output of which fed the control unit of a Granville-Phillips automatic pressure regulator (series 216) which controlled the flow rate of the major component in the high pressure source. The instrument was modified to allow it to operate in the negative ion mode.

Literature Cited.

1. Field, F.H., MTP Int. Rev. of Sic., Series One, Physical Chemistry, Vol. 5 (Ed. A. Maccoll), Ch. 5, Butterworths, London, 1972.
2. Ferrer-Correia, A.J.V. and Jennings, K.R., Int. J. Mass Spectrom Ion Phys., (1973), 11, 111.
3. Drewery, C.J., Goode, G. C., and Jennings, K.R., Int. J. Mass Spectrom Ion Phys., (1976), 22, 211.
4. Drewery, C.J. and Jennings, K.R., Int. J. Mass Spectrom. Ion Phys., (1976), 19, 287.
5. Ferrer-Correia, A.J.V., Sen Sharma, D.K. and Jennings, K.R., Org. Mass Spectrom., (1976), 11, 867.
6. Drewery, C.J., Goode, G.C., and Jennings, K.R., Int. J. Mass Spectrom Ion Phys., (1976), 20, 403.
7. Jennings, K.R., Mass Spectrometry, Vol. 4, Specialist Periodical Report, Chemical Society, London, Ed. R.A.W. Johnston, 1977, p. 209.
8. Goode, G.C. and Jennings, K.R., Adv. in Mass Spectrom., (1974), 6, 797.

9. Dawson, J.H.J., and Jennings, K.R., J. Chem. Soc. Faraday Trans. II, (1976), 72, 700.
10. Harrison, A.G. and Jennings, K.R., J. Chem. Soc. Faraday Trans. I, (1976), 72, 1601.
11. Bruins, A.P., Ferer-Correia, A.J.V., Harrison, A.G., Jennings, K.R. and Mitchum, R.K., Adv. in Mass Spectrom. 7, (in press).

RECEIVED December 30, 1977

10

Selectivity in Biomedical Applications

CATHERINE FENSELAU

Department of Pharmacology and Experimental Therapeutics, Johns Hopkins University School of Medicine, Baltimore, MD 21205

Often in pharmacology, clinical studies, toxicology and studies of environmental residues, the sample is a minor component of a very complex mixture. Figure 1 shows an example of a typical mixture extracted from urine. An analogously obtained extract from blood, sweat or cerebral spinal fluid would contain even more compounds, because these physiologic fluids have a higher proportion of lipophillic endogenous components. If one is performing qualitative or quantitative assays of a specific component of such a mixture, it is easy to see how interference from other compounds may lessen the reliability of the assay, degrade the sensitivity of the assay, and interfere with quantitation.

Documentation of the loss in sensitivity and also in precision of mass spectral analyses of samples from physiologic fluids is presented in Tables 1 and 2. Table 1 indicates that, while 100 picograms of bis-2-chloroethylamine (a metabolite of the antitumor drug cyclophosphamide) can be quantitatively detected as

Table 1. Quantitation of Bis-(2-chloroethyl)amine

Pure	1×10^{-10} g
from urine	3×10^{-8} g
from plasma	9×10^{-8} g
Signal/Noise = 30/1	

the trifluoroacetyl derivative in a particular gas chromatograph-mass spectrometer system, samples which had been extracted from urine and derivatized could only be assayed reliably down to about 30 nanograms

© 0-8412-0422-5/78/47-070-188$10.00/0

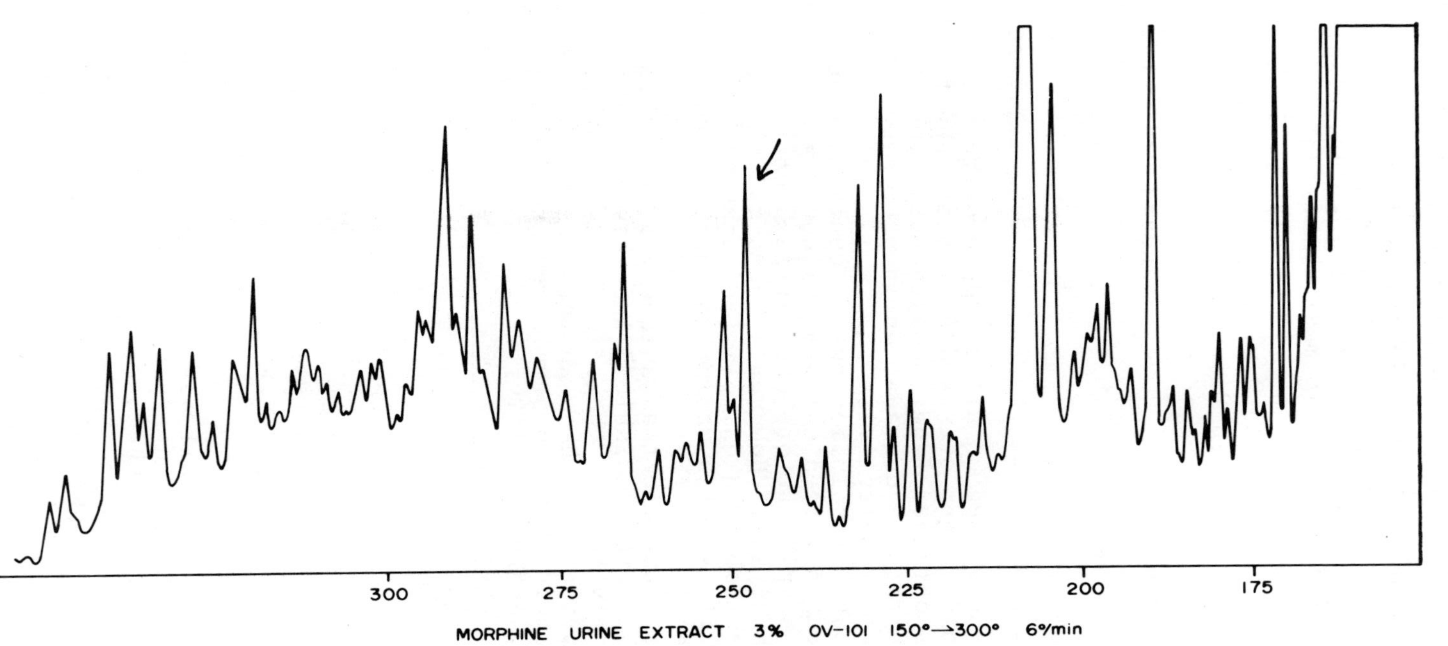

Figure 1. Gas chromatogram of urine extract from a morphine-treated rabbit

using the same protocol. In moving from urine to plasma, another three-fold loss in sensitivity was experienced. In Table 2, standard deviations are presented of isotope ratio measurements of C^{13}-containing carbon dioxide from two sources. Here measurements performed on urine samples were found to have a three-fold larger standard deviation than those made on samples obtained from (homogeneous) limestone (1). This illustrates a degradation in precision.

Table 2. Isotope Ratio Measurements

Sample	Standard Deviation
CO_2 from limestone	± 0.05 %
CO_2 from urine	± 0.15 %

Enhancement of Selectivity

The easiest solution to the contamination problems encountered in biomedical samples would be, of course, to purify samples before one carries out the mass spectral analysis. However, biomedical samples are usually available in only limited amounts, and in practice, it is often easier to employ selective methods of analysis, that is, to utilize mass spectral techniques that select as uniquely as possible for the sample of interest. Table 3 indicates five areas in which one may enhance the selectivity of analyses by mass spectrometry. These approaches can be used in various combinations to provide enhanced reliability, precision, and/or sensitivity.

Table 3. Approaches for Enhancing Selectivity

1) Sample preparation
2) Ion production
3) Ion separation
4) Ion detection
5) Data processing

The most important aspect of selective sample preparation is, of course, isolation and purification. This is an art form which will not be addressed in this essay, except to point out that gas chromatography provides a very powerful technique for purification or separation of the compound of interest. All of the subsequent illustrations will be measurements made by combined gas chromatography-mass spectrometry.

Selective derivatization procedures should also be evaluated, for example the formation of boronate diesters from cis diols. However, many compounds in a mixture are usually derivatized by any chemical reaction, and derivatization reactions usually contaminate a sample further, rather than selecting for the compound of interest. One useful function of derivatization is to increase the molecular weight of the sample beyond the mass range of most interfering ions.

Another example of selective sample preparation might be the generation of isotope clusters. In Figure 2 are shown the structures of propoxyphene and its d_7-analog. The major mode of fragmentation is also indicated. A 1:1 mixture of these two compounds was administered to an animal, and urine was collected and extracted. The drug related metabolites were picked out of the many components detected by gas chromatography-mass spectrometry on the basis of isotopic clusters at m/e 91 and m/e 98. These peaks represented the benzyl and d_7-benzyl ions indicated in the Figure. This approach permitted at least 8 metabolites to be located and subsequently identified (2).

Although selective ionization, the second category in Table 3, has historically denoted photoionization and low energy electron impact ionization, the low sensitivity of these techniques has prevented their widespread application. Ionization techniques which currently appear to offer the greatest potential for enhancing the selectivity of analyses include positive and negative chemical ionization (3) and field ionization and desorption, all addressed by other authors in this volume. The utility of chemical ionization is illustrated below.

The third category is selectivity in ion separation in the ion analyser. Certainly high resolution analysis comes to mind as a method for reducing the ambiguity in an analysis. This was illustrated quite early in a pharmacologic application of Bommer and Vane (4). Collisional activation with two stages of ion separation, and some of the defocussed metastable techniques are also examples of selectivity in ion separation. In all these cases, however, reliability of the analysis is increased at the expense of sensitivity.

The fourth category in which one can increase the selectivity of a mass spectral analysis is by selective detection. Here the preeminent example is selected ion monitoring. In this technique, one monitors the intensities of preselected, characteristic ions as a function of time. As an example, Figure 1 contains the gas chromatogram detected by flame ionization of an extract of urine from a morphine-dosed rabbit. In this investigation in the author's laboratory, the glucuronide metabolite of morphine was sought. Its approximate elution time is indicated by the arrow. Figure 3 shows the selected ion records of several ions characteristic of morphine glucuronide. All of these profiles are simpler than the gas chromatogram in Figure 1. In addition, it may be observed that when higher masses are monitored, fewer components of the urine extract are detected. Needless to say, selected ion monitoring can be used in combination with other selective techniques, most notably gas chromatography and chemical ionization.

Ion profiles analogous to those provided by selected ion monitoring can also be reconstructed from repetitively scanned conventional spectra, best done with a computer. In this case, all of the ions detected in each scan are stored in the computer so that any of them may be called out and profiled. However, the overall technique is less sensitive than selected ion monitoring by a factor as great as 10^4. This reflects primarily the greater sampling time for the selected ion technique and also the fact that sample is not wasted while uninformative areas of the spectra are being scanned and during analyzer reset periods.

In Table 4, the practical lower detection limits are presented for various kinds of gas chromatographic detectors. Total ion current monitoring with the mass spectrometer has a sensitivity roughly comparable to

Table 4. Sensitivity of GC Detectors

Total Ion Current Monitoring	10^{-8} g
Flame Ionization	10^{-9} g
Electron Capture	10^{-12} g
Selected Ion Monitoring	10^{-12} g
Mass Chromatograms	10^{-8} g

flame ionization detection, while selected ion monitoring with the mass spectrometer is clearly competitive with the best electron capture detectors. Selected ion monitoring is of course much more generally applicable than electron capture detection. Thus, selected ion monitoring has been reported to be used with sensitivity in the picogram range with electron

91 98

Propoxyphene d_7-Propoxyphene

Figure 2. Structures and major fragmentation of propoxyphene and d_7*-propoxyphene*

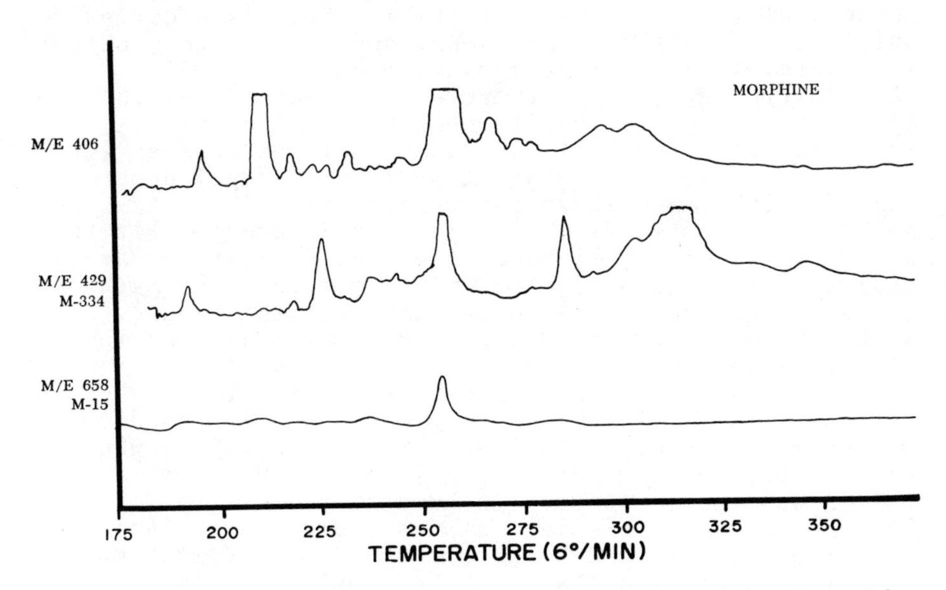

Figure 3. Selected ion records of the urine extract chromatographed in Figure 1 (re-injected)

impact ionization, with positive ion chemical ionization, and also with negative ion chemical ionization. The author submits that this is indeed a category of high performance mass spectrometry.

Analysis of Drug Metabolites

The discussion of selected ion monitoring above serves as an introduction to an account of work at Hopkins in which mass spectrometry has been used in the selected ion monitoring mode to provide enhanced sensitivity and reliability in the analysis of metabolites of cyclophosphamide in urine and blood. Figure 4 shows the structure of cyclophosphamide (I) and of its major metabolites as they are now understood. Cyclophosphamide is a widely used antitumor drug, which is not itself biologically active. Rather, it has been known for some years that cyclophosphamide is converted to cytotoxic active metabolites by oxidative metabolism in the hepatic microsomes. Although this was understood early on, nonetheless, the drug was used clinically for more than 10 years before the structures of any of its hepatic metabolites were elucidated, and in fact, successful identification of the metabolites occurred only after mass spectrometry was applied to the problem in several laboratories, including our own.

In Figure 5, the structure is presented for the cytotoxic active metabolite phosphoramide mustard, which was first isolated and characterized from an *in vitro* incubation of radioisotope labelled cyclophosphamide with mouse liver microsomes (5). The metabolite was treated with diazomethane, converting it to a mixture of mono-, di- and trimethyl derivatives as is indicated in Figure 5. The major fragmentation path of these derivatives are also indicated in the Figure. Although the group at Hopkins was very pleased at the time to have identified the first active metabolite of cyclophosphamide, nonetheless this metabolite had been obtained from an *in vitro* experiment, and it was deemed necessary to examine blood and urine of patients who were receiving the drug to confirm that such a metabolite was formed *in vivo* in humans.

Because much smaller samples would be available from the patients and because these would be much more highly contaminated than was the sample from the *in vitro* microsomal experiment, selected ion monitoring was employed to look for this metabolite.

In order to use this highly selective detection technique, one must know what the conventionally scanned mass spectrum of the compound looks like. In Figure 6 is the mass spectrum of the dimethyl derivative of phosphoramide mustard. Ions were chosen from

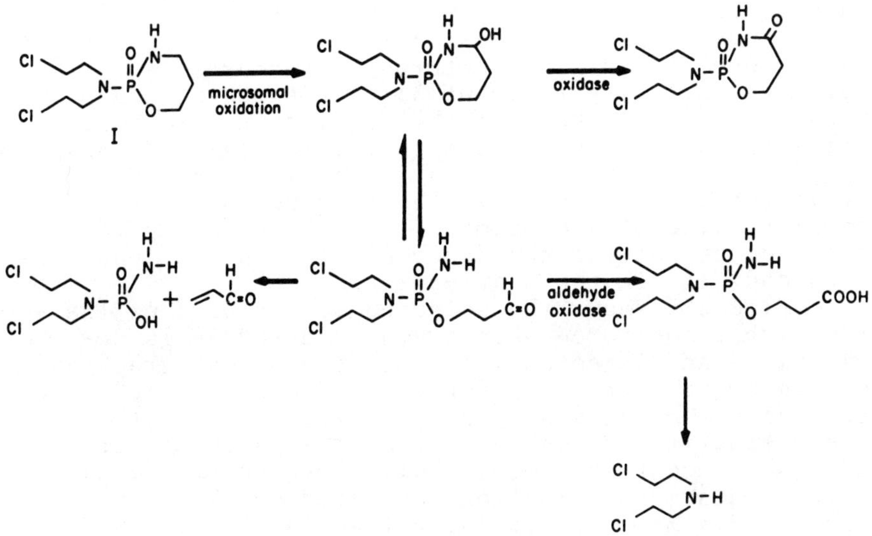
microsomal oxidation
oxidase
aldehyde oxidase
COOH
I

Figure 4. Cyclophosphamide I and its major human metabolites

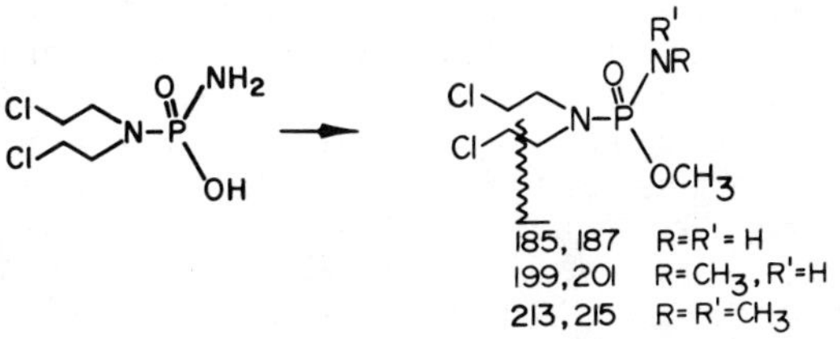
185, 187 R=R' = H
199, 201 R= CH3, R'=H
213, 215 R= R'=CH3

Figure 5. Structures and fragmentation of the derivatives of phosphoramide mustard treated with diazomethane

this spectrum for selected ion monitoring which are, on the one hand, highly characteristic of the substance being sought, and on the other, sufficiently intense to provide good sensitivity. From this spectrum m/e values 199 and 201 were chosen for monitoring. Because these peaks represent ions containing chlorine isotopes, one of their characteristic features is their relative intensity ratio of 3:1. Figure 7 contains the mass spectrum of the trimethyl derivative of phosphoramide mustard. From this spectrum the pair of peaks at m/e 213 and 215 was selected for monitoring. Here again these peaks occur in the spectrum with an intensity ratio of approximately 3:1, reflecting the presence at natural abundance of Cl^{35} in the lower mass ions and Cl^{37} in the heavier ions.

Figure 8 shows the selected ion records of these four ions, monitored when a standard sample is derivatized and injected into the gas chromatograph and when a urine extract is derivatized and injected into the gas chromatograph mass spectometer. From this experiment it was concluded that the patient's urine did indeed contain the metabolite phosphoramide mustard. This identification was based on the comparable retention times of material extracted from the urine and the standard compound, on the formation of the characteristic fragment ions with their appropriate relative intensities, and on the characterization of all three derivatives of the metabolite. (The monomethyl derivative was detected in a separate injection.)

In order to provide a frame of reference for the selectivity of this detection method, the total ion current gas chromatogram of the same urine extract is shown in Figure 9. The arrows on the axis indicate the points at which the di- and trimethyl derivatives of the metabolite were detected by selected ion monitoring. One of these arrows falls on the trailing edge of the larger peaks in the mixture. A conventional scanned spectrum was obtained at this point, which is shown in Figure 10. This is the spectrum of a methylalkyl phthalate, a member of a family of contaminants often encountered in clinical samples. All peaks contributed to the spectrum by the metabolite of cyclophosphamide had intensities less than 1% relative to the intensity of the base peak of the contaminant. Nonetheless, the selective detection method completely ignores this gross contaminant, because it does not form ions of the masses being monitored.

In Figure 11, selected ion profiles are presented, obtained by monitoring the chromatographic separation

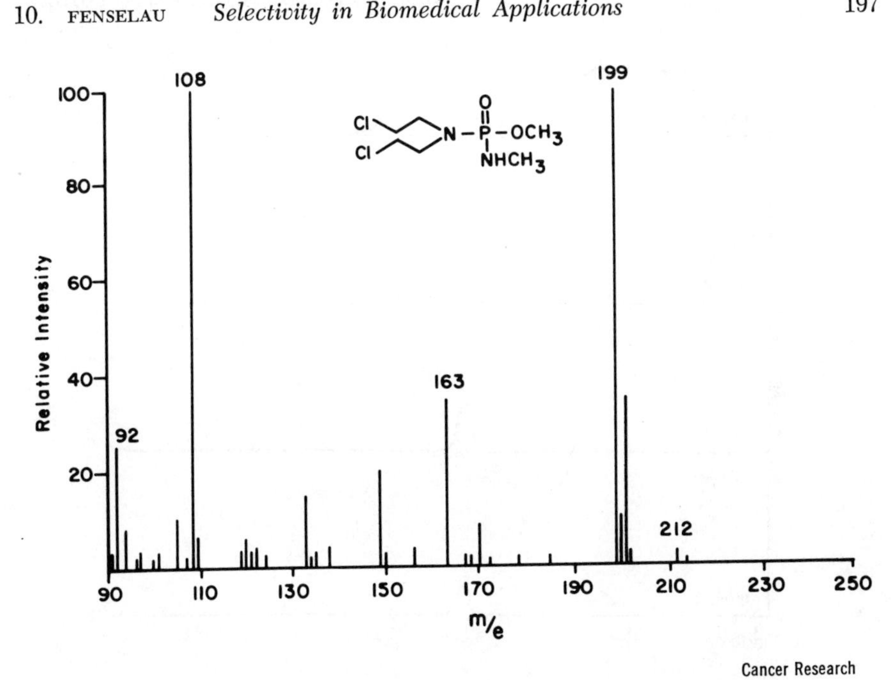

Figure 6. Electron impact mass spectrum of dimethyl phosphoramide mustard

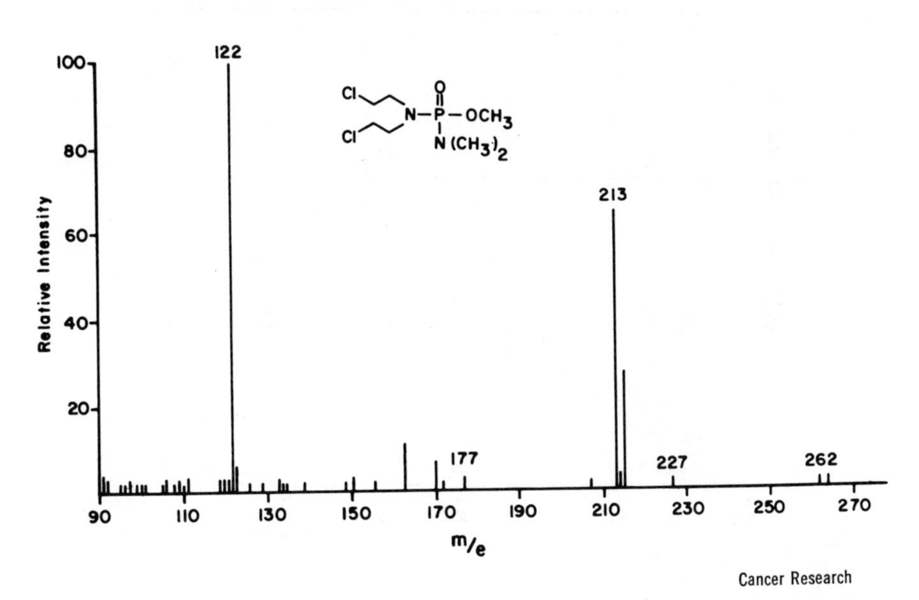

Figure 7. Electron impact mass spectrum of trimethyl phosphoramide mustard

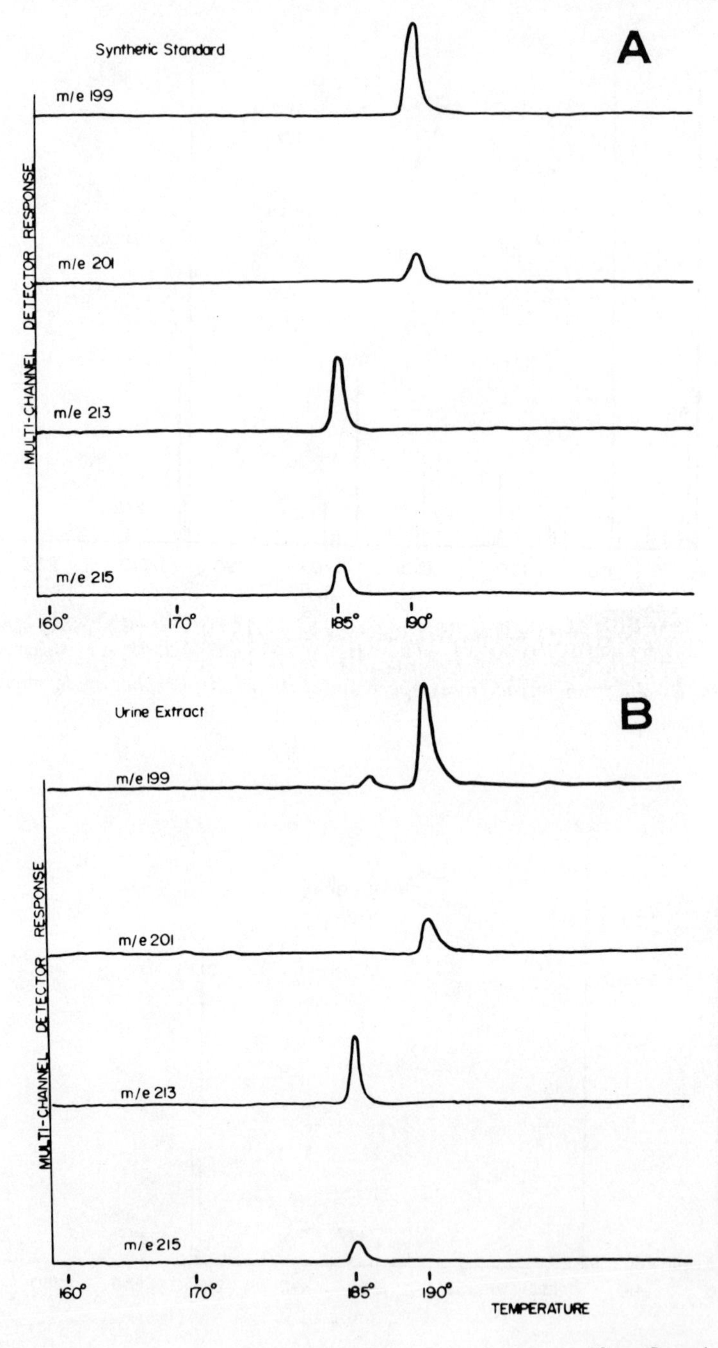

Cancer Research

Figure 8. Selected ion records of urine extract from a patient receiving cyclophosphamide and of authentic derivatized phosphoramide mustard

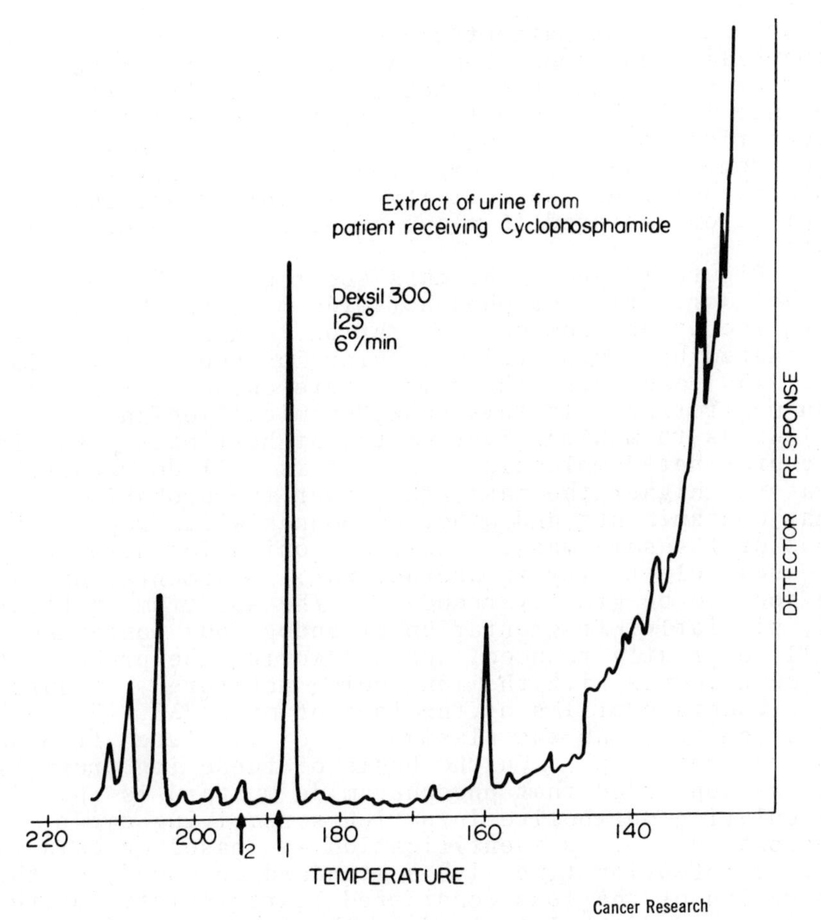

Figure 9. Gas chromatogram of the urine extract profiled in Figure 9

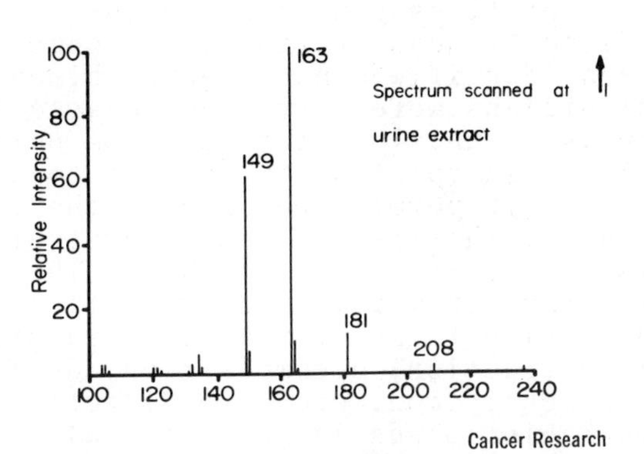

Figure 10. Electron impact mass spectrum taken at ↑₁ in Figure 10

of an extract of patient plasma. In this case, many lipophilic endogenous contaminants fragment to form ions of the masses being monitored, and the interference is high enough to render the metabolite's identification ambiguous. Accordingly, the mass spectrometer had to be employed in such a way as to obtain additional selectivity. In this case, chemical ionization was used in place of electron impact ionization.

Figure 12 shows the mass spectrum of the monomethyl derivative of phosphoramide mustard obtained with isobutane chemical ionization. This spectrum is dominated by protonated molecular ion species (m/e 235, 237, 239) and these three ions were chosen for selected ion monitoring. In this case, chemical ionization allows us to monitor ions of the highest mass possible, the protonated molecular ion. It is well documented that the higher the mass, the lower the probability that contaminants and other compounds will fragment to ions of the same mass. Chemical ionization also increases selectivity in another way. Fragmentation can be seen to be greatly reduced in the spectrum in Figure 12; similarly, fragmentation of endogenous contaminants will be greatly reduced, again lowering the probability of coincidence with the ions being monitored. Figure 13 presents profiles of the ions of mass 235, 237 and 239, monitored as the plasma extract is eluted from the gas chromatograph. On the basis of these ion profiles, it was concluded that phosphoramide mustard is indeed a circulating metabolite in patients receiving cyclophosphamide. This identification was based on coincident retention time with a standard compound, on the formation of the ions considered characteristic of the derivatized metabolite, and on the formation of these ions in the appropriate intensity ratios of about 9:6:1.

The analog profiles generated by the selected ion monitor are directly quantitatable, and once the metabolite had been identified in patients' blood and urine, the clinicians were very keen to assay its concentration as a function of time. One particular advantage which mass spectrometry brings to assays of this sort is that it permits the use of the ideal internal standard, that is the compound itself labelled with heavy isotopes, H^2, C^{13}, O^{18}, N^{15}. These internal standards will have chemical characteristics most nearly approximating those of the assay sample itself, and losses in extraction, decomposition, derivatization, column absorption, etc. are directly compensated by comparable losses in the internal standard.

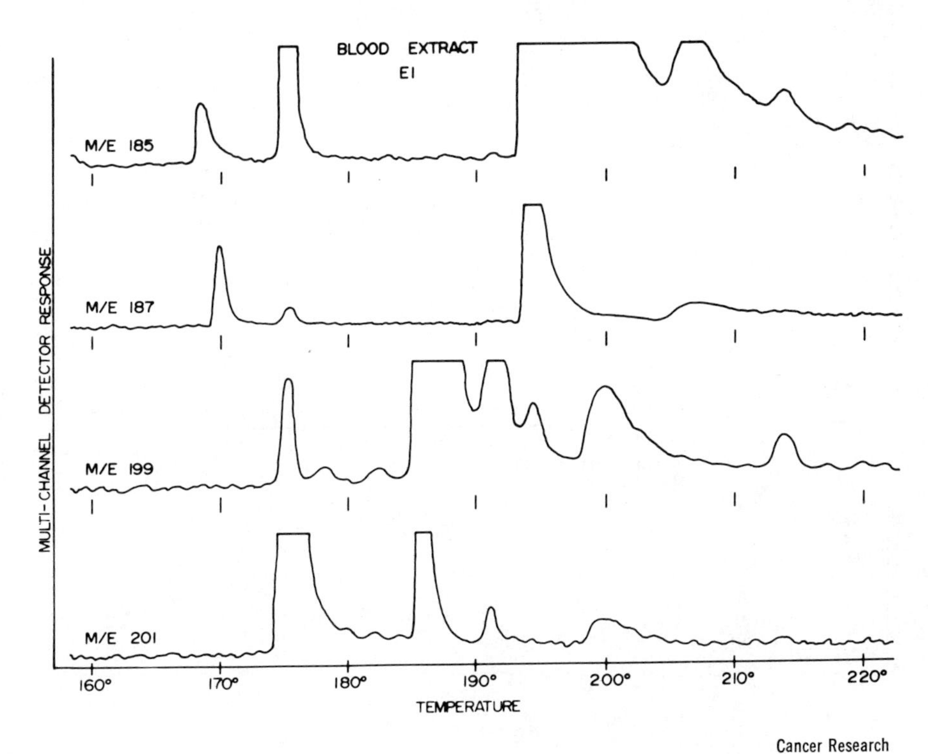

Cancer Research

Figure 11. Selected ion records obtained by electron impact of plasma extract from a patient receiving cyclophosphamide

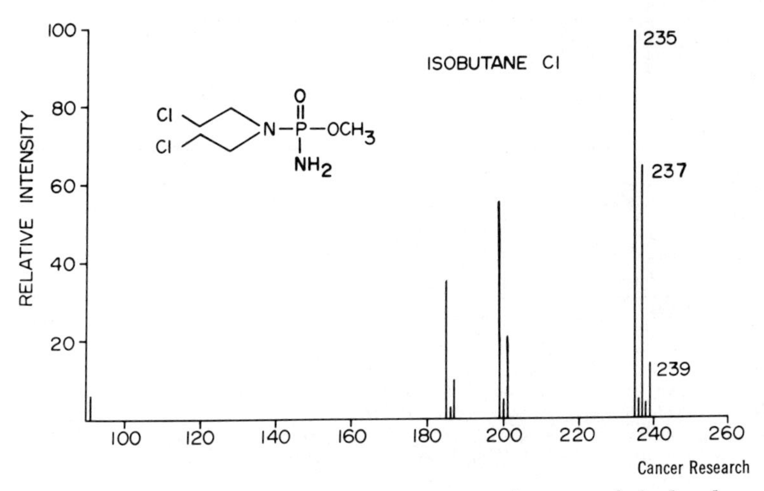

Cancer Research

Figure 12. Chemical ionization mass spectrum of monomethyl phosphoramide mustard

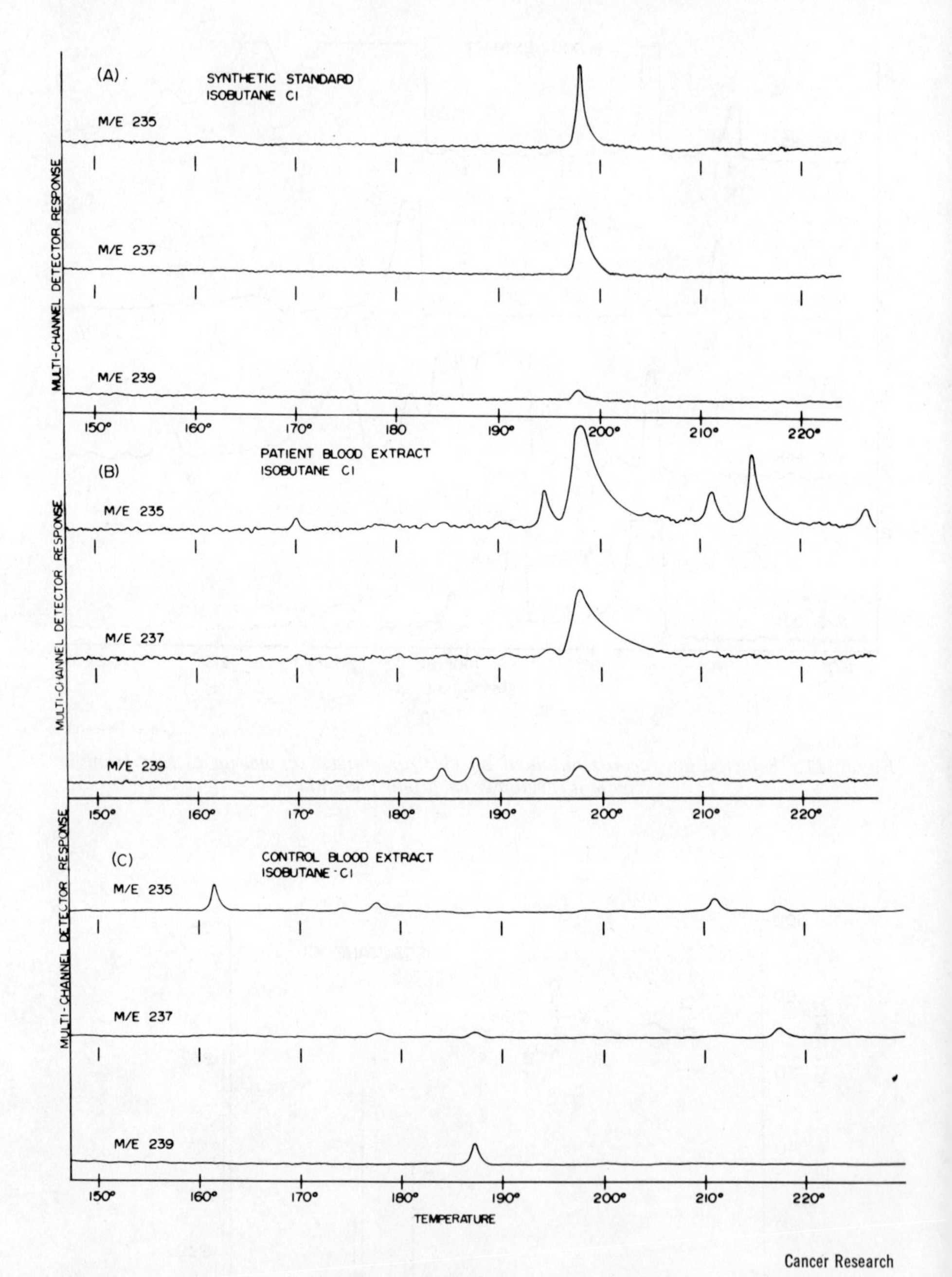

Cancer Research

Figure 13. Selected ion records obtained by chemical ionization of: (A) authentic derivatized phosphoramide mustard, (B) plasma extract from a patient receiving cyclophosphamide, (C) control plasma extract.

The first step in the development of an assay technique for phosphoramide mustard was the synthesis of a $\underline{d}_4$-analog to be used as an internal standard. Part of the synthetic scheme is indicated in Figure 14. Lithium aluminum deuteride reduction is an ideal step for introducing quantitative amounts of deuterium into specific locations in the molecule. The product of the reaction with thionyl chloride, nornitrogen mustard, is a general building block present in all the metabolites of cyclophosphamide and also a number of other antitumor alkylating drugs. $\underline{d}_4$-Nornitrogen mustard was elaborated to $\underline{d}_4$-phosphoramide mustard and $\underline{d}_4$-cyclophosphamide by literature methods, preparatory to undertaking clinical assays of the corresponding unlabelled compounds. $\underline{d}_4$-Nornitrogen mustard and $\underline{d}_4$-phosphoramide mustard were also used in studies of mustard alkylation mechanisms (6).

Figure 15 presents scans of the molecular ion regions of $\underline{d}_0$- and $\underline{d}_4$-phosphoramide mustard trimethyl derivatives. Each compound contains two chlorine atoms, and thus three protonated molecular ions are detected in each spectrum. From these two sets of [M + H] ions four peaks were selected for monitoring. Although some sensitivity is sacrificed in monitoring four peaks rather than two, additional reliability is gained since the intramolecular intensity ratio can be checked as an assurance that the correct metabolite is being detected, and that interference is minimal. Some flexibility is also gained, because the intermolecular ratios of metabolite to internal standard may be calculated in several different combinations. Figure 16 presents the selected ion records from a plasma sample containing $\underline{d}_0$-phosphoramide mustard and $\underline{d}_4$-phosphoramide mustard. It is important to know that cross talk is minimal, that there is little or no signal contributed by the $\underline{d}_4$-compound to masses monitored for the $\underline{d}_0$-compound, and vice versa. Figure 17 presents a calibration curve for phosphoramide mustard extracted from plasma. A log-log scale is used so that the measurements may be presented through two orders of magnitude. All points except the lowest were determined with experimental uncertainties between 1 and 5%.

Figure 18 presents the profiles of the concentrations in patient plasma of nornitrogen mustard, phosphoramide mustard and cyclophosphamide as a function of time through 24 hours. The selectivity offered by the mass spectral assay compares favorably with the alternative nonspecific spectrophotometric assay based on the formation of colored products from any alkylating compounds present. In addition, the mass spec-

$HN(CH_2CO_2C_2H_5)_2 \xrightarrow{LiAlD_4} HN(CH_2CD_2OH)_2 \xrightarrow{SOCl_2} HCl.HN(CH_2CD_2Cl)_2$

Figure 14. Synthesis of d_4*-nornitrogen mustard*

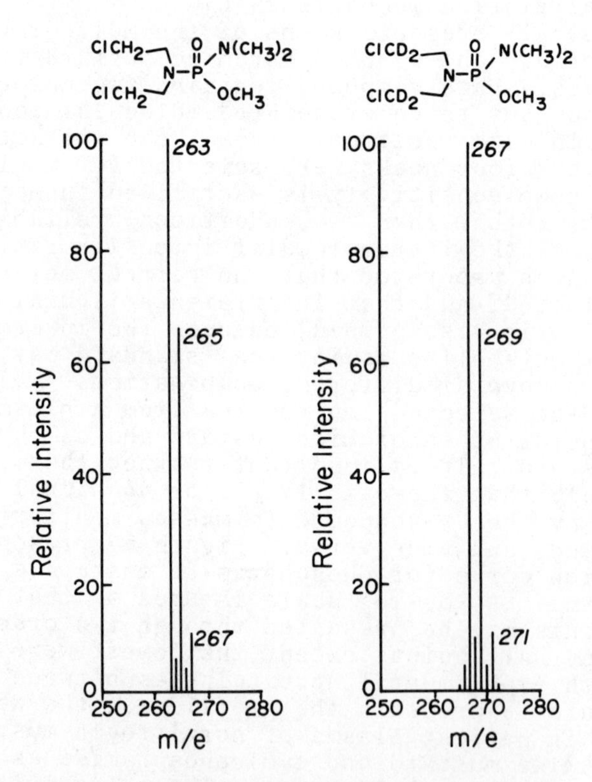

Figure 15. Protonated molecular ion region (chemical ionization) of phosphoramide mustard and d_4*-phosphoramide mustard*

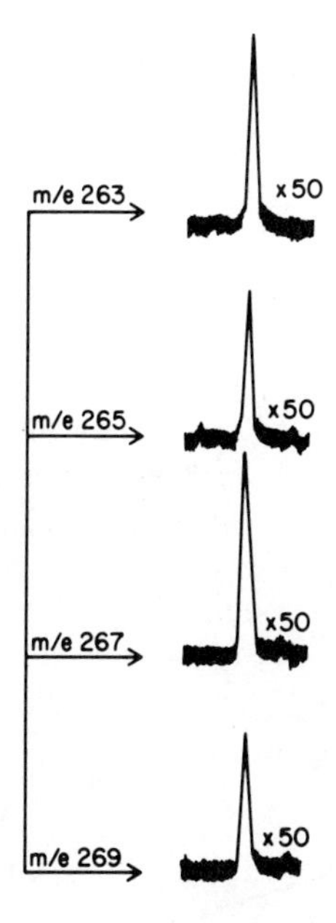

Figure 16. Selected ion records of an extract of plasma containing phosphoramide mustard and d_4-phosphoramide mustard

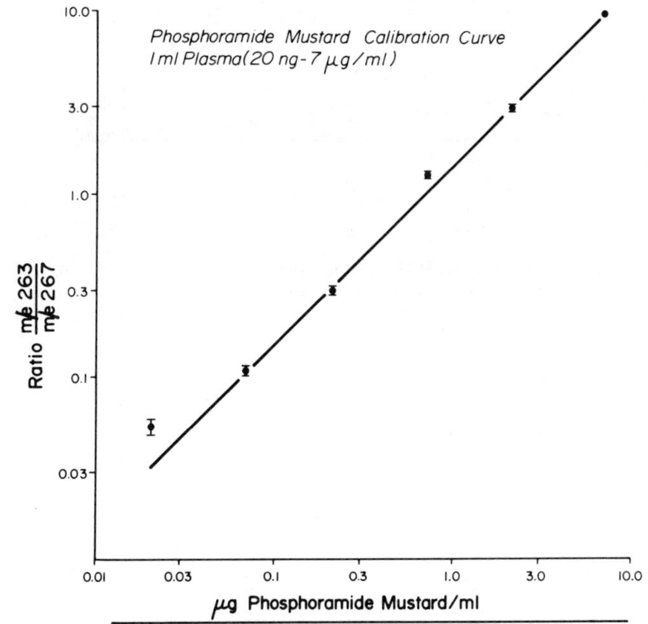

Cancer Research

Figure 17. Ratio of intensities of peaks at m/e *263 and 267 as a function of the amount of phosphoramide mustard added to 10* μg d_4*-phosphoramide mustard cyclohexylamine salt in one mL plasma*

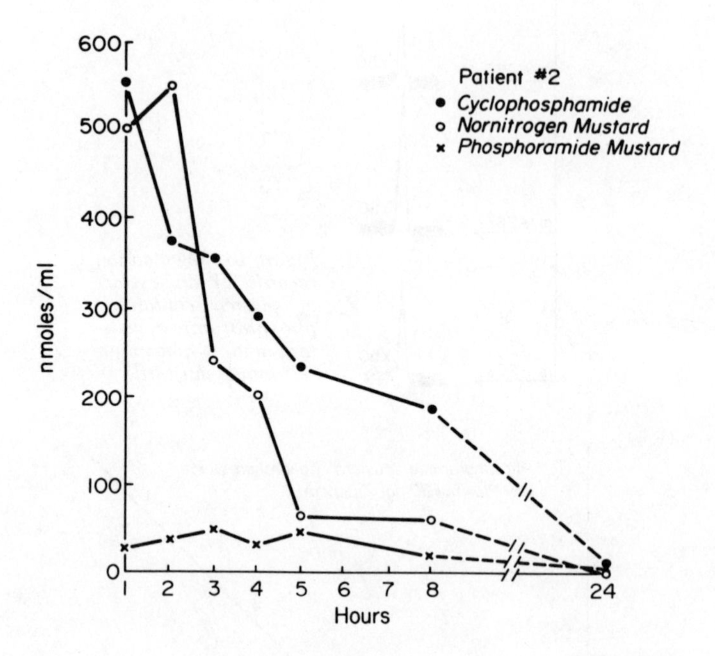

Cancer Research

Figure 18. Concentration of cyclophosphamide nornitrogen mustard as a function of time in plasma of a patient receiving 75 mg/kg in a 1 hr infusion

tral values are uniformly higher than those obtained by spectrophotometry. In part this is because decomposition of these unstable compounds during workup and derivatization is compensated by the stable isotope labelled internal standard. The mass spectrometric assay is highly reliable, highly specific, relatively expensive and tedious.

The assay whose development and application is described above employs selectivity from a number of the categories in Table 3. Selectivity is used in sample preparation. First of all the gas chromatographic inlet system was used. Secondly heavy reliance was placed on isotope clusters, both from chlorine and from deuterium/hydrogen. Selectivity in ion production has been incorporated by the use of isobutane chemical ionization. Selectivity in ion detection was introduced by the use of selected ion monitoring. Some sensitivity was traded for reliability and for capability to do quantitation by monitoring four ions instead of only one.

At intervals of increasing frequency, claims are presented in the literature for increased sensitivity, either associated with a new technique or a particular instrument. Can increased sensitivity be used in biomedical applications? Yes! Methodology or instruments with improved sensitivity can make important contributions. However, it must be remembered that even working with samples present in nanogram amounts sufficient to be detected by conventional scanning, analyses are stymied as the sample is lost in the normal noise or interference of the other components of a physiologic fluid. This is further compounded by ions contributed from column bleed and instrument background. For increased sensitivity to be utilizable and useful, it must be applied in conjunction with methodology which provides appropriate selectivity as well.

Literature Cited

1. G. E. Von Unruh, D. J. Hauber, D. A. Schoeller and J. M. Hayes. Detection Limits of Carbon-13 Labelled Drugs and Metabolites. Biomed. Mass Spectrom (1974) 1 345-349.
2. S. L. Due, H. R. Sullivan and R. E. McMahon. Propoxyphene: Pathways of Metabolism in Man and Laboratory Animals. Biomed. Mass Spectrom (1976) 3 217-225.

3. H. P. Tannenbaum, J. D. Roberts and R. C. Dougherty. Negative Chemical Ionization Mass Spectrometry - Chloride Attachment Spectra. Anal. Chem. (1975) 42 49-54 and references therein.
4. P. Bommer and F. Vane. The Use of Fragmentation Patterns of 1,4-benzodiazepines for the Structure Determination of Their Metabolites by High Resolution Mass Spectrometry. Fourteenth Annual Conference on Mass Spectrometry and Allied Topics, Dallas, Texas, 1966; M. A. Schwartz P. Bommer and F. Vane. Diazepam Metabolites in the rat: Characterization by High Resolution Mass Spectrometry and Nuclear Magnetic Resonance. Arch. Biochem. (1967) 121 508-516.
5. M. Colvin, C. A. Padgett and C. Fenselau. A Biologically Active Metabolite of Cyclophosphamide. Cancer Res. (1973) 33 915-918.
6. M. Colvin, R. B. Brundrett, M. N. Kan, I. Jardine and C. Fenselau. Alkylating Properties of Phosphoramide Mustard. Cancer Res. (1976) 36 1121-1126.
7. C. Fenselau, M. N. Kan, S. Billets and M. Colvin. Identification of Phosphorodiamidic Acid Mustard as a Human Metabolite of Cyclophosphamide. Cancer Res. (1975) 35 1453-1457. (Reprinted by permission of Cancer Research, Inc.)
8. I. Jardine, C. Fenselau, M. Appler, M. N. Kan, R. B. Brundrett and M. Colvin. The Quantitation by Gas Chromatography Chemical Ionization Mass Spectrometry of Cyclophosphamide, Phosphoramide Mustard and Nornitrogen Mustard in the Plasma and Urine of Patients Receiving Cyclophosphamide Therapy. Cancer Res. Reprinted by permission of Cancer Research, Inc.

RECEIVED December 30, 1977

11

Prognosis for Field Desorption Mass Spectrometry in Biomedical Application

CHARLES C. SWEELEY, BERND SOLTMANN, and JOHN F. HOLLAND*

Department of Biochemistry, Michigan State University, East Lansing, MI 48824

Field Desorption

Field desorption mass spectrometry is a method of increasing importance to the biological and medical sciences. It is reported by many investigators to be a panacea for the analysis of polar, non-volatile compounds, compounds that are simply not amenable to other types of mass spectrometry. Conversely, it is claimed by others to be a black art that is more mystical than scientific, with successful analyses resulting more from good fortune than from clever experimental design. The truth is obviously somewhere between these divergent views. But the potential of the method for revealing definitive information on molecular species makes a better understanding of its processes a concern of all scientists who are engaged in efforts to determine the structures of molecules of biological importance. Some typical successes of the method to date include carbohydrates and saccharides (1-3), glycosides (6), nucleosides and nucleotides (5), polypeptides (6), hormones (7) and drugs and drug metabolites (8-10). In spite of this impressive list, the potential of the method is just beginning to be realized. A vexing problem persists, however, in that the number of failures remains disturbingly high in many investigations.

In field desorption mass spectrometry, the sample to be analyzed is physically placed upon the ion source anode when it is external to the vacuum chamber. Subsequently, the anode is placed in the ion source and the vacuum restored. The sample is heated in the presence of a very large voltage field. At a specific temperature, called the best anode temperature (BAT), ions will be formed, many of which are accelerated and focused into the mass spectrometer for dispersion and

*Author to whom correspondence should be addressed.

detection. This method of ionization, from its inception by Beckey (11), has been successfully applied to a number of compounds that are polar and non-volatile and, in general, not amenable to analysis using other types or ionization. The ions produced by FD have very low internal energy, and the resulting mass spectra normally have large peaks related to their molecular ions, with few or no fragment ions present. Cluster ions are often observed, but the spectra produced are much less complex than those obtained by electron impact ionization. For example, riboflavin gives a complex electron impact spectrum that contains a very small molecular ion, M^{+}. However, when observed by FD, M^{+} is dominant and the only major ion in the spectrum. It is clear that the ionization processes in FD, whatever they may be, yield ions with little internal energy and fragmentation is minimized. Let us examine ways in which this may occur.

A theoretical model for field desorption was originally developed from field ionization theory, based upon mathematical derivations by Mueller (12) and Gomer (13). As shown in Fig. 1, this model proposes that the potential energy curves for the electrons of a molecule on the surface of a conductor are greatly altered in the presence of strong electric fields (~ 10^{8} V/cm). Voltage gradients of this magnitude are obtainable in the vicinity of very sharp points since applied voltages exhibit large, non-linear gradient distortions in the vicinity of small radii of curvatures, with the field rising asymptotically as the surfaces are approached (14).

Field image theory predicts that ions may be formed in one of two manners in these very large fields. Firstly, as seen from Fig. 1, the energy of ionization has been greatly reduced. The possibility of thermal energy providing the stimulus for the electron to escape increases as the energy well becomes shallower. Secondly, if the portion of the energy line that lies below the level of the energy in the well becomes close to the well itself, the probability for electron tunneling will increase and ions can be generated by this quantum mechanical mechanism. Indeed, this is considered to be the dominant mode in field ionization mechanisms. Unfortunately, in practice, the field image theory has provided the analyst with little intuitive feeling for the processes involved as they relate to experimental observations.

As users, we, like many others, were very frustrated by the large number of seemingly random failures. On the other hand, successful analyses were often

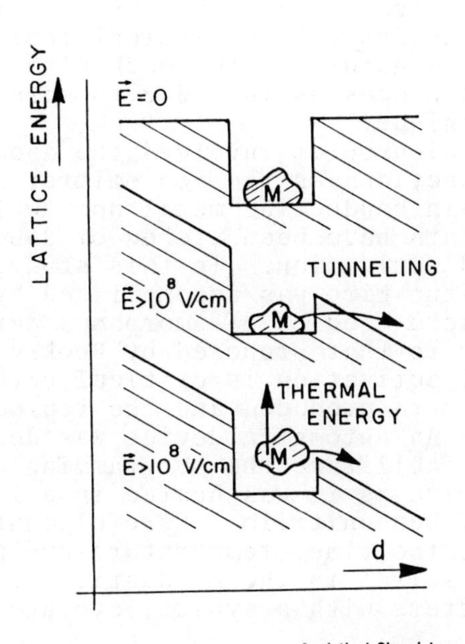

Analytical Chemistry

Figure 1. The effect of a strong field on a potential energy curve. The ordinate is energy and the abcissa is distance from the emitter surface.

obtained with no knowledge as to what was done right or, for that matter, often with no knowledge of anything being done differently. The only meaningful observation was that chemical parameters were, in all cases, influential to the results of analyses. It was this dilemma that encouraged our investigation into field desorption. This investigation has centered on two major objectives: to control the variables of analysis in order to gain analytical reproducibility; and to study the nature of the mechanism involved in the ionization processes in order to more intelligently apply the technique.

The initial problem involved the anode itself. This device functions as the ion emitter and is comprised of a thin conducting metal upon which sharp needles or points have been etched or deposited in a process called activation. In this study, activation of a 10μ tungstun wire was accomplished by the deposition of micro-needles of amorphous carbon in a method similar to that proposed by Beckey et al. (15). The process of activation is critical both to the quality of spectra produced and the reproducibility of the analysis. An automatic device was designed and fabricated to facilitate the programming of the temperature of the wire as it was heated in a low pressure environment of benzonitrile. Careful control and duplication of the time, temperature and pressure parameters have resulted in the production of uniform activated emitters with a typical average needle length of 30μ.

The two most commonly used methods of sample addition to the activated emitters are dipping and microsyringe extrusion (16-18). In the dipping method, the emitter is inserted into a solution of the sample and the amount adhering to the micro- needles after removal is allowed to dry prior to analysis. A variation of this method involves the freezing of the sample on the emitter surface after insertion into the sample solution (19). This process can be repeated, allowing the accumulation of larger amounts of sample on the emitter. Dipping methods are convenient, but a more reproducible quantity of sample can be applied using a microsyringe. In this method, a known volume of sample solution is extruded from a calibrated syringe (usually in the range of 200 nanoliter to one microliter) and allowed to adhere to the micro-needles on the emitter surface. A convenient variation of this method places the extruded drop into the region between two parallel activated emitters as shown in Fig. 2. After addition, the emitters are slowly separated, during which time

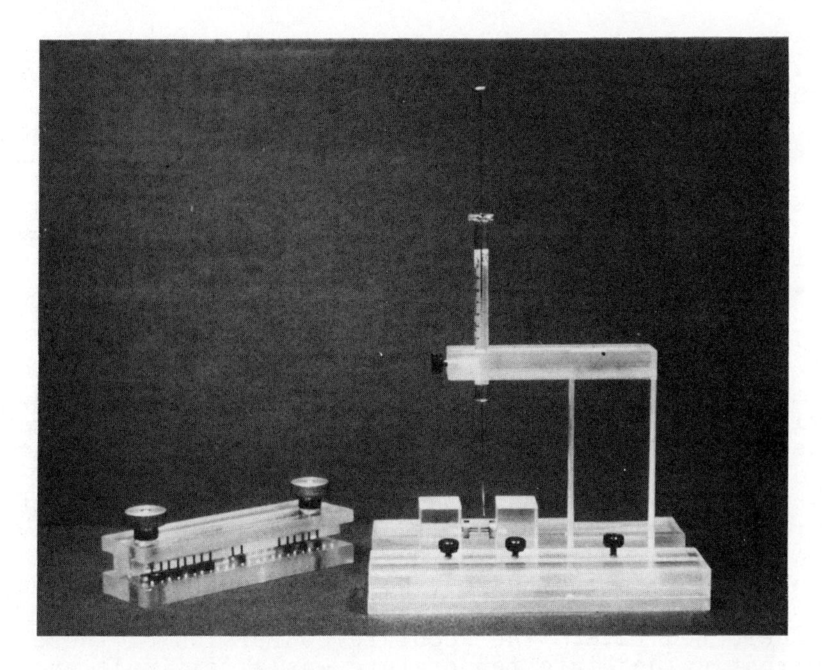

Figure 2. Microsyringe sample addition using adjacent parallel emitters. Separation will result in the entire sample being sorbed to either one of the emitters.

the total drop will cling to one or the other emitter. This approach greatly increases the possibility of a drop being applied to and held by an emitter (20).

The greatest uncertainty in the initial studies in field desorption mass spectrometry was the absence of an accurate knowledge of the temperature during analysis. Heat is applied to the sample by passing a current through the emitter wire; however, all of the original methods involved the manual setting of an uncalibrated dial. Even under these primitive conditions, the need for exact temperature control was not critical until reproducible emitters were available. At that time an appealing alternative presented itself: namely, to calibrate the current passing through the emitter against temperature. The emitter current can then be used as the means of temperature measurement and control (21-23). This was accomplished in our laboratory by the design and fabrication of an emitter current programmer (ECP) (22). This solid state device enables the regulation of current at any value between 0 and 85 ma to the nearest 0.1 ma and makes possible a linear program between any two current values at a preselected rate. The accurate conversion of these current values into temperature was a difficult and time consuming task. Figure 3 illustrates the nature of the current-temperature relationship for the uniform activated emitters used in this study. The upper values on this curve were obtained via an optical pyrometer which viewed the emitters thru a special salt window mounted in an experimental ion source constructed for research objectives. The middle part of the curve was determined by using resistance measurements of the emitter wire to extrapolate between the upper and lower regions. The lower part of the curve was obtained by melting a series of known compounds upon the emitter surface.

Table 1 illustrates the reproducibility achievable with the ECP on multiple analysis using the same emitter and on successive analyses using different emitters. This device, for the first time, allowed control of the temperature of the sample during analysis and provided a reasonable estimate of its actual value. An additional benefit of the ECP is realized when a linear temperature scan is performed with the mass spectrometer being cyclically scanned by the data system. The resulting stored data files can output mass chromatograms in formats similar to those of gas chromatography-mass spectrometry with the exception that the abscissa is temperature and not elution time. Mixtures whose components desorb at different temperatures can be resolved in time, and co-desorbing compounds can be resolved by mass differentiation. Figure 4 illustrates the type of output obtained from a linear thermal scan of the nucleoside, adenosine. The points on the current *vs*. intensity curve of the total ion intensity (TII) are created by summing all of the ion intensities for each scan. In the insert in Fig. 4, this quantity is plotted against the current axis along with the ion intensities of *m/e* 267 (M^+) and *m/e* 268 $(M+1)^+$ in a format similar to that of mass chromatograms. Although we have no preference, some authors refer to this type of output as a thermogram (21,23). The mass spectrum shown in Fig. 4 was taken at the BAT, scan number 23. Notice the dominance of the molecular ion cluster in the spectrum. This is very typical of FD spectra.

Table 1. REPRODUCIBILITY OF EMITTER CURRENT PROGRAMMER (ECP). Reprinted with permission from Anal. Chem., 48, 428 (1976). Copyright American Chemical Society.

Wire#	Current (mA)	# of Measurements
1	19.2 ± 0.4	5
2	21.3 ± 0.4	5
3	20.3	1
4	21.0	1
5	20.2	1

Figure 5 contrasts the electron impact (EI) spectrum of a synthetic nucleoside with its FD spectrum. Again the dominance of the molecular ion species in the FD spectrum is observed while the same species is hardly detectable in the EI spectrum. Figure 6 compares the EI and FD spectra of glutamic acid. The major FD peak for this compound is the M+1 ion, a situation not uncommon with other amino acids. Tyrosine, however, as shown in Fig. 7, exhibits no molecular ions in the EI spectrum and a very large M^+ ion in the FD spectrum. Protonation tendencies would not predict this observation. Quite possibly the nonbonding electron pair on the phenolic hydroxyl is involved in the ionization of this species *via* a truly field dependent quantum mechanical mechanism.

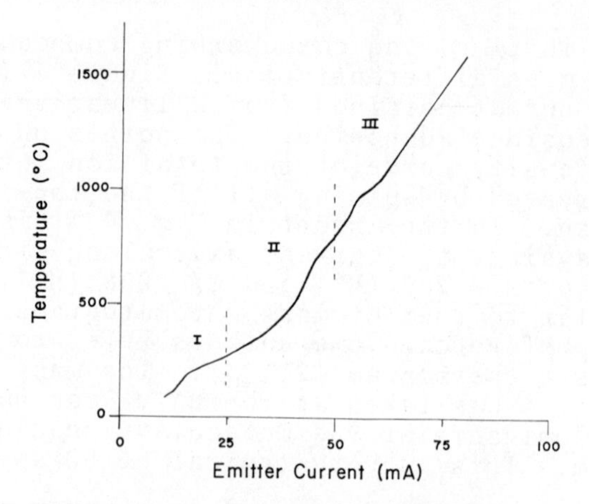

Figure 3. Temperature vs. the emitter current for the activated emitters used in this study (10 μ tungsten wire—30 μ average needle length). Region I obtained by melting of known samples and cross calibration to direct probe heated; Region II obtained by resistance measurements on activated emitters; Region III obtained by resistance and optical pyrometer measurements.

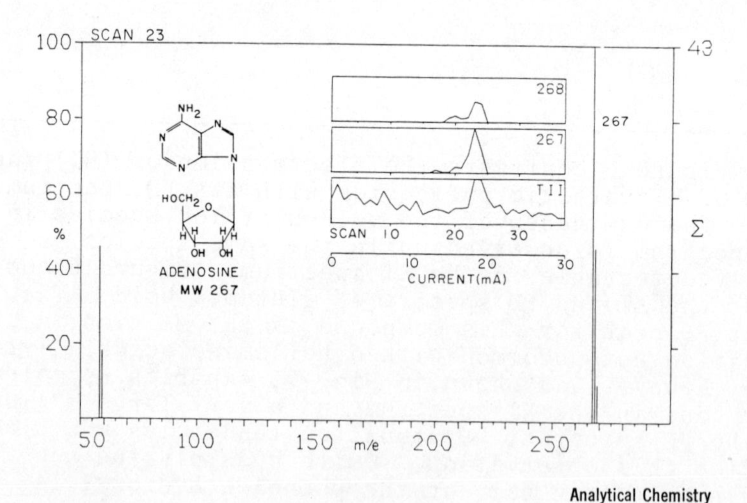

Figure 4. Field desorption spectrum of adenosine obtained by emitter current programming

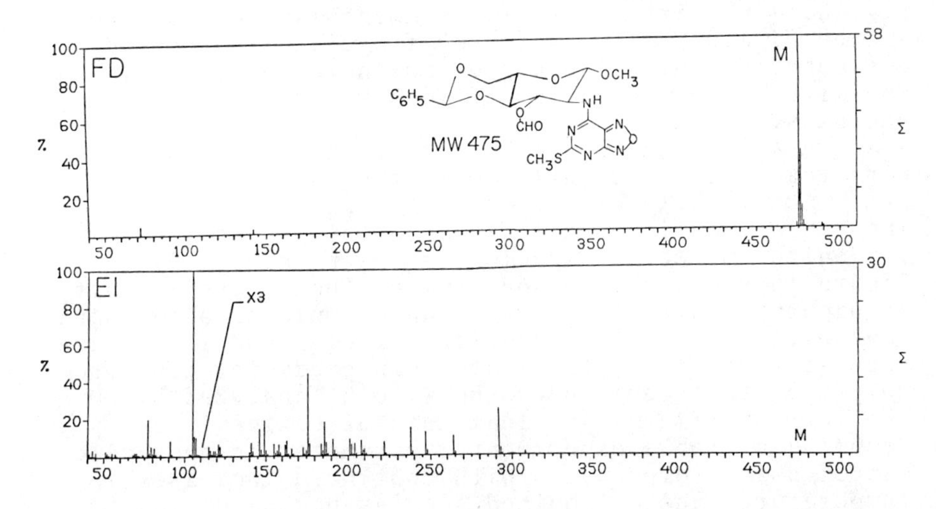

Figure 5. Comparison of field desorption and electron impact spectra of a synthetic nucleoside analog. Sample furnished by E. C. Tayler and L. E. Crane of Princeton University, Princeton, NJ.

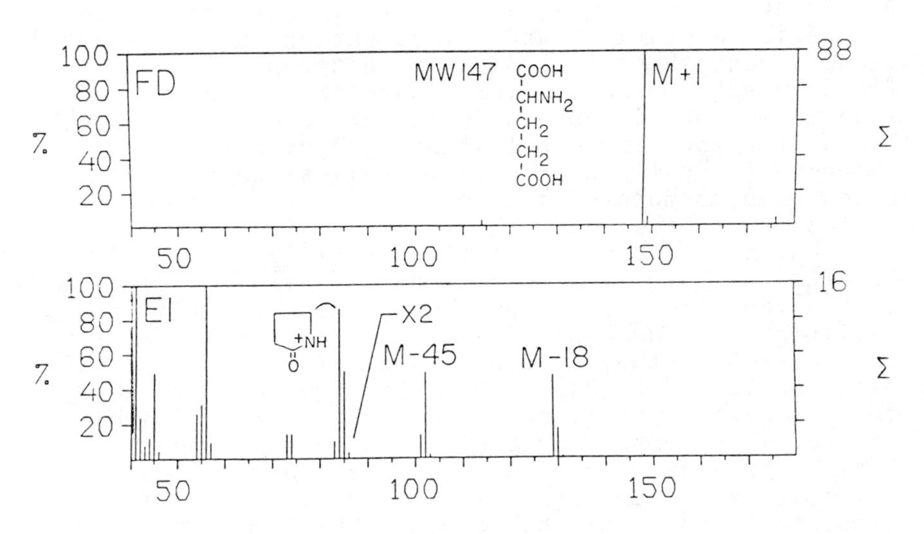

Figure 6. Comparison of field desorption and electron impact of glutamic acid

FD spectra of galactitol and sucrose are shown in Fig. 8. In neither case was a molecular ion observed in the EI mode. The increased fragmentation of the disaccharide indicates either an increased thermal lability or the formation of ions with higher internal energies.

In all of the above examples, the spectra shown were taken at or slightly below the mid-point of the BAT. At slightly higher temperatures, the presence of inorganic impurities will lead to the production of cationized molecular species as shown in Fig. 9. Figure 10 shows the FD spectrum of the potassium salt of naphthyl sulfate. This is an example of a compound that will yield only cationized species and illustrates the usefulness of cationization in producing ions from molecules that could not otherwise be analyzed by FD. The introduction of small amounts of cation impurity is becoming a viable experimental parameter in the analysis of polar compounds. With continued increases in temperature, the cationized species become dominant. If there are significant levels of impurity, the inorganic cation itself becomes the major species detected.

The analysis of field desorption spectra has indicated a ubiquitous dependence of the nature of the ions produced upon emitter temperature, the solvent used, and the presence of small cations. These parameters strongly suggest a chemical involvement in the ionization mechanism, and investigations were conducted to determine the exact role of the strong electric field in this method. Field image theory would predict than an ion should desorb in the presence of a strong field at temperatures below those required in the absence of the field. In fact, this situation has often been postulated and was generally accepted as fact prior to this study.

Electron Impact Ionization of Field Desorbed Samples. Figure 11 shows the results of an investigation which compares the BAT for uridine with large variations of the applied voltage field. Additional data for this study were obtained using a novel technique called EI/D (24). This technique employs the thermal desorption of the sample from the activated emitter with subsequent ionization by the electron beam. The geometric changes of the source necessary for this technique are shown in Fig. 12. This method generates ions after the compound has left the emitter and the temperature of maximum ion detection is called best emitter temperature (BET) in a nomenclature analogous to BAT. For all the compounds tested to date, BET

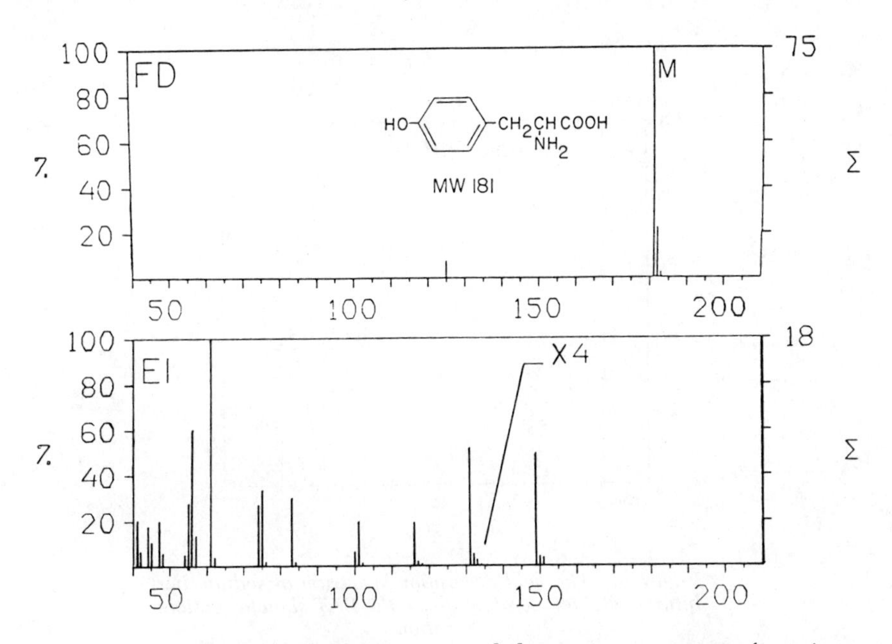

Figure 7. Comparison of field desorption and electron impact spectra of tyrosine

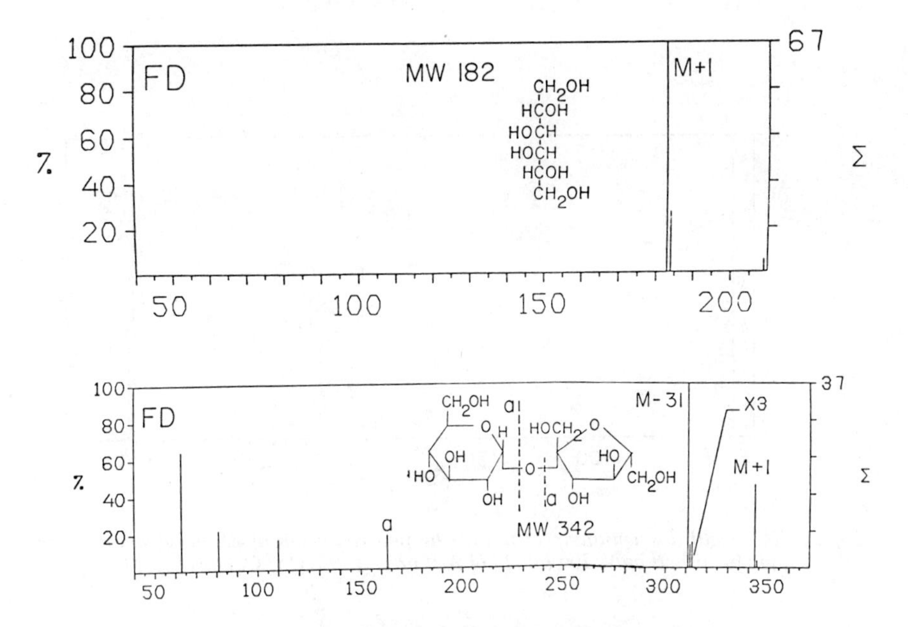

Figure 8. The field desorption spectra of galactitol and sucrose

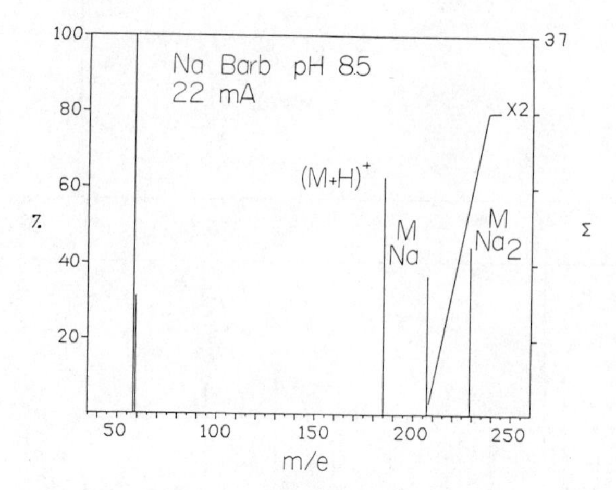

Figure 9. The field desorption spectrum of sodium barbiturate at a temperature above the BAT showing cationization

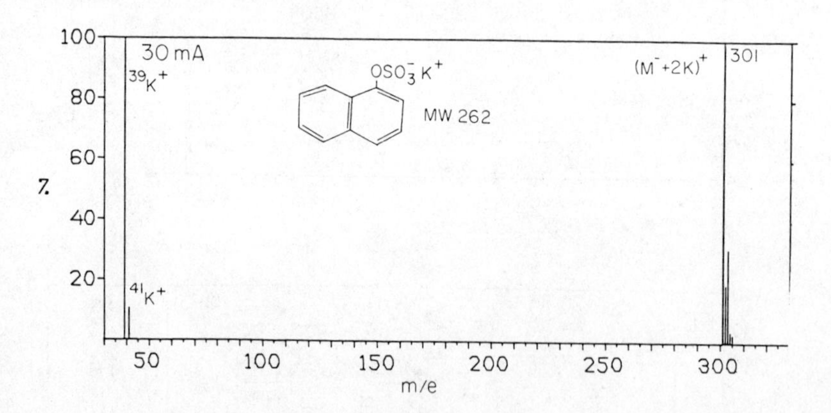

Figure 10. Field desorption spectrum of the potassium salt of naphthyl sulfate. Sample furnished by V. Rheinhold of Arthur D. Little Co., Cambridge, MA.

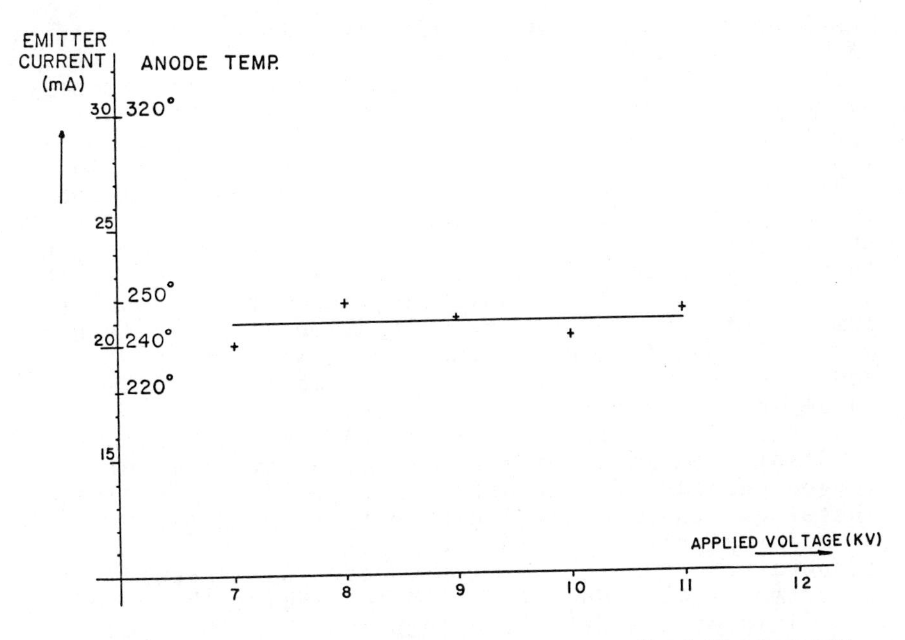

Biomedical Mass Spectrometry

Figure 11. The dependence of the BAT on the applied electric field

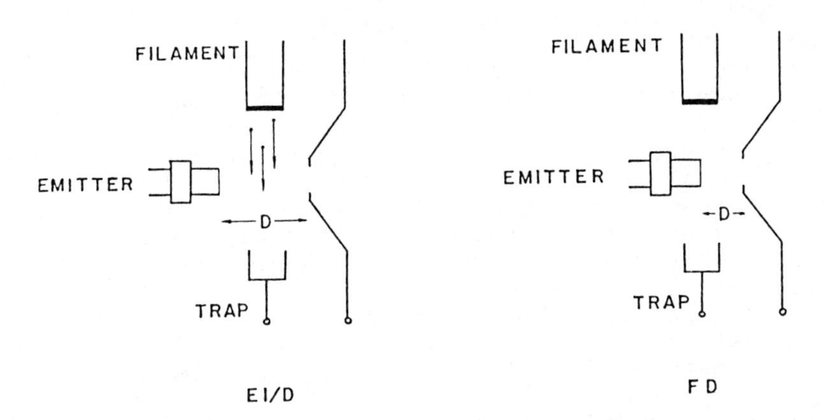

Figure 12. Ion source geometries for field desorption (FD) and electron impact ionization using a field desorption emitter (EI/D)

has been the same as BAT, and they both are independent of the applied fields throughout the range from 14KV to 3KV. The 3KV acts as a minimum for these experiments since this is the accelerating voltage for the mass spectrometer being used, a Varian CH-5-DF. This presents a dilemma, of course, since the voltage fields that are needed to collect, accelerate and focus the ions in a magnetic sector mass spectrometer may also be involved in ion production. This is a situation akin to a person attempting to find out what goes on in the dark. In order to see anything, he must somehow illuminate the region during which time it is no longer dark. Ion generation in the much lower fields of the source of a quadrupole mass spectrometer would be an ideal way to extend these studies.

Below the threshold of the BAT, measurable amounts of desorption were not found to occur even for prolonged periods of time and with high applied fields. These results strongly indicate that the field is not of major significance in the determination of the temperature at which desorption occurs. A series of experiments was conducted then to resolve the effect of the field on the rate of desorption. In these experiments, the rate of desorption, with the FD field on, was compared with the rate of desorption with the fields of the mass spectrometer off. Figure 13 illustrates one type of experiment from this series. Figure 13A shows the normal thermal desorption curve obtained by a linear temperature scan under standard FD conditions. In the remaining curves, the thermal scan was repeated with the same amount of sample on the emitter; however, all of the voltage fields of the mass spectrometer were left off until the points marked were reached, I for B, II for C, and III for D. The scans were then completed as in curve A. This experiment indicates that the sample is leaving at a rate that is independent of the applied fields. Within experimental error, no effect of the field was observed for any of the compounds examined. In summary, neither the temperature at which desorption occurred nor the rate of mass transfer from the emitter wire into the spectrometer vacuum were influenced by the high electric field. Clearly the field image theory does not accommodate these observations.

In another series of experiments conducted at a constant emitter temperature, the ratio of ion intensities between M^+ and MH^+ remained invarient over the entire applied field range; and the ion patterns of the spectra produced also remained unchanged within the reproducibility of the system. On the other hand,

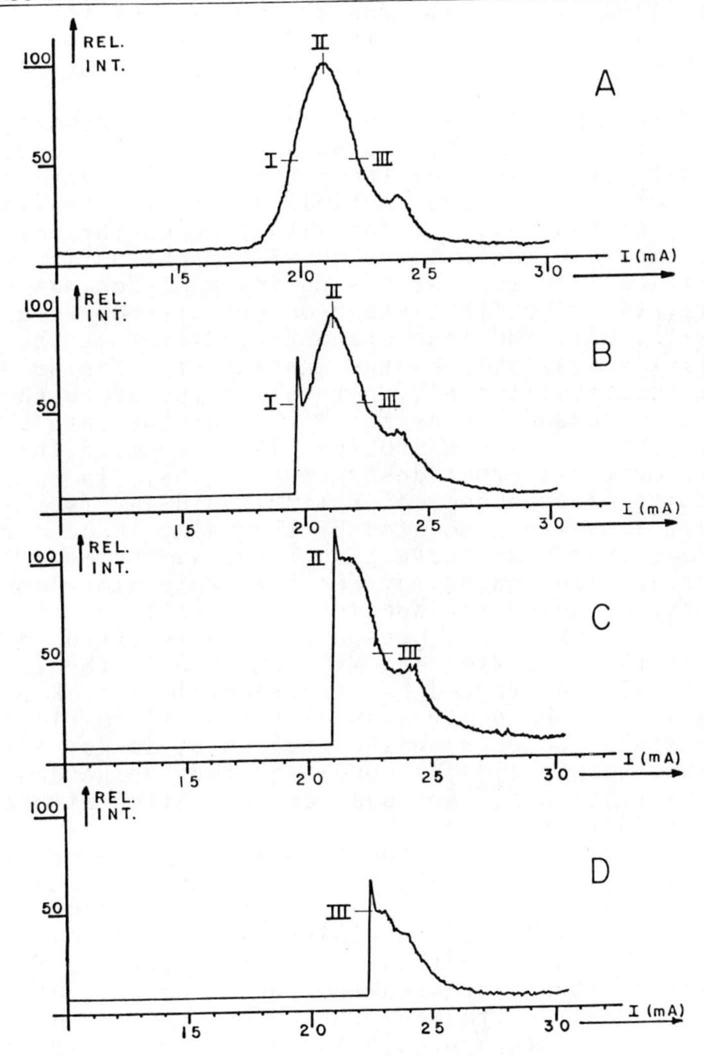

Biomedical Mass Spectrometry

Figure 13. Measurement of mass transfer in the presence of and the absence of applied voltages

changes in emitter temperature, solvent used in preparation, and the presence of small amounts of inorganic cations significantly altered the nature of the ions produced. Throughout these studies, changes in chemical parameters altered the results of the analysis while changes in the applied field only varied the absolute intensities of the ions detected but not their ratios.

Based upon the analyses of our experimentation and the observations of many others, we have proposed an alternate mechanism for the generation of many of the ions that are observed in field desorption analysis (25). For convenience, the situation on the emitter for a single organic compound of typical purity can be classified into four stages during analyses, as shown in Fig. 14. The first stage occurs at temperatures below the BAT. No ions are observed even at the highest fields available by the instrument. The second stage occurs at the BAT. At this temperature the organic compound begins to melt, changing into a liquid mobile phase. Exposed to the high vacuum of the mass spectrometer, thermal desorption begins. In this expanding mixed phase, of solid-liquid-gas (semi-fluid), ions are generated by thermally induced chemical reactions. At these temperatures, the mobility of the organic component may generate molecular ions and clusters of molecular species. If small amounts of inorganic cations are present, which is often the case, ions of the type $M_2\oplus^+$ and $M\oplus^+$, where $\oplus$ is the inorganic cation, will be formed by electrophilic attack of the cations upon the neutral species. At these elevated temperatures, electrophilic attachment is not adiabatic but endothermic (26), a condition that coincides with the observation of increased cationization at higher emitter currents.

During state III, the temperature is elevated to the point where the cation mobility is sufficient to allow the detection of only cationized molecular species. Ion clusters, containing two or more cations, begin to appear. At this higher temperature, fragmentation often increases and the appearance of a larger number of fragment ions occurs.

If there are appreciable amounts of inorganic salts, the spectrum may become dominated by the highly mobile cation. This stage occurs at the high side of the BAT and the organic species is rapidly thermally desorbed into the vacuum.

In the final stage, encompassing the highest temperatures, ions and cluster ions arising from the inorganic lattice appear. These ions are very similar

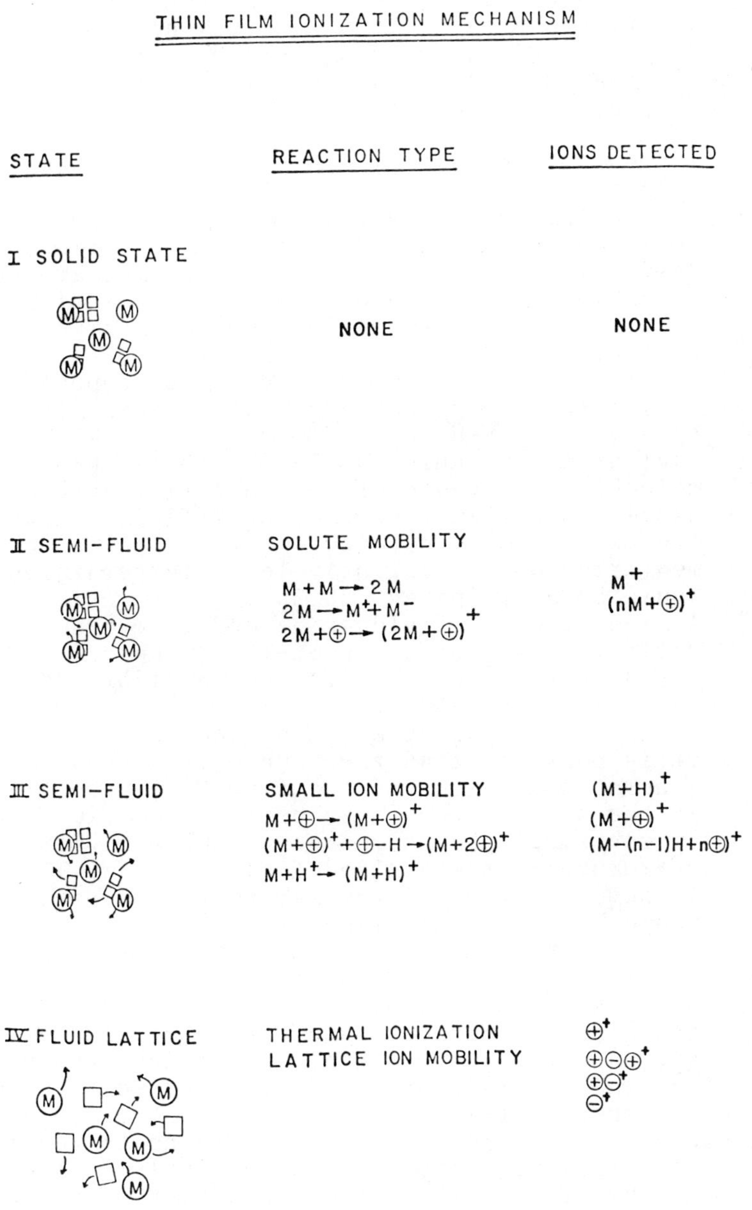

THIN FILM IONIZATION MECHANISM

STATE	REACTION TYPE	IONS DETECTED
I SOLID STATE	NONE	NONE
II SEMI-FLUID	SOLUTE MOBILITY $M + M \rightarrow 2M$ $2M \rightarrow M^{+} + M^{-}$ $2M + \oplus \rightarrow (2M + \oplus)^{+}$	M^{+} $(nM + \oplus)^{+}$
III SEMI-FLUID	SMALL ION MOBILITY $M + \oplus \rightarrow (M + \oplus)^{+}$ $(M + \oplus)^{+} + \oplus - H \rightarrow (M + 2\oplus)^{+}$ $M + H^{+} \rightarrow (M + H)^{+}$	$(M+H)^{+}$ $(M+\oplus)^{+}$ $(M-(n-1)H+n\oplus)^{+}$
IV FLUID LATTICE	THERMAL IONIZATION LATTICE ION MOBILITY	$\oplus^{+}$ $\oplus\ominus\oplus^{+}$ $\oplus\ominus^{+}$ $\ominus^{+}$

Biomedical Mass Spectrometry

Figure 14. A proposed ionization mechanism for temperature-dependent chemical generation of ions observed in field desorption spectra

to those obtained by a conventional Knudsen cell (27). The temperature attainable by emitter heating may exceed 1800°C and in the final stage of the analysis, these inorganic species may be driven off with the ions produced in accordance with classical Langmuir evaporation.

The essential parts of this proposed mechanism may be condensed as follows:

(1) Active thermal desorption must procede or be concomittant with ion formation and detection.

(2) Most, if not all, of the ions observed are generated by thermally-induced chemical reactions occurring in the semi-fluid phase on or near the emitter surface.

(3) The relative organic and inorganic compound mobilities are paramount in determining the ionic species observed.

(4) Most of the organic sample desorbs as neutral molecules. Because of the thin film nature of the sample on the emitter and the efficient heating process, desorption occurs with a minimum of fragmentation and at slightly lower temperatures than would be anticipated (28,29).

(5) A high electric field is needed to extract and collect ions from the semi-fluid phase.

(6) The high field will shift the equilibria of rapid thermally generated chemical reactions to favor formation of positive ionic species.

(7) It is possible that the high energy needed for the generation of ions of the type M^{+} may be attained in the surface chemistry of the semi-fluid phase by classical thermodynamic parameters with little influence of the applied field.

(8) In summary, the emitter may be viewed as a high temperature microchemical reactor.

Whether the model, particularly items 2 and 7, is essentially proven to be correct or not, it has already been of considerable value. It has focused attention upon the chemical parameters of the analysis and, thereby, has been instrumental in the decrease of an original failure rate of 50-70% to a present failure rate of approximately 5%. Inorganic cations, the presence of which were originally a bane to the method, are now simply reagents to be selectively applied where they may be of benefit in the production of cationized species of molecules. In addition, the model introduces a degree of chemical intuition into the method and, in so doing, relieves many of the anxieties engendered by attempts to correlate experimental results with classical theory.

This model strengthens the use of field desorption as an analytical method for polar, non-volatile compounds and expands its potential for the analysis of complex biological molecules. Clearly, the role of the field must continue to be a subject of investigation until the exact nature of ion formation can be defined. Once this occurs, the method should pass from the present predominantly artisan stage to that of an applied science.

References

1. Moor, J., and Waight, E. S., Org. Mass Spectrom., (1974), 9, 903.
2. Krone, H., and Beckey, H. D., Org. Mass Spectrom., (1969), 2, 427.
3. Schulten, H. R., Beckey, H. D., Bellel, E. M., Foster, A. B., Jarman, M., and Westwood, J. H., J. Chem. Soc. Chem. Commun., (1973), 13, 416.
4. Schulten, H. R., and Games, D. E., Biomed. Mass Spectrom., (1974), 1, 120.
5. Schulten, H. R., and Beckey, H. D., Org. Mass Spectrom., (1973), 7, 861.
6. Winkler, H. U., and Beckey, H. D., Biochem. Biophys. Res. Commun., (1972), 46, 391.
7. Adlercreutz, H., Soltmann, B., and Tikkanen, M., J. Steroid Biochem., (1974), 5, 163.
8. Games, D. E., Jackson, A. H., Millington, D. S., and Rossiter, M., Biomed. Mass Spectrom., (1974), 1, 5.
9. Maurer, K. H., and Rapp, U., Biomed. Mass Spectrom., (1975), 2, 307.
10. Sammons, M. C., Bursey, M. M., and Brent, D. A., Biomed. Mass Spectrom., (1974), 1, 169.
11. Beckey, H. D., J. Mass Spectrom. and Ion Phys., (1969), 2, 500.
12. Muller, E. W., Phys. Rev. (1956), 102, 618.
13. Gomer, I., J. Chem. Phys., (1959), 31, 341.
14. Eyring, C. F., MacKcown, S. S., and Millikan, R. A., Phys. Rev. (1928), 31, 900.
15. Beckey, H. D., Hilt, E., Schulten, H. R., J. Phys. E., (1973), 6, 1043.
16. Beckey, H. D., Heindricks, A., and Winkler, H. U., Int. J. Mass Spectrom. and Ion Phys., (1970), 3, 9.
17. Barofsky, D. F., and Barofsky, E., Int. J. Mass Spectrom. Ion Phys., (1974), 14, 3.
18. Schulten, H. R., Beckey, H. D., Boerboom, A. J. H., and Meuzelaar, H. I. C., Anal. Chem., (1973), 45, 2358.

19. Olson, K. L., Cook, J. C., Rinehart, K. L., Biomed. Mass Spectrom., (1974), 1, 358.
20. Soltmann, B., 2nd Annual Field Desorption Workshop, Urbana, 1975.
21. Kummler, D., and Schulten, H. R., Org. Mass Spectrom., (1975), 10, 813.
22. Maine, J. W., Soltmann, B., Holland, J. F., Young, N. D., Gerber, J. N., and Sweeley, C. C., Anal. Chem., (1976), 48, 427.
23. Moor, J., and Waight, E. S., Org. Mass Spectrom., (1974), 9, 903.
24. Soltman, B., Sweeley, C. C. and Holland, J. F., Anal. Chem., (1977), 49, 1164.
25. Holland, J. F., Soltmann, B., and Sweeley, C. C., Biomed. Mass Spectrom., (1976), 3, 340.
26. Hiraoka, K., and Kebarle, P., #48L 7th Int'l Conf. on Mass Spec., Florence, Italy, Sept. 1976.
27. Cherpha, W. A., and Ingham, U. G., J. Phys. Chem., (1955), 59, 100.
28. Beuhler, R. J., Flanigan, E., Greene, L. J., and Friedman, L. Biochem., (1974), 13, 5060.
29. Beuhler, R. J., Flanigan, E., Greene, L. J., and Friedman, L., J. Am. Chem. Soc., (1974), 96, 3990.

RECEIVED December 30, 1977

12

Mass Spectrometry Applications in a Pharmaceutical Laboratory

D. A. BRENT, C. J. BUGGE, P. CUATRECASAS, B. S. HULBERT, D. J. NELSON, and N. SAHYOUN

Wellcome Research Laboratories, Burroughs Wellcome Co., Research Triangle Park, NC 27709

To discuss mass spectral research applications in the pharmaceutical area, we have selected a series of representative problems taken from work carried out at the Wellcome Research Laboratories. At this facility, the mass spectrometers are located in a central laboratory servicing all departments involved in pharmaceutical research and production. The instruments used are the Varian MAT CH5-DF and 731 double focussing mass spectrometers. Both instruments are interfaced to a Varian SS100 data system and a Varian C-1024 time averaging computer. Each instrument can be coupled to a gas chromatograph, has defocussed metastable ion capability, and can measure accurate masses.

Analysis of Individual Components in A Complex Mixture By Selected Ion Monitoring

Mass Spectrometry has been a powerful tool in the analysis of complex mixtures (1-3). When coupled to a gas chromatograph, a full mass spectrum can be taken of each peak eluting from the chromatograph. When these peaks contain more than one component, the mass spectrometer can be used as a specific detector to quantitate each component under a single peak (4). This is accomplished by sequentially monitoring ions that are characteristic of each component. The area of the envelope

© 0-8412-0422-5/78/47-070-229$05.00/0

created by monitoring the intensity of an ion with time is proportional to the concentration of the compound being measured. In addition, single or multiple ion dection is more sensitive than scanning methods. This type of analysis can be applied to complex solids, e.g. tyramine in freeze dried brain tissue (5). Further specificity in the monitoring of complex mixtures can be accomplished by carrying out the measurement at high resolution (6). Allopurinol, oxipurinol, hypoxanthine, xanthine, and uric acid were directly quantitated in freeze dried muscle tissue at the 10-100 ppm level with an accuracy of ± 20% (6). A resolution of 20,000, 10% valley definition, was used to quantitate these purines and purine analogs. At this resolution some integer masses were resolved into as many as five fractional masses. One of the fractional masses at each integer mass selected is specific for the compound to be measured.

We have applied the technology of analyzing a mixture by single ion monitoring at high resolution to patent infringement cases (7). In Argentina, Austria, Czechoslovakia, Hungary, Mexico, Poland, Portugal, Spain, Switzerland, Uruguay, Venezuela, Yugoslavia, inter alia, newly developed drugs can only be protected by patents covering synthetic routes rather than by patents covering the compound itself or its usage. Therefore, it is valuable to have a method to determine the synthetic route used to monitor possible infringement cases. A compound cannot be manufactured absolutely pure. Some of the impurities present may characterize a single synthetic route or a particular class of routes. The samples are analyzed for the presence of characteristic impurities. The material to be analyzed (along with binding agents, excipients, and other drugs) is ground by mortar and pestle to a fine powder. About 3 mg of a sample is introduced into the mass spectrometer via the direct probe inlet. The sample is examined under low resolution conditions at a variety of probe temperatures for ions characteristic of known impurities. If these ions exist, the same experiment is carried out at 10,000 resolution, 10% valley definition. For example, one of the synthetic routes (8) to the drug trimethoprim, (1), involves an intermediate acetal, (2). Ions used for the identification of the impurity must be resolved from ions in the spectrum of the drug, binder, or excipients. One of the ions choosen for (2) is its molecular ion. One reason for this choice is that (1) cannot interfer because it is of lower molecular weight. Figure 1 is an oscillographic trace taken of the molecular ion of (2) at a concentration of .0005% in trimethoprim. The sample was heated from ambient to 110° (left to right in the figure). The m/e 293 and m/e 295 regions were alternately brought into focus using the peak matcher accessory on the mass spectrometer. Ions from the chemical mass marker perfluorokerosene (PFK) are used as reference peaks to determine the appropriate position for the

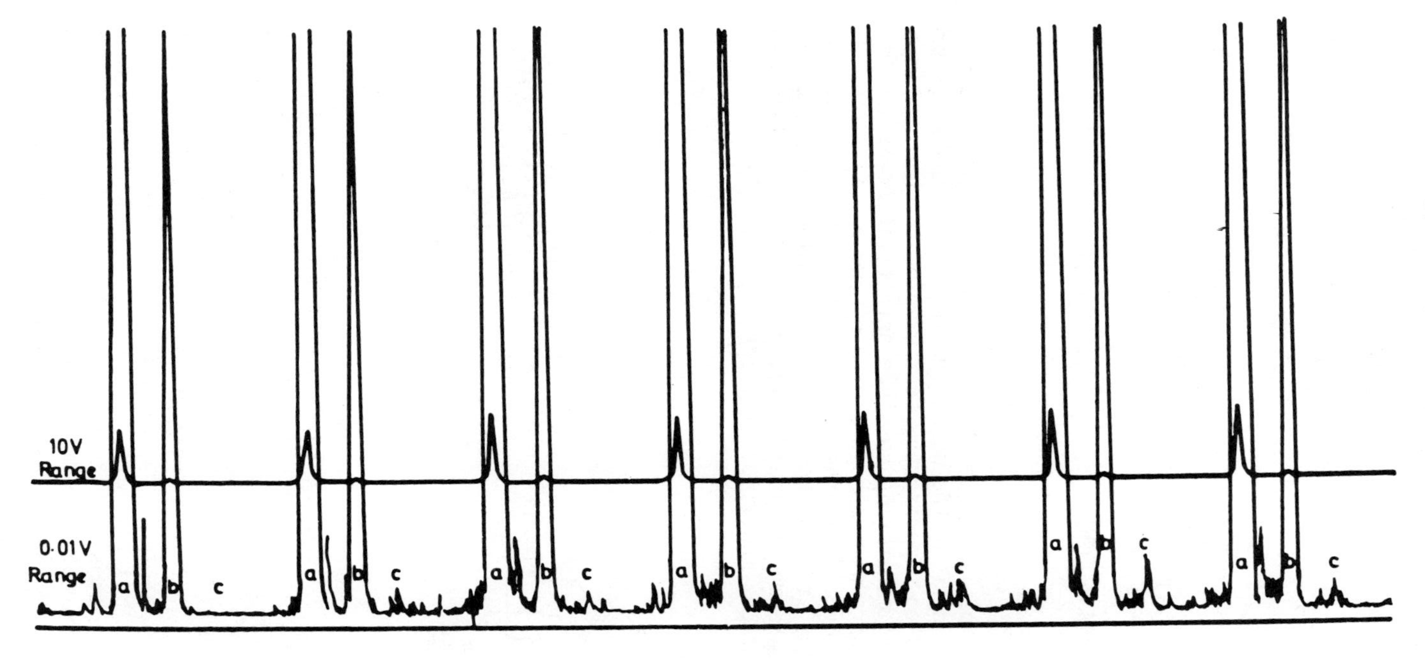

Figure 1. Sequential single-peak monitoring of the molecular ion of (2). The sample was heated from ambient to 110°C (left to right on the chart); (a) m/e 292.9824 from PFK, (b) m/e 294.9970 from PFK, (c) m/e 295.1420, the molecular ion of (2).

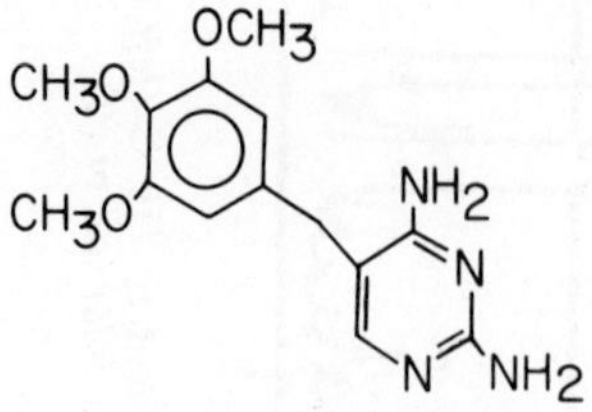

molecular ion of (2). The PFK is introduced into the source via a heated inlet at the same time the direct insertion probe is being used to vaporize the test sample. We have employed this method of monitoring possible patent infringements for about five years with a high degree of success in defense of our patents covering synthetic routes to trimethoprim.

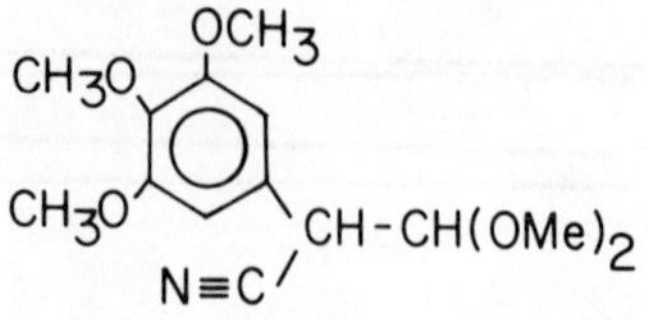

A similar strategy was used to determine whether or not some of the pharmacological activity of a codeine, (3), sample was due to trace impurities of morphine, (4).

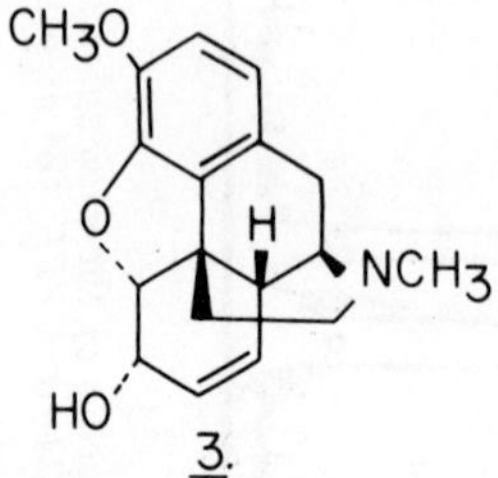

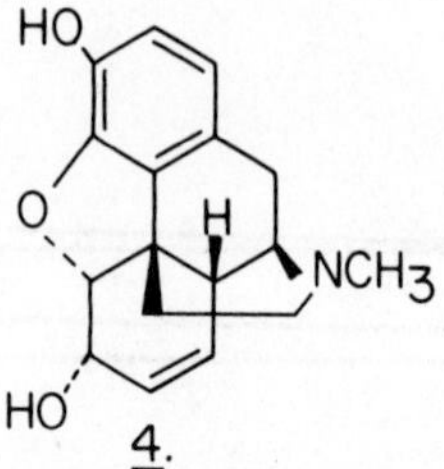

Both morphine and codeine have intense molecular ions and small fragment ions. The molecular ion, m/e 299, of codeine fragments by the loss of a methyl group to give an ion at m/e 284. The ^{13}C isotope of this fragment has the same integer mass as the molecular ion of morphine. They differ in compositions by a ^{13}C vs ^{12}CH. Therefore their accurate masses differ slightly (285.1320, ^{13}C isotope peak from the M-15 of codeine and 285.1365, the M of morphine. The resolution necessary to separate these ions with a 10% valley is 63,775 ($M/\Delta M$).

This problem can be approached two ways using mass spectrometry. The first is to calculate the intensity of the ^{13}C

isotope peak expected and compare this intensity with an averaged intensity observed experimentally. There are several problems involved with this approach. Accurate measurement of peak heights is difficult, and small variations exist in abundances of naturally occurring isotopes. A low concentration of morphine will lead to only a small perturbation of the isotopic ratio. Even if a difference were observed, it is not certain that morphine is the cause.

A second approach is to resolve the potential m/e 285 doublet at >60,000 resolution, obtain the percentage of peak height of m/e 285 due to an ion whose accurate mass corresponds to morphine, and calculate an approximate concentration of morphine in codeine by a direct comparison of their molecular ions. The latter approach was selected. At first, we had difficulties in tuning the resolution of our Varian MAT 731 to the level needed for this experiment. We sent a sample of the codeine to Drs. Gross and Lyon at the University of Nebraska - Lincoln to attempt the experiment on their A.E.I. MS-50. The m/e 285 peak is 0.95% of the base peak, m/e 299. The A.E.I. MS-50 set to a resolution of >70,000 showed a doublet at m/e 285. The intensity of the ions was too weak to get a good peak height measurement. At the time Drs. Gross and Lyon had no method of integrating this signal. After careful source cleaning and alignment we were able to obtain a resolution of 62,500 with the Varian MAT 731 and integrated the signal by taking 264 scans into a time averaging computer, Figure 2.

The experiment was run over a twenty degree probe temperature range to limit possible distillation effects. In making the calculation the following assumptions were made:

1) No change in concentration was observed due to fractional distillation.

2) The ionization cross sections of morphine and codeine are the same.

3) The same amount of ion current is contained in molecular ions of morphine and codeine at 70 eV.

4) The molecular ion of codeine does not fragment by the loss of CH_2.

It is felt that these approximations are reasonable for the accuracy needed in the pharmaceutical evaluation. The experiment indicated a level of 0.7 parts per thousand of morphine in the codeine sample. This level was too small to account for the pharmacological effects observed.

Mixture Analysis By Metastable Ion Methods

Metastable methods are discussed more fully in other sections of this volume. Of the several choices to examine metastable ions, the defocused metastable ion techniques of accelerating voltage scans keeping the magnetic and electric sector constant and mass analyzed ion kinetic energy methods have been most profitable to our research. The defocused

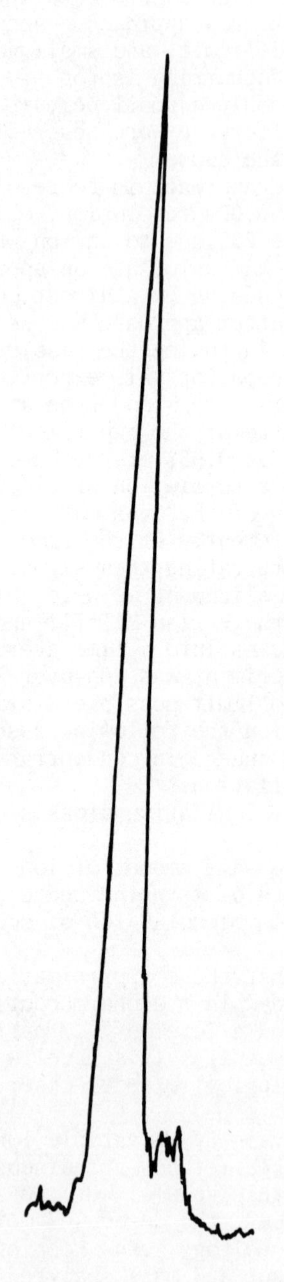

Figure 2. The average of 264 scans over the m/e 285 doublet at 62,500 resolution

metastable work was carried out on the Varian MAT CH5-DF mass spectrometer. This instrument has a reverse geometry, i.e. the source is followed by the magnetic sector, electric sector, and electron multiplier. In this configuration no metastable ions are seen in the primary ion spectra. However, accelerating voltage scans keeping the magnetic and electric sectors constant can be used to observed metastables formed in the first field free region. Mass analyzed ion kinetic energy spectra (MIKES), or direct analysis of daughter ions (DADI) spectra can be used to observe metastable ions formed in the second field free region (9,10).

The accelerating voltage scan experiment is carried out by focusing a daughter ion on the electron multiplier at 1 or 2 kV accelerating voltage. For the transition $M_1^+ \rightarrow M_2^+ + (M_1 - M_2)$, the accelerating voltage is increased until a daughter ion from a specific parent/daughter decomposition is observed. The mass of the parent is calculated by the equation $M_1 = M_2(Vm/Vo)$ where $M_1 \equiv$ mass of the parent ion; $M_2 \equiv$ mass of the daughter ion; $Vo \equiv$ accelerating voltage required to focus the primary ion daughter; and Vm is the accelerating voltage required to focus the metastable ions daughter. Thus, all parents of a given daughter ion can be determined within a parent/daughter ratio that is limited by the maximum accelerating voltage of the instrument, e.g. a Varian MAT CH-5DF has a maximum ratio of 3 and a Varian MAT 731 has a maximum ratio of 5.

In the MIKES or DADI mode, a parent is focused on the electron multiplier of a reverse geomety instrument. The daughter ions from the decomposition in the second field free region are brought into focus by decreasing the potential on the electric sector. The mass of the parent ion of these daughters has been selected by adjustment of the magnetic sector. Thus all daughters from a given parent over the complete mass range can be determined. The equation used is $M_2 = M_1(E/Eo)$; where Eo is the electric sector's potential while focusing the parent and E is the electric sector potential needed to observe the daughter ion.

An example of a problem that can be solved by defocused metastable techniques is given below. In the reaction of (5) with an equivalent of base, it is not certain whether the first methylation occurs at the phenolic site or on the pyrimidine section of the molecule (11). The crude reaction mixture was examined by low resolution electron impact mass spectrometry using 70 eV of ioning energy. The spectrum (Figure 3) appears to indicate a mixture. The masses at m/e 277, 291, 305 and 319 are molecular ions of (5) and mono-, di-, and trimethylated (5). Reducing the electron voltage from 70 eV to 20 eV reduces the fragmentation. Therefore, most of the ion current in the spectrum is contained in the molecular ions, Figure 4. From the 20 eV spectrum the following ratio of concentrations was

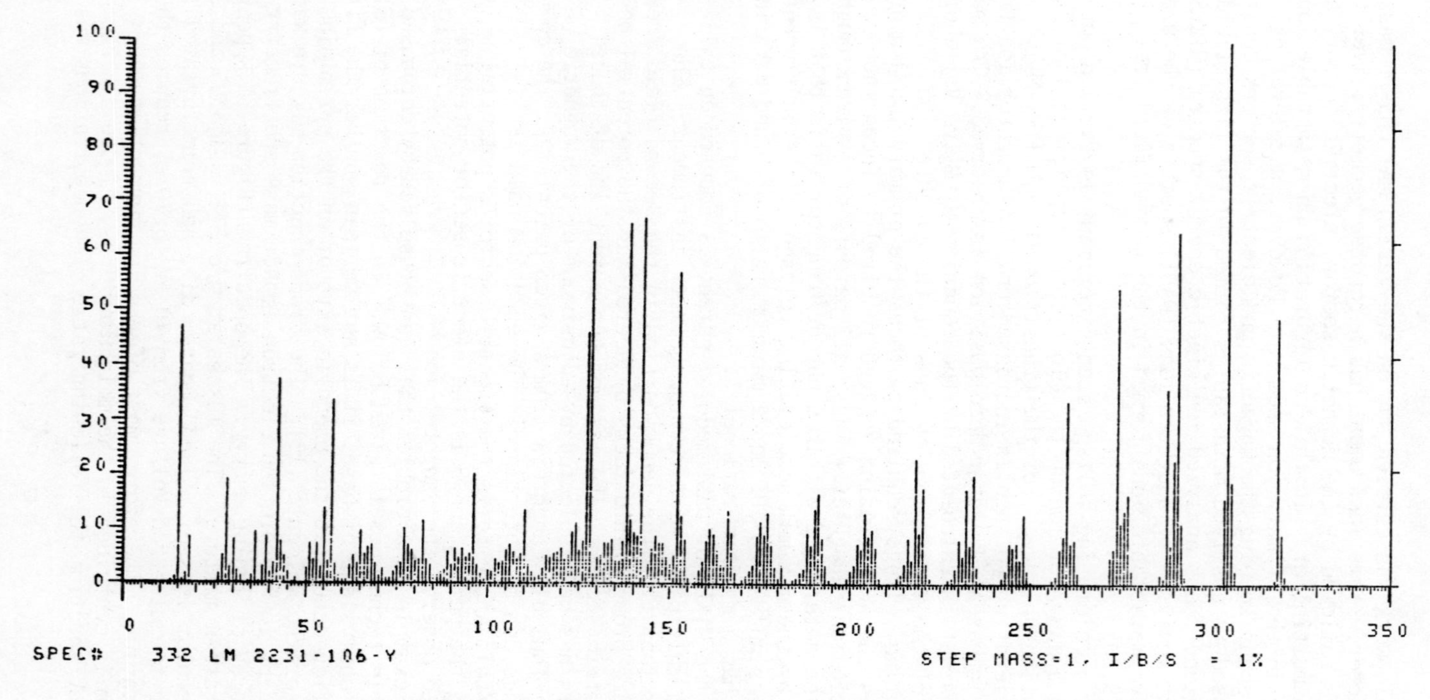

Figure 3. Mass spectrum of the products of the reaction of (5) with methyl iodide; 70 eV.

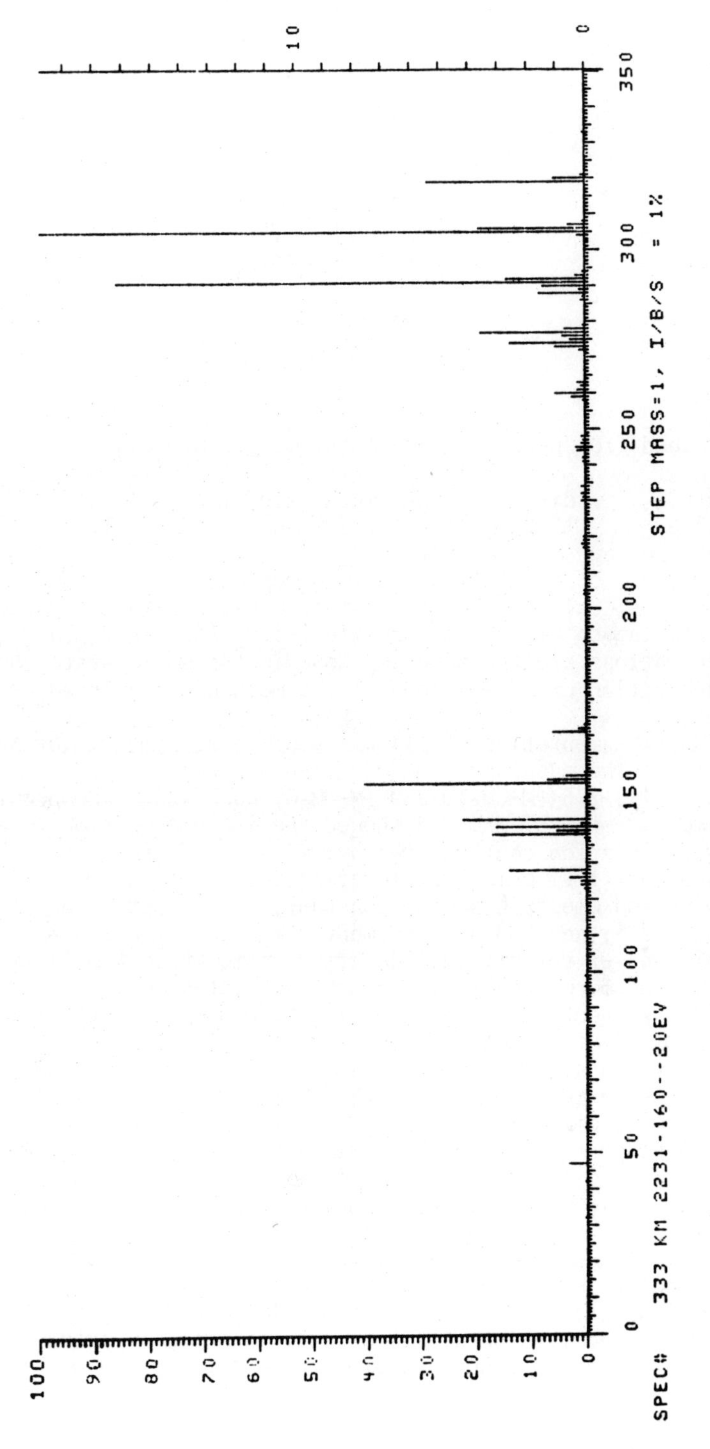

Figure 4. Mass spectrum of the products of the reaction of (5) with methyl iodide; 20 eV.

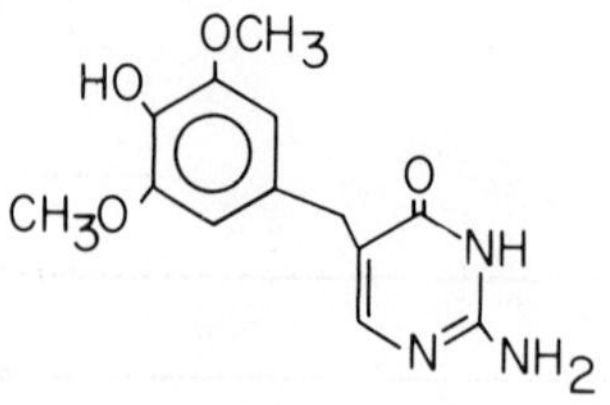

5. M.W. 277

determined from peak heights of the molecular ions:

tri-	di-	mono-methylated (5)	(5)
2.8	10.0	8.2	1.7

The accuracy of this measurement is dependent on the following assumptions: all the ion current is contained in the M ions; no distillation effect has occurred, i.e. the vapor phase concentrations are the same as those in the solid state; and the ionization cross sections of the molecules analyzed are the same.

The fragmentation of (1) was studied as a model for methylated (5). The mass spectrum of (1) has prominent peaks at m/e 290 (M), 275 (M-CH_3); 259 (M-CH_3O), and 123. The accurate mass measurement of m/e 123 showed the ion composition to be $C_5H_7N_4$. This ion contains the pyrimidine and benzylic methylene portions of (1). Using accelerating voltage scans with the magnetic and electric sector constant, it was determined that m/e 290, 275, and 259 all fragmented to yield m/e 123.

To answer the original question as to where the first methylation occurs on (5), the monomethylated species were examined without purification using defocused metastable ions, (the DADI or MIKES method). All monomethylated products have a molecular ion at m/e 291. Since these ions are known to fragment losing the benzene portion of the molecule, those molecular ions that fragment yielding m/e 138 are methylated on the pyrimidine portion of the molecule and those molecular ions that yield m/e 124 are methylated on the benzene portion of the molecule. The compositions of m/e 291, 138 and 124 were confirmed by accurate mass measurement. One cannot make a conclusion about the origin of m/e 124 and 138 without the use defocused metastable ions or by examining only purified monomethylated species. For example m/e 124 would be generated from the molecular ion of (5) as well as monomethylated (5), and m/e 138 can be generated from the molecular ion of either

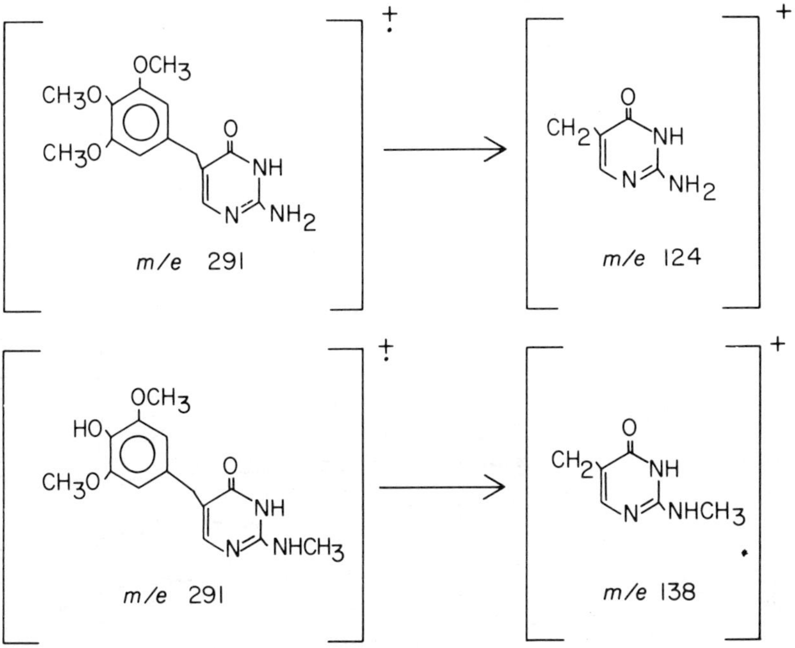

m/e 291 → m/e 124

m/e 291 → m/e 138

di- or monomethylated (5). The DADI spectrum is given in Figure 5. The DADI spectrum of m/e 291 shows a transition, m/e 291 → m/e 138 but not m/e 291 → 124. Within the limits of detection of the technique, it appears that of the monomethylated portion of the crude mixture, methylation occurs only on the pyrimidine section of the molecule.

The Use Of Defocused Metastable Ions In Defining Structure

The use of metastable ions to define a parent/daughter relationship can also be useful in structure problems. For example, we were asked if we could distinguish (6) and (7) by mass spectrometry. The electron impact mass spectrum of the compound analyzed gave a molecular ion and fragmentation pattern which would fit either (6) or (7) with the expection of M- $C_4H_5NO_2$. The elemental composition of this ion was confirmed by accurate mass measurement. Structure (7) could lose $N{\equiv}C\text{-}CO_2Et$ directly to yield M- $C_4H_5NO_2$. However, (6) could give the same ion by sequential loss of $CH_2{=}CH_2$ by a McLafferty rearrangement yielding a $\text{-}CO_2H$ which loses CO_2 followed by the loss of HCN. Although there are some exceptions, it is felt that metastable transitions are one step losses and usually occur in one section of the molecule (12,13). The molecular ion was examined by DADI and shown to fragment directly to M- $C_4H_5NO_2$, thus favoring sturcture (7). However, other than the possible errors already

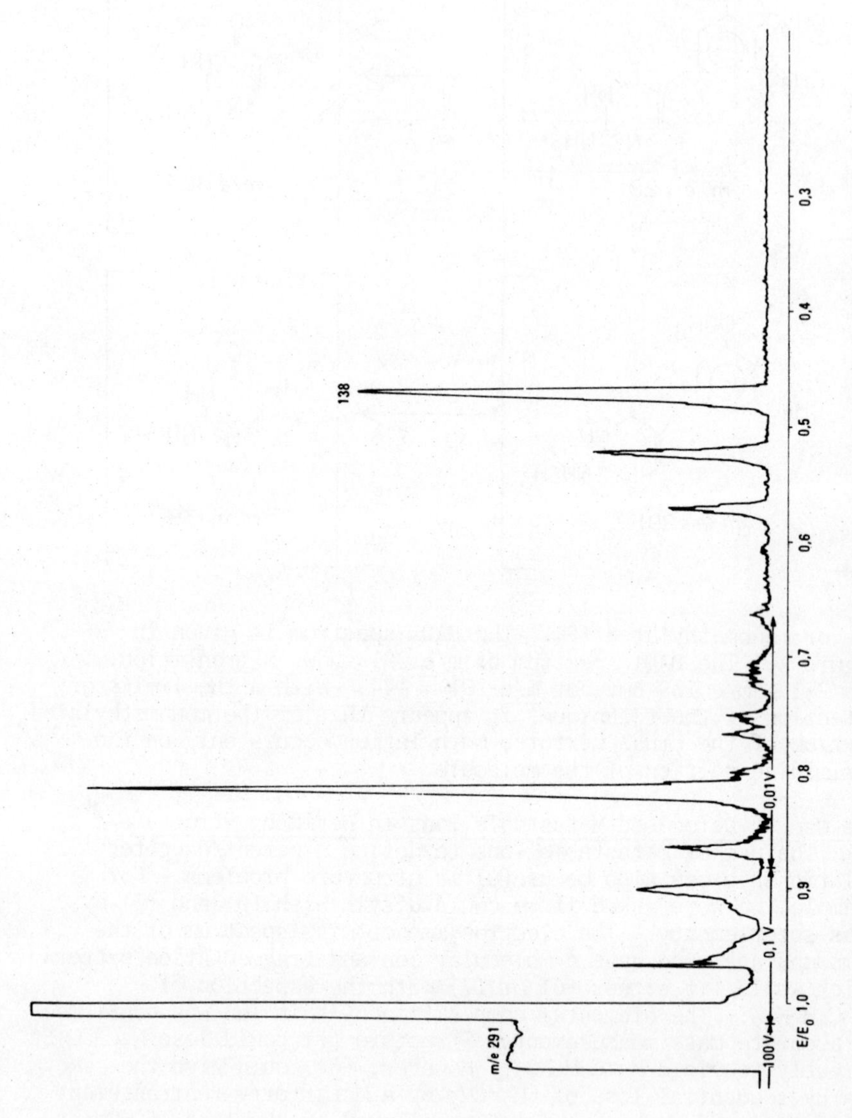

Figure 5. DADI spectrum of m/e 291 *from the mixture of methylated products*

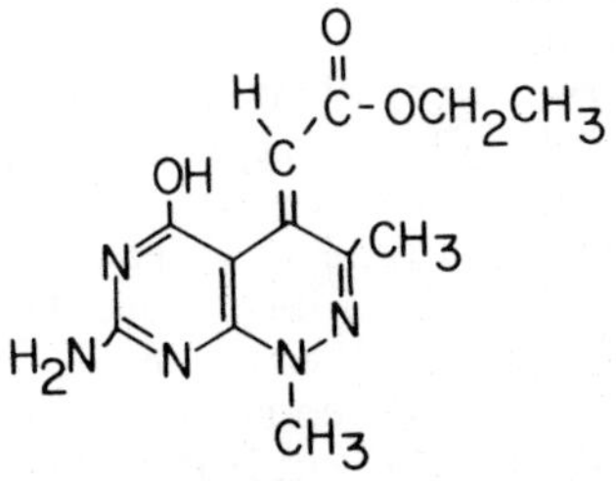

6.

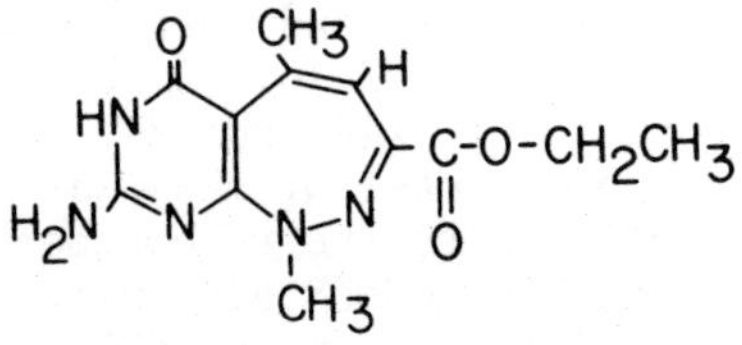

7.

mentioned, molecules may rearrange thermally during volatilization or after ionization to a new structure from which the metastable ion is formed. So although this data was an early indication for structure (7), other physical and chemical evidence were required. Subsequent physical and chemical evidence supported the mass spectrometeric results.

Analysis Of A Mixture Via GC-MS

A protocol for inducing liver microsomal activity was established using Aroclor® 1254 lot RCS9999 as the inducing agent. When a new lot of this material, RCS088, was purchased it was noted that the physical properties were different. RCS9999 was yellow and slightly viscous; RCS088 was clear and very viscous. The latter being more typical of Aroclor® 1254. The problem was to determine the difference between these two lots of material and if possible recreate lot RCS9999. The biological activity data could not be used conveniently because these tests were too time consuming. It was feared that the two materials would differ in their capacity to induce liver microsomal activity. Therefore a physical method was sought to distinguish any differences in the two lots of material.

Aroclor® is a mixture of polychlorinated hydrocarbons. The first two digits of the numbers following the name refer to the numbers of carbons in the system which are chlorinated and the last digits refer to the percentage of chlorine in the mixture (14). A literature procedure was followed to separate Aroclor® into its components by gc (15). The more highly chlorinated components elute last. The column used was 6 ft x 2 mm 3% OV-1 on Gas Chrom. Q 100/120 mesh. The column was held at 100° for 8 min and brought to 200° at 4°/min and held. The chromatogram of RCS 088 was clean for the first few minutes. At higher temperatures, at least 15 components eluted. RCS9999 had the same high temperature profile but had over 10 components in the low temp section of the chromatogram. These peaks could arise from trichlorobenzenes (a common diluent of Aroclor®), less highly chlorinated biphenyls, or other materials. RCS9999 was examined by gc-ms and a combustion analysis was carried out for percent chlorine.

Via gc-ms, the peaks which eluted first were identified as mono-, di-, tri-, and tetra-chlorinated biphenyls. The combustion analysis indicated 46.55% chlorine. The gas chromatograms of Aroclor® 1254, 1248 and 1242 were examined. A blend of 2 parts of 1242 and 1 part of 1254 was most like RCS9999. These results were returned to the researcher initiating the problem. The blend and lot RCS088 are currently being tested to compare their activity in inducing liver microsomes to the activity of lot RCS9999.

Metabolite Analysis Using Mass Spectrometry

Mass spectrometry plays an important role in structure elucidation of drug metabolites. The small amounts of material isolated in these studies are still usually within the capabilities of a mass spectrometer. The metabolites are often monitored by some tag throughout their isolation and purification, *e.g.* radio label, stable isotope label, specific chromaphore, *etc.* Drug biotransformation and biochemical conjugations are normally known processes (16). The compound being examined is usually known to be a metabolite, and often the gross structural features of the parent drug are still present. The mass spectrum of the parent drug is studied carefully by low and high resolution mass spectrometry and defocused metastable ion techniques. The fragmentation of the parent is used for a model for the metabolite.

The reinvestigation of a tlc system used to isolate the metabolites of trimethoprim (1) led to the discovery of a new metabolite (17). The compound had a fluorescence on tlc similar to trimethoprim 1-oxide, a known metabolite. The mass spectrum of the material indicated a molecular weight 16 Daltons greater than the parent drug. The elemental composition of parent plus oxygen was further supported by accurate mass measurement of

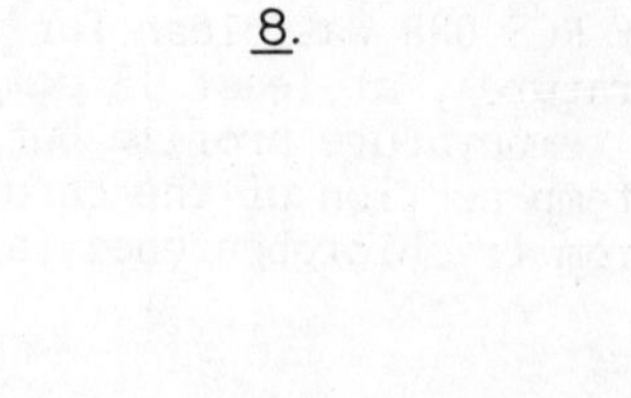

8.

m/e 306. The molecular ion fragmented by losing oxygen, and the remaining spectrum looked like that of the parent drug. The loss of oxygen from the molecular ion was indicative of the oxygen residing on N and not on carbon. The lack of an M-2 appeared consistent with N→O and not -NH-OH. Since the 1-oxide was a known metabolite which had a different R_f on tlc than this product, the metabolite was likely the 3-oxide, (8). The ir, uv, mass spectrum, and chromatographic properties of the metabolite were identical with synthetic (8).

Analysis Of Natural Products Using Field Desorption Mass Spectrometry

Field desorption mass spectrometry (fdms) is a gentle method of ionizing molecules from the solid state (18). A solution or suspension of the compound to be analyzed is coated on an emitter. The emitter is a 10 μM tungsten wire on which carbon dendrites are grown. The emitter is introduced into the source on the end of a direct probe. Most of the solvent evaporates during the insertion of the direct probe into the source. Several thousand volts are placed on the emitter. In addition a small heating current can be applied. Under the influence of the high field and gentle heating, molecules are ionized by the loss of an electron to the emitter surface or by attachment of a cation (19). This technique has extended the range of compounds which can be examined by mass spectrometric methods to include thermally labile and/or non- volatile molecules (18). For many samples, fdms eliminates the need for derivatizing the target molecule.

Field desorption mass spectrometry, nuclear magnetic resonance (nmr) spectroscopy, and high pressure liquid chromatography (hplc) were used to determine the structure of a compound isolated from toad erythrocyte ghosts which inhibited adenylate cyclase. This product gave one uv absorbing peak on a DEAE-cellulose ion-exchange column (20). Its molecular weight was estimated to be approximately 500 Daltons by Bio-Gel P-4 chromatography. Digestion with RNA-ase, DNA-ase, phospholipase A and C, papain, typsin, chymotrypsin, 5'-nucleotidase, spleen and snake venom phosphodiesterases, beef and heart cyclic-nucleotide phosphodiesterase did not inhibit the activity of this compound. Some peptidases and alkaline phosphatase produced diminished activity. Therefore, nucleotides and peptides were suspected to be possible subunits of the biologically active material. An early experiment with electron impact mass spectrometry had produced no infotmation. An hydrolysis followed by amino acid analysis indicated the presence of α-aminoadipic acid.

Since electron impact mass spectrometry had not worked, field desorption mass spectrometry was attempted. A summary of the ions seen is given below. Only integer mass values were obtained, and the isotope peaks were too weak or complicated by

$[M+H]^+$ ions to use to determine elemental composition. Also when this work was carried out we had no means of adding spectra to improve the signal to noise ratio. Therefore we manually counted the number of times an ion appeared in spectra taken at one milliamp increments of emitter heating current above the first evidence of ions as observed on the total ion current monitor. (For a more detailed explanation of the mechanics involved in obtaining an field desorption mass spectrum, the authors suggest reading reference 18.) Compound assignment was based on integer mass values fitting known amino acid or nucleotide molecular weights.

Table I

Field Desorption Mass Spectrum of the Adenylate Cyclase Inhibitor

m/e	No. of times the ion was seen.	Comment
90	(6)	Alanine + H
98/99	(3)	$[H_3PO_4]^+ / [H_4PO_4]^+$
122	(8)	Cysteine + H
135	(4)	Adenine
144	(1)	?
251, 252, 253	(4)	?
268	(1)	Adenosine
303/304	(2)	?
314/315	(2)	?
332	(2)	?
348	(2)	Adenosinemonophosphate
354	(4)	?

No α-amino adipic acid was seen. A repeat of the hydrolysis/ amino acid analyzer experiment showed an error; alanine was misassigned as α-aminoadipic acid. The possibility of the presence of cysteine was not confirmed by the hydrolysis experiment. The presence of nucleosides and-tides was supported by the field desorption experiment.

A fourier transform proton nmr was run on the material. Two different substituted adenines yielding a single proton at δ8.14 were seen. A triplet at δ6.48 and a doublet at δ6.02 were observed. These signals were thought to be due to the 1' CH of a 2'-deoxysugar and the 1'CH of a 2'-oxysugar. Looking back at the field desorption results, some unassigned peaks could fit 2'-deoxy AMP, i.e. m/e 251, 252, 253 and 332. The nmr signals at δ6.48 and δ6.02 were approximately of equal intensity.

Since the unknown contained a mixture of nucleotides, the sample was subjected to hplc using a Partisil-10-SAX anion exchange column. Elution with phosphate buffer was known to separate nucleoside monophosphates with good resolution. The hplc results indicated two large 254 nm absorbing peaks of equal intensity eluting in the monophosphate region (both of which had u.v. spectra identical to 5'-AMP) and a small peak eluting at the region of ATP. Mixing experiments showed the first peak to be AMP. This component was destroyed by reaction with periodate. The second peak cochromatographed with 2'-deoxy AMP and was not affected by periodate oxidation. These peaks were collected and tested for biological activity. The inhibitory activity remained with the "2'-deoxy AMP" peak. At this point synthetic 2'-deoxy-5'-AMP was tested and found to be inactive. A 5'-nucleotide would be contrary to the undiminished activity after treatment with 5-nucleotidase. However, the reason that the 5'-nucleotidase experiment may have been in error is that the enzyme might have special requirements such as the presence of a 2'-hydroxy group.

Work then branched in two directions. Enzyme data still pointed to the possibility of amino acids or peptides linked to the active molecule. The nucleotide was examined for amino acids by hydrolysis followed by chromatography on an amino acid analyzer or trimethylsilylation, then gc or gc-ms. In addition more enzymes were tested on the peak isolated by hplc. It was found that rye grass 3'-nucleotidase diminished the peak height on hplc yielding deoxyadenosine. Therefore, synthetic 2'-deoxy-3'-AMP was tested. It had the same inhibitory activity against adenylate cyclase and the same chromatographic properties as the material isolated from toad erythrocyte ghosts. The compound in question is 2'-deoxy- 3'-AMP. The only piece of structural uncertainty is its reactivity with peptidase. This piece of evidence, likely caused by impure enzyme, delayed the structure assignment by giving a false lead. However, if it weren't for the other enzyme data, the problem would be orders of magnitude more difficult.

Summary

We have not attempted to survey the entire scope of mass spectrometry as applied to the pharmaceutical industry. Rather we have taken a cross section of problems to illustrate the utility of high performace mass spectrometry to problems found in the industry. The use of high resolution mass spectrometry, defocussed metastable ion techniques, coupled gas chromatography-mass spectrometry, and the new ionization technique of field desorption have been highlighted. In all cases successful solution to a problem depended on good communication between researchers to establish the history of the problem, define the goals, and to plan the experiments which might lead to a solution. In all cases, although mass spectrometry was emphasized, other physical methods as well as wet chemistry played vital interactive roles in the problem's solution. The high performance mass spectrometer may have been circumvented in many of these problems, but this would be at the expense of sample and time. The savings of time and the ability to work with extremely small quantities of sample are the keys to the impact of modern instrumentation.

Acknowledgement

The authors wish to acknowledge the helpful assistance on several problems referred to in the text by Drs. R.W. Morrison, D. Clive, and J.W. Findlay and Mr. R.F. Butz.

References

1. Biemann, K., "Mass Spectromety", McGraw Hill Book Co., Inc., New York, 1962.
2. Waller, G.R., "Biomedical Applications of Mass Spectrometry", Wiley-Interscience, New York, 1972.
3. Junk, G.A., Int. J. Mass Spectrum. Ion Phys. (1972) 8, 1.
4. Sweeley, C.C., Elliot, W.H., Fries, I. and Ryhage, R., Anal. Chem., (1966) 38, 1549.
5. Boulton, A.A. and Majer, J.R., Can. J. Biochem. (1971) 49, 993.
6. Snedden, W. and Parker, R.B., Anal. Chem. (1971), 43, 1651.
7. Brent, D.A. and Yeowell, D.A., Chem. Ind. (1973), 190.
8. Cresswell, R.M. and Mentha, J., U.S. Patent 3513185, 19 May, 1970; Hoffer, M., U.S. Patent 3341541, 12 September, 1967.
9. Beynon, J.H., Cooks, R.G., Amy, J.W., Baitinger, W.E. and Ridley, T.Y., Anal. Chem., (1973), 45, 1023A.
10. Cooks, R.G., Beynon, J.H., Caprioli, R.M. and Lester, G.R., "Metastable Ions", Elsevier Scientific Pub. Co., Amsterdam, New York, 1973.
11. Brent, D.A. and Rouse, D.J., Varian CH5DF/MAT 311 Application Note No. 12, (1973).

12. Jennings, K.R., Chem. Commun., (1966), 283.
13. Letcher, R.M. and Eggers, S.H., Tetrahedron Lett. (1967), 3541.
14. Armour, J.A., J. Chromatogr. (1972), 72, 275.
15. Hirwe, S.N., Borchard, R.E., Hansen, L.G. and Metcalf, R.L., Bull. Environ. Contam. Toxicol. (1974), 12, 138.
16. LaDu, B.N., Mandel, H.G. and Way, E.L., eds., "Fundamentals of Drug Metabolism and Drug Disposition", The Williams and Wilkins Co., Baltimore, Md., 1971.
17. Sigel, C.W. and Brent, D.A., J. Pharm. Sci., (1973) 62, 674.
18. Brent, D.A., J. Assoc. Off. Anal. Chem. (1976) 59, 1006.
19. Holland, J.F., Soltmann, B. and Sweeley, C.C., Biomed. Mass Spectrom. (1976) 3, 340.
20. Sahyoun, N., Schmitges, C.J., Siegel, M.I. and Cuatrecasas, P., Life Sci. (1976) 19, 1961.

RECEIVED December 30, 1977

13

Detection and Identification of Minor Nucleotides in Intact Deoxyribonucleic Acids by Pyrolysis Electron Impact Mass Spectrometry

J. L. WIEBERS

Department of Biological Sciences, Purdue University, West Lafayette, IN 47907

It has been demonstrated previously that intact DNA or polydeoxyribonucleotides can be subjected to mass spectrometric analysis without prior derivatization or chemical or enzymatic hydrolysis (1,2). When a polydeoxyribonucleotide is exposed to pyrolysis electron impact mass spectrometry (Py-EI-MS), the primary fragmentation process involves cleavage at the phosphodiester bonds linking the nucleotide residues. Subsequent fragmentations result in ion products which are diagnostic for the common nucleotide components of the polynucleotide (3). Ribooligonucleotides are not susceptible to Py-EI-MS and remain nearly inert under such conditions. Presumably, this different behavior is due to the presence of the 2'-hydroxyl group in the ribo compounds. In the original scheme (1) for the cleavage of deoxyribopolynucleotides, it is proposed that the fragmentation involves extrusion of the 3'- and 5'-phosphodiester entities from the polynucleotide, and that extrusion of the oxygen functions in the 3'- and 5'-positions is more energetically favored in deoxyribo compounds than in ribo compounds. This appears to be a reasonable if not a complete explanation of the process.

Similar types of fragmentations of nucleic acids have been observed in pyrolysis field desorption mass spectrometry (Py-FD-MS) (4,5,6). These studies demonstrated that DNA and RNA can be analyzed by Py-FD-MS, and reaction mechanisms involved in the pyrolysis process were proposed. High resolution analysis permitted the assignment of probable structures to the pyrolysis products of the sugar and phosphate moieties as well as the nucleic acid components. In more recent investigations (7,8), high resolution (HR) Py-FD-MS yielded structural information on the ion products from unprotected ribo-dinucleotides. Evidence for the formation of nucleoside cyclophosphates from the ribodinucleotides permitted the differentiation of dinucleotide sequence isomers. The application of the HR-Py-FD-MS method to the base sequence analysis of oligonucleotides has been proposed (7,8). Low resolution pyrolysis electron impact mass spectrometry (LR-Py-EI-MS) has been successfully employed for the sequence determination of di-, tri-, and tetra-deoxyribo-oligonucleotides by applying computerized pattern recognition analysis to the spectral results (3,9). An intriguing study on the analysis of the pyrolysis products of DNA by combined low energy electron impact and collisional activation spectrometry allowed the assignment of the structures of the different components in the complex mixture from the pyrolysis without prior separation (10).

The investigations cited above indicate that both structural and sequence information about nucleic acids can be obtained *via* pyrolysis mass spectrometry. One aspect of such information is the detection of modified nucleotides in DNA and is the subject of this paper. Modified components are thought to have functional significance for the DNA molecule, and, the detection and location of such components in viral genomes is expected to contribute to a better understanding of host-viral relationships.

Pyrolysis Electron Impact Mass Spectral Analysis of DNA.

Methodology. Solutions of DNA or polydeoxyribonucleotides containing 0.01 to 1.0 A_{260nm} units are introduced into capillary sample tubes and taken to dryness *in vacuo* in a desiccator containing P_2O_5. Samples are introduced into the spectrometer (DuPont 21-490 B) by direct probe, slowly heated to about 250° and the analysis is carried out at 70 eV, 3 x 10^{-6} Torr, source temperature = 200°. Spectra are recorded at the point at which the maximum number of ions are generated as indicated by the ion monitor.

Fragmentation Pattern. The general scheme proposed for the mass spectral fragmentation of DNA together with the suggested structures of the ion products are outlined in Figure 1. Species *a* is considered to be the first volatile product that is released from the polynucleotide chain and, thus, is submitted to electron-impact ionization, which fragments the molecule to the purine or pyrimidine base and to methylfuran. Subsequently, ions *b* and *c* are

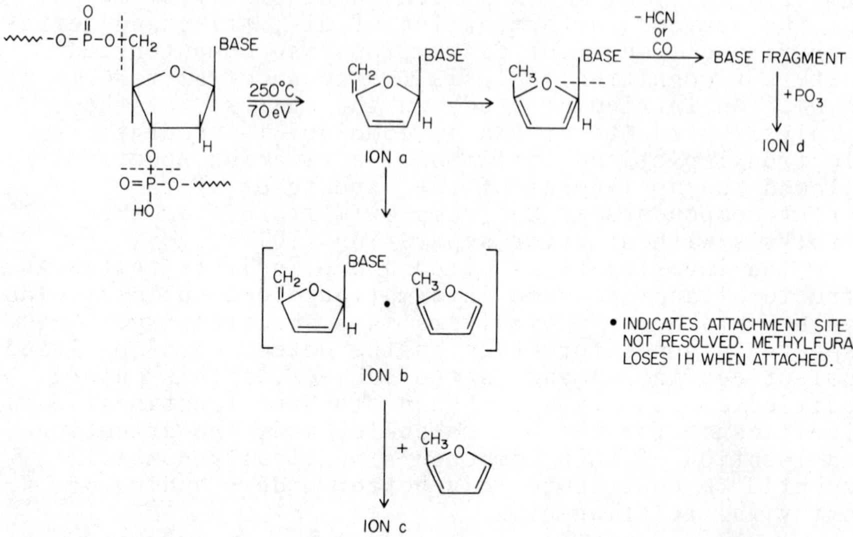
BASE
250°C
70 eV
CH2
CH3
H
O
HO
O=P-O
-HCN
or
CO
BASE FRAGMENT
+PO3
ION a
ION b
ION c
ION d
• INDICATES ATTACHMENT SITE NOT RESOLVED. METHYLFURAN LOSES 1H WHEN ATTACHED.

Nucleic Acids Research

Figure 1. Proposed scheme for the electron impact and pyrolytic fragmentation of polydeoxyribonucleotides

formed through the attachment of a PO_3 moiety to the exocyclic amino group of the base (presumably through a phosphoramidate linkage. This ion only appears in spectra of nucleotides that contain a free exocyclic amino group (3). The purine or pyrimidine base fragment and the ion types a, b, c, and d, are specific for each of the common deoxyribonucleotide residues found in DNA. The mass values of these ion types for the common residues are listed in Table I. Their appearance in the mass spectrum of salmon sperm DNA is shown in Figure 2.

Table I. Mass Values (m/e) of Diagnostic Ions Derived from Residues in DNA and Polydeoxyribonucleotides.

Ion Type	Nucleoside Residue					
	dA	dC	dG	dT	d-m^{5}C	d-hm^{5}C
Base + H	135	111	151	126	125	141
a	215	191	231	206	205	221
b	295	271	311	286	285	301
c	375	351	391	366	365	381
d	186	162	202	---*	176	192

* Ions belonging to the d class are derived from PO_3 attachment to exocyclic amino groups of the bases and do not appear when such groups are either absent or alkylated.

Detection of Modified Nucleotides in DNAs.

The same types of ion products documented above can be used to detect modified nucleotide residues present in DNAs. For example, 5-methyldeoxycytidine appears in small amounts in many DNAs. A mass spectrum of 5-methyldeoxycytidine-5'-phosphate (Figure 3) shows ions at m/e 176, 205, 285, and 365. These values are analogous to the values of the ion types a, b, c and d, for common DNA components (Table I) and are diagnostic for the presence of 5-methylcytidine residues in DNA. A spectrum (Figure 4) of wheat germ DNA

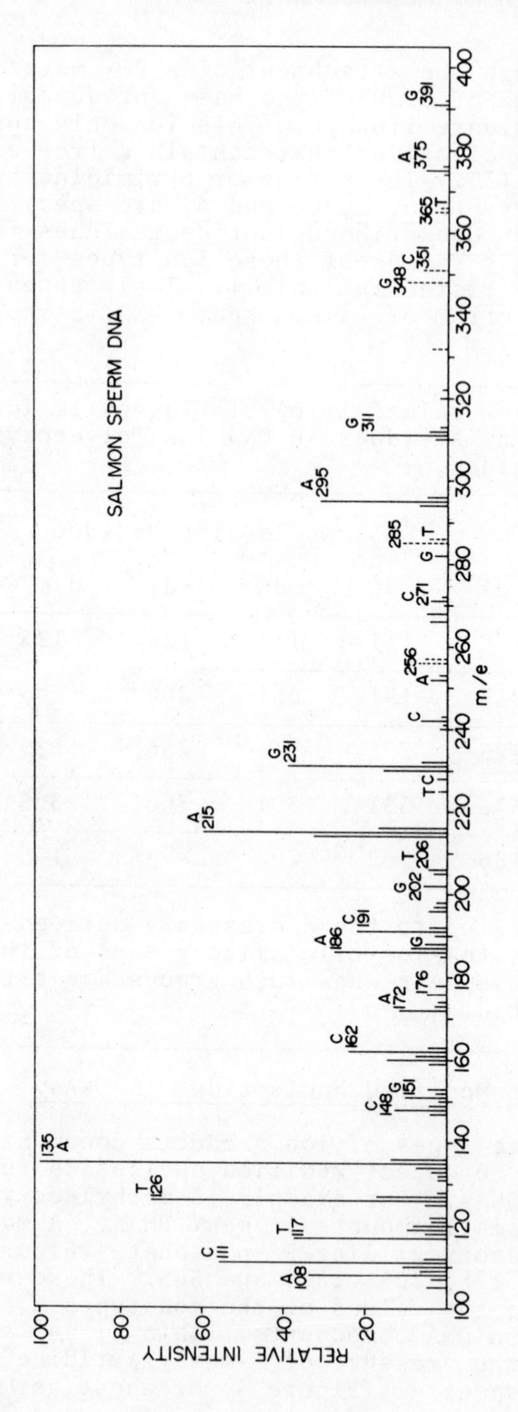

Figure 2. Mass spectrum of salmon sperm DNA. A, G, C, and T, indicate the ions that arise from deoxyadenosine, deoxyguanosine, deoxycytidine, and thymidine residues, respectively. In Figures 2–8, peaks indicated by broken lines are drawn at 10 times their actual relative intensity.

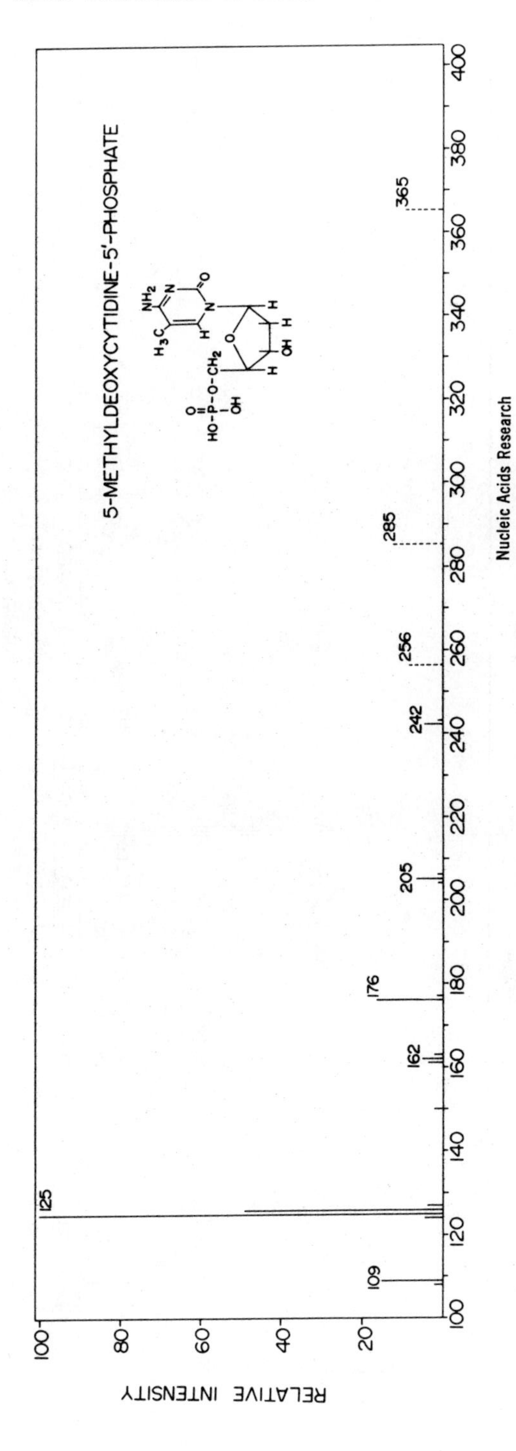

Figure 3. Mass spectrum of 5-methyldeoxycytidine-5′-phosphate

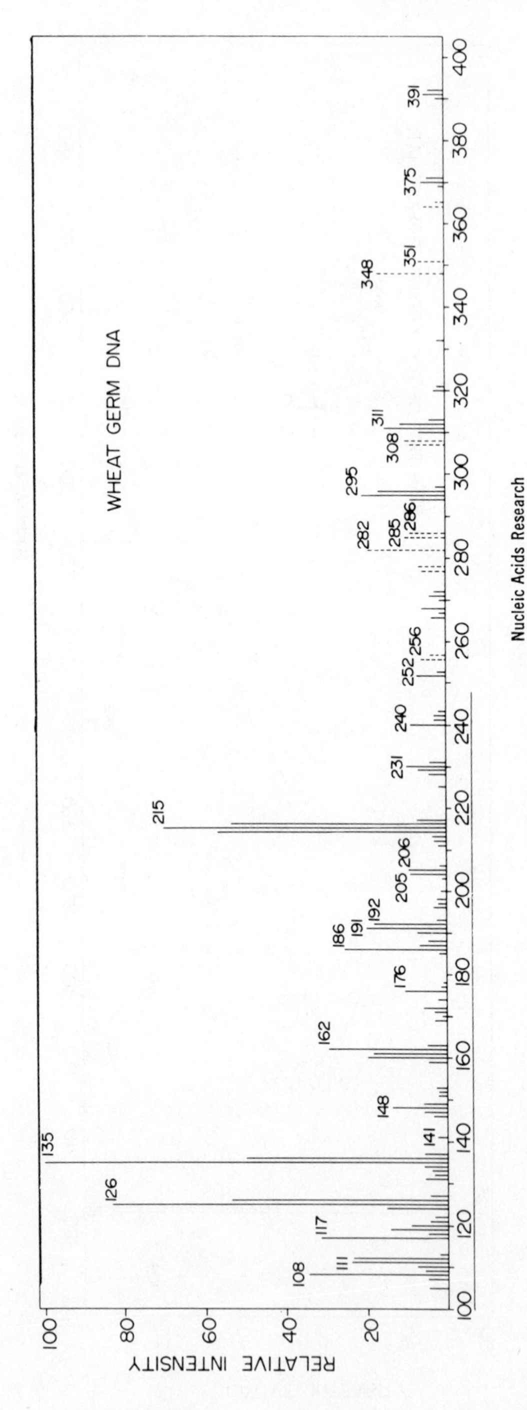

Nucleic Acids Research

Figure 4. Mass spectrum of wheat germ DNA

[which is known to contain 5.6 moles of 5-methyldeoxy-cytidine per 100 gram atoms of DNA phosphorus (11)] clearly shows the expected ion types at m/e 176, 205, 285, and 365 for the 5-methyldeoxycytidine entity.

Some DNAs have one of the common components completely replaced by modified residues. For example, bacteriophage T2 DNA is known to have all of its deoxycytidine residues hydroxymethylated (12). The spectrum of T2 DNA (Figure 5) exhibits none of the diagnostic ions that originate from deoxycytidine residues (Table I), but rather, shows ions with the expected values for products deriving from 5-hydroxymethyldeoxycytidine residues (Table I). These results from this particular DNA emphasize the validity of the mass spectral detection method since they qualitatively confirm the findings of the chemical composition studies on this DNA. As a further control for the method, mass spectral analysis of the synthetic polynucleotide poly(dA·dT) (Figure 6) yielded only the expected ion products deriving from dA and dT, and no extraneous ions were observed.

Mass spectra of bacteriophage DNAs are more complex than those of the bacterial hosts that they infect. For example, the spectrum of *E. coli* DNA (Figure 7) displays diagnostic ions for 5-methyldeoxy-cytidine, whereas, the spectrum (Figure 8) of the DNA from bacteriophage ØX-174 (which infects *E. coli*) indicates the presence of 5-methyldeoxycytidine residues as well as several different minor purine and pyrimidine components. A recent study (13) of four bacteriophage DNAs yielded spectra which showed several series of ion products that correspond to the ion types *a*, *b*, *c*, and *d* and have mass values which could be indicative of the presence of methylated deoxyadenosines and of substituted deoxyuridines.

Although the function(s) of modified components in DNAs has not been resolved, it is of great interest to molecular biologists to determine not only which modified nucleotides occur in bacteriophages, but also, to determine their location in the whole DNA molecule. The strategy that we are using in our laboratory to help to answer these questions is (a) to detect and identify the modified constituents in the intact DNA molecule by mass spectral analysis, and (b) to treat the DNA with appropriate endonuclease restriction enzymes. This technique permits the cleavage of the molecule at specific sites yielding specific fragments whose sequential order can be reconstructed. The fragments are separated and collected by gel electrophoresis and, subsequently, are analyzed by

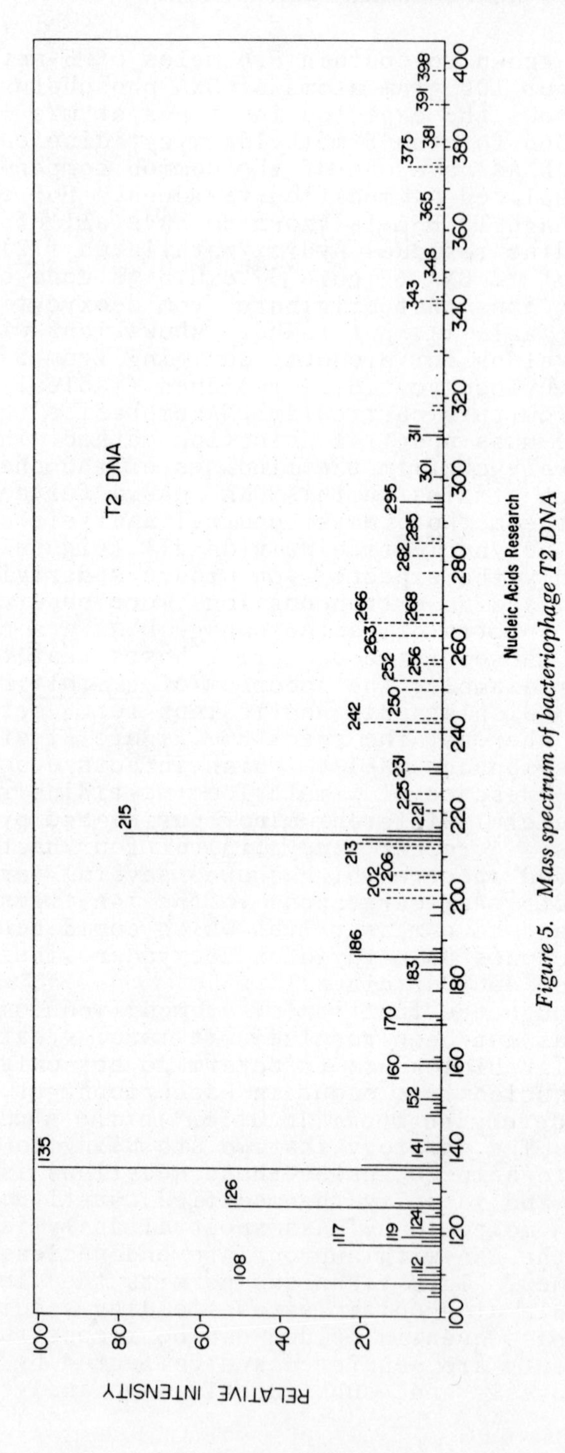

Figure 5. Mass spectrum of bacteriophage T2 DNA

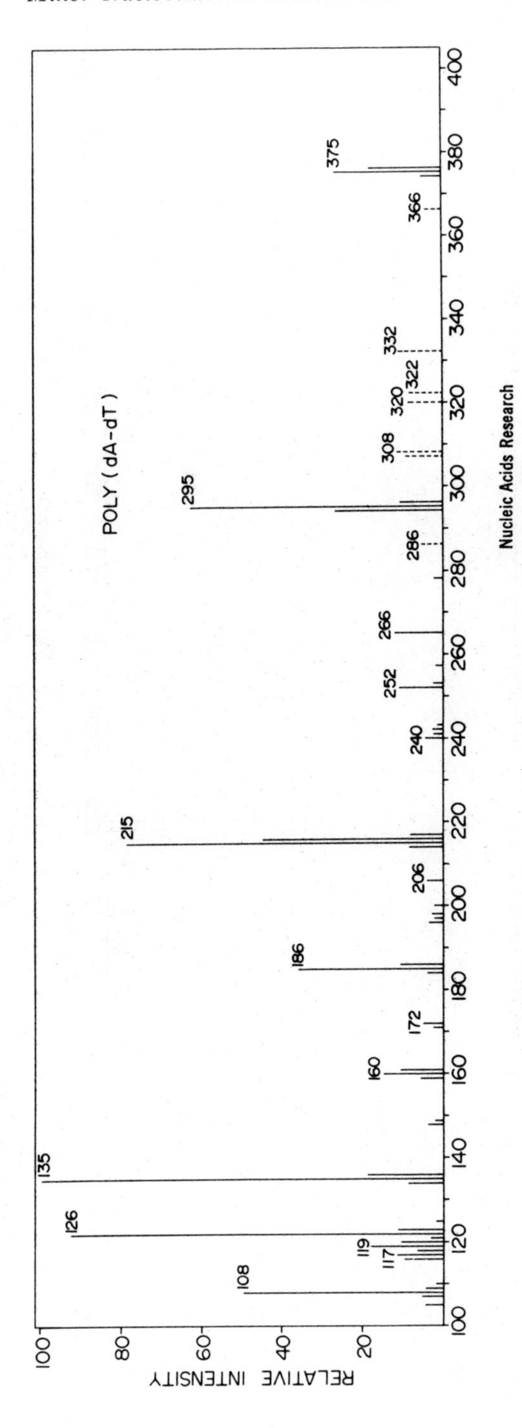

Figure 6. Mass spectrum of synthetic polynucleotide poly(dA · dT)

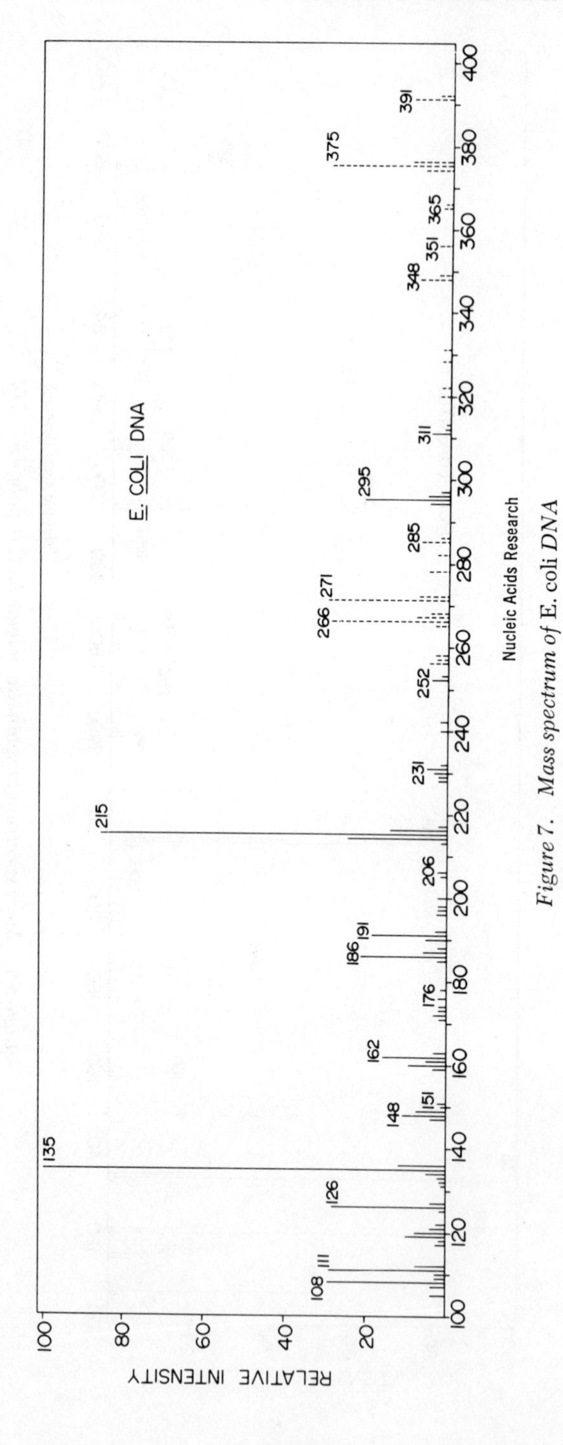

Figure 7. Mass spectrum of E. coli *DNA*

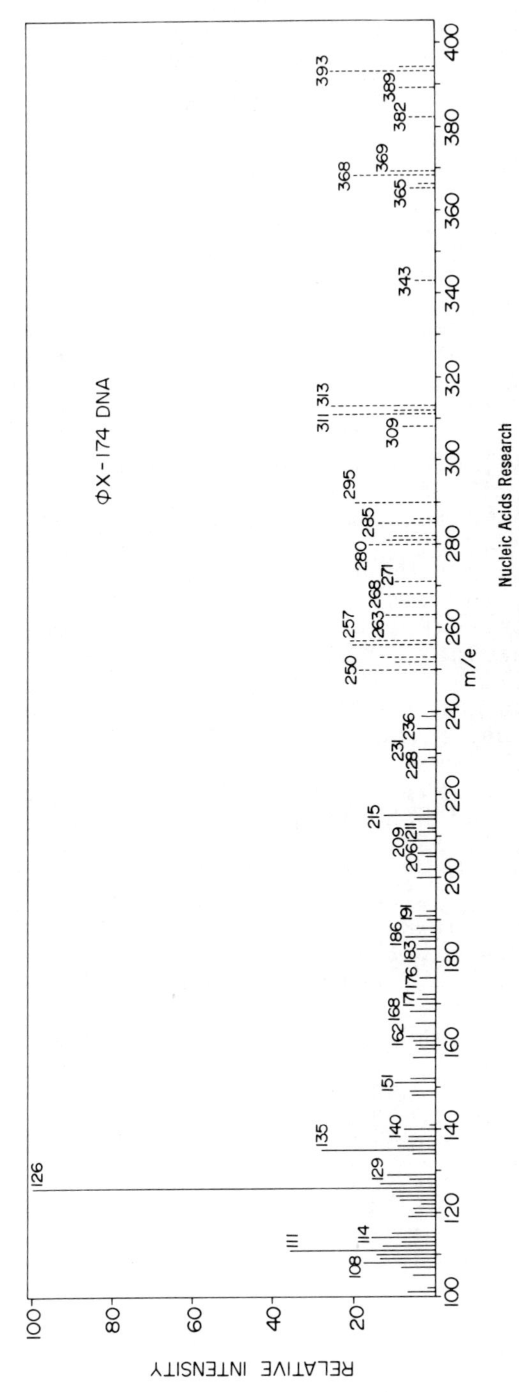

Nucleic Acids Research

Figure 8. Mass spectrum of bacteriophage ØX-174 DNA

mass spectrometry for the existence of modified residues. This process is repeated with other endonuclease restriction enzymes that cleave the DNA molecule at different sites, yielding smaller and different fragments, which then are submitted to mass spectral analysis. Ultimately, this procedure is expected to permit the localization of the unusual residues in specific regions of the molecule. At the present time there is no other method for locating very small amounts of minor nucleotides in DNA. Thus, the sensitivity and the definitiveness of the mass spectrometer is being exploited in the solution of a currently significant problem in molecular biochemistry.

Literature Cited.

1. Charnock, G.A. and Loo, J.L. Anal. Biochem., (1970), 37, 81.
2. Wiebers, J.L., Anal. Biochem., (1973), 51, 542.
3. Wiebers, J.L. and Shapiro, J.A., Biochemistry, (1977), 16, 1044.
4. Schulten, H.R., Beckey, H.D., Boerboom, A.J.H., and Meuzelaar, H.L.C., Anal. Chem., (1973), 45, 2358.
5. Meuzelaar, H.L.C., Kistemaker, P.G., and Posthumus, M.A., Biomed. Mass Spectrom., (1974), 1, 312.
6. Posthumus, M.A., Nibbering, N.M.M., Boerboom, A.J.H., and Schulten, H.-R. Biomed. Mass Spectrom. (1974), 1, 352.
7. Schulten, H.-R. and Schiebel, H.M., Z. Anal. Chem. (1976), 280, 139.
8. Schulten, H.-R. and Schiebel, H.M., Nucleic Acids Res., (1976), 3, 2027.
9. Burgard, D.R., Perone, S.P., and Wiebers, J.L., Biochemistry, (1977), 16, 1051.
10. Levsen, K. and Schulten, H.R. Biomed. Mass Spectrom., (1976), 3, 137.
11. Spencer, J.H. and Chargaff, E. Biochim. Biophys. Acta, (1963), 68, 18.
12. Wyatt, G.R. and Cohen, S.S. Biochem. J., (1953), 55, 774.
13. Wiebers, J.L. Nucleic Acids Res., (1976), 3, 2959.

RECEIVED December 30, 1977

14

Ultra-High Resolution Mass Spectrometry Analysis of Petroleum and Coal Products

H. E. LUMPKIN and THOMAS ACZEL

Exxon Research and Engineering Co., Analytical Research Laboratory, Baytown, TX

The detailed characterization of petroleum in terms of molecular composition and advances in mass spectrometry have proceeded abreast for about 30 years. From the analysis of gases and low boiling liquids in the 1940's and gasoline type analysis in the early 1950's (6), the frontiers moved toward higher boiling fractions as heated inlet systems evolved (9,11,15). But inlet systems mutated and evolved so rapidly that the mass separation or resolving power of the basic instruments could not keep apace. Components of the very complex material we call "rock oil" were vaporized into MS ion sources and the resulting spectra were partially or wrongly interpreted. Distillation, thermal diffusion, and solid-liquid elution chromatography were used to separate petroleum into fractions, but the fractions were still very complex. Meanwhile, work in gas chromatography was beginning to show excellent separations in the lower boiling ranges and for specific mixtures of components. However, GC has not been of major value in separations of the higher boiling ranges because of its lack of resolution.

In the 1960's high resolution (5,13) and in the mid-1970's ultra high resolution (14) mass spectrometers became available and were used in the characterization of petroleum and, more recently, coal liquids (1). When ion production methods which simplify the spectra such as low ionizing voltage (7), field ionization, and chemical ionization, are used along with the inherent power of these instruments, even such intransigent mixtures as petroleum and coal products begin to yield. This paper reviews the work and summarizes the results of using the low ionizing voltage electron impact technique with high and ultra high resolution mass spectrometers for the quantitative analysis of petroleum fractions and coal products. The instruments used were the AEI MS9 and MS50 mass spectrometers.

For a method to be quantitative it must be based on spectra which are reproducible. In addition, there should be a linear response to concentration, and the relative sensitivities among all the components to be determined should be known. We have used calibration data (relative or absolute instrument response to a given amount of a component) derived from pure compounds and separated concentrates. Extrapolation of calibration data into regions where direct instrumental measurements on calibrants are not available have been made by considering aromaticity, structure, and molecular weight (4,10,12).

The low ionizing voltage method has both advantages and disadvantages with respect to methods using higher ionizing voltages (Table I). Obviously we believe that for aromatic and polar compound mixtures, the low voltage technique is superior. The major advantage is the simplification of the spectra to the point that only molecular ions are produced. Moreover, there is no interference among components, and calibration data are more simply derived. The major disadvantage is the loss in sensitivity - a factor to 5 to 20.

TABLE I. Advantages and Disadvantages of High Resolution - Low Voltage Technique.

Advantages	Disadvantages
Molecular Ion Spectra. General Applicability - regardless of sample origin, processing, boiling range. Data on individual homologs. Extrapolated calibration data.	Limitation to Double Bond Compounds - Olefins, Aromatics, Polars. Relatively low sensitivity and reproducibility.

Nomenclature and High Resolution.

In the following discussion, the chemical names of the compounds and compound types will be used interchangeably with the molecular ion series, where the series is derived from the empirical formula expression C_nH_{2n+z}. The z-number is a measure of hydrogen deficiency (Table II). Thus butylbenzene, $C_{10}H_{14}$, is the C_{10} member of the C_nH_{2n-6} series of alkylbenzenes. Dibenzofuran is the first member and nucleus of the $C_nH_{2n-16}O$ or -16/O series. The nucleus determines the series, and alkyl substitution extends the carbon number but does not affect the empirical formula. Unfortunately for the analytical chemist, the carbon number of almost all series continue on and on in both petroleum and coal extracts to as high molecular weight compounds as can be vaporized. Of course, as the boiling point of the fraction is increased, more and more compound types and series types are found.

TABLE II.
Hydrogen Deficiency Nomenclature

Compound Type	Example	Formula	Z in C_nH_{2n+z}
Paraffins	n-Hexadecane	$C_{16}H_{34}$	+2
Naphthenes	MeCyclohexane	C_7H_{14}	0
Alkylbenzenes	Butylbenzene	$C_{10}H_{14}$	-6
Alkylnaphthalenes	Naphthalene	$C_{10}H_8$	-12
Phenanthrenes/ Anthracenes	DiMePhenanthrene	$C_{16}H_{14}$	-18
Dibenzofurans	Dibenzofuran	$C_{12}H_8O$	-16 O
Benzodithiophenes	MeBenzodithiophene	$C_{11}H_8S_2$	-14 S_2

If the materials to be examined were only hydrocarbons, a resolving power of about 7000 would be adequate because this is sufficient to separate the hydrocarbon series up to a mass of about 650. But both petroleum and coal-derived materials are made up of compounds containing the elements O, N, and S as well as hydrocarbons. These heteroatom components lead to increased complexity because now many compound types have molecular ions differing slightly in mass at the same nominal molecular weight. Thus multiplets are found throughout the spectra, and resolving power requirements are dictated by the heteroatoms encountered and the molecular weight. Of course, the more resolution one has available, the higher the molecular weight at which

TABLE III.
Mass Doublets and Resolving Power

Doublet	ΔMass	Example		Max. Mass at RP[a] 10K	80K	150K
H_{12}-C	.094	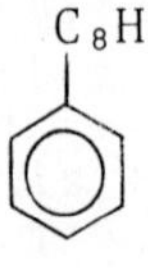 $C_{14}H_{22}$ 190.1721	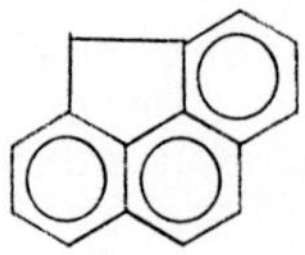 $C_{15}H_{10}$ 190.0782	940	7520	14000
C_2H_8-S	.091	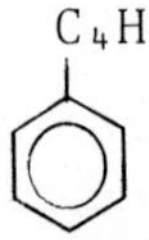 $C_{10}H_{14}$ 134.1095	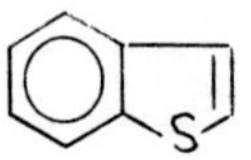 C_8H_6S 134.0190	910	7280	13500
CH_4-O	.0364	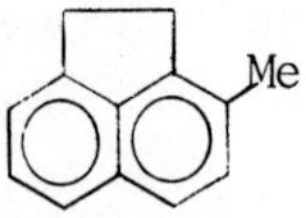 $C_{13}H_{12}$ 168.0939	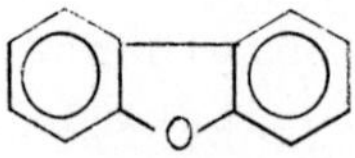 $C_{12}H_8O$ 168.0575	364	2910	5460
CH_2-N	.0126	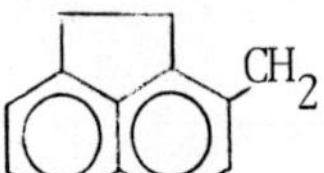 $C_{13}H_{11}$ 167.0861	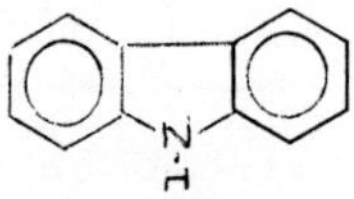 $C_{12}H_9N$ 167.0735	126	1010	1890
SH_4-C_3	.0034	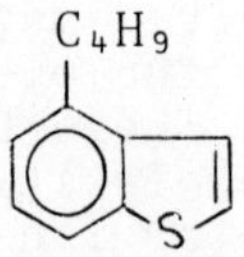 $C_{12}H_{14}S$ 190.0816	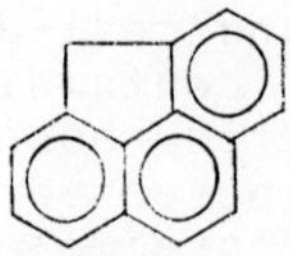 $C_{15}H_{10}$ 190.0782	34	270	500

TABLE IV.
Resolving Power Required to Separate Members of Possible Multiplets at m/e 300 and 301

Formula	Mass	ΔMass	RP Required
$C_{22}H_{36}$	300.2817	-	-
$C_{21}H_{32}O$	300.2453	0.0364	8,000
$C_{20}H_{28}O_2$	300.2089	0.0364	8,000
$C_{20}H_{28}S$	300.1912	0.0177	17,000
$C_{23}H_{24}$	300.1878	0.0034	88,000
$C_{19}H_{24}SO$	300.1548	0.0330	9,000
$C_{22}H_{20}O$	300.1514	0.0034	88,000
$C_{24}H_{12}$	300.0939	0.0575	5,000
$^{13}CC_{22}H_{24}$	301.1912	-	-
$C_{22}H_{23}N$	301.1830	0.0082	37,000
$^{13}CC_{21}H_{20}O$	301.1548	0.0282	11,000
$C_{21}H_{19}NO$	301.1466	0.0082	37,000

one can separate a doublet (Table III). The closest lying doublet requiring the greatest resolution in the samples we analyze is the SH_4-C_3 doublet (Table IV).

Data Acquisition and Computer Programs.

High resolution analyses of petroleum and coal products produce such a volume of data that an on-line data acquisition system is required. We currently use a Columbia Scientific Industries MS readout system to sense peak tops and slope changes and to furnish digital data on peak height and occurence time to a magnetic tape recorder. Data from the magnetic tape are read and handled by computer programs which perform a number of sophisticated functions (3,8) as outlined in Table V. A major function is mass measurement. From knowledge of the occurrence times and masses of peaks from reference compounds in a time-intensity spectrum and the times of the sample peaks, the masses of the sample peaks

TABLE V.

Data Acquisition and Reduction

Purpose:	Determine formulas and intensities of 100-5000 peaks and calculate percent abundance.
Functions:	Convert analog to digital signals. Measure peak heights or areas. Measure peak positions in time units. Separate partially resolved multiplets. Reject noise spikes. Recognize standard reference peaks. Calculate masses and assign formulae. Calculate quantitative analysis and print out input, detailed data, and summaries.

can be calculated. The accuracy of the mass measurements is a function of the electronic and mechanical stability of the mass spectrometer. The MS50 is a very stable instrument leading to accurate mass measurements. When masses are known to sufficient accuracy, the computer assigns unique empirical formulae to peaks and ultimately the appropriate series in the general C_nH_{2n+z} nomenclature. The quantitative analysis is calculated in a separate program using the measured peak heights, the series assignments, and an extensive set of calibration data. Examples of the form of data given in the quantitative program will be discussed later.

Resolution and Mass Measurements.

Good mass measurement even at a resolving power of 20,000 are insufficient to separate possible multiplets in complex petroleum and coal mixtures. For the data in Table VI, the computer has offered two or three possible formulae for several of the peaks, together with the respective error in mass measurement. The low lying peak at mass 296 is $C_{18}H_{16}S_2$, based on mass measurement, while the mass 296.157 peak is probably composed of unresolved ions from the indicated hydrocarbon and sulfur compounds. At mass 298 the mass measurements are poorer, but one would select the -28/S and -16/SO rather than the -38 and -26/O types when their mass measurements are compared to that for the -24 series.

TABLE VI.
Coal Extract (~20,000 Resolution)

Intensity	Meas m/e	Error MMU	Formula	(Series)
2,138	296.070	7.7	$C_{24}H_8$	($_{-40}$)
2,138	296.070	4.3	$C_{21}H_{12}S$	($_{-30}$S)
2,138	296.070	0.9	$C_{18}H_{16}S_2$	($_{-20}S_2$)
25,136	295.121	0.5	$C_{22}H_{16}O$	($_{-28}$O)
17,744	296.157	1.0	$C_{23}H_{20}$	($_{-26}$)
17,744	296.157	-2.4	$C_{20}H_{24}S$	($_{-16}$S)
6,868	298.084	6.0	$C_{24}H_{10}$	($_{-38}$)
6,868	298.084	2.6	$C_{21}H_{14}S$	($_{-28}$S)
5,708	298.142	6.1	$C_{22}H_{18}O$	($_{-26}$O)
5,708	298.142	2.7	$C_{19}H_{22}SO$	($_{-16}$SO)
15,832	298.175	2.9	$C_{23}H_{22}$	($_{-24}$)

Coal liquefaction products contain relatively less sulfur types and more nitrogen and oxygen types than some petroleum fractions. Less resolution is required for these latter heteroatoms than for sulfur, and R=40,000 appears adequate to resolve many constituents without resorting to any prior separative steps. For example, 464 components in

106 different compound types were found for the coal product reported in Table VII. Ultra high resolution (80,000) dynamic scanning of a high sulfur petroleum fraction taxes the capabilities of the mass spectrometer, the data acquisition system, and the computer programs because of the large number of peaks as shown in Table VIII. We had to increase the dimensions of our arrays from 3500 to 5000 for scans of this nature. In Table IX is shown the data at selected masses for only the hydrocarbon and sulfur compound

TABLE VII.
Unseparated Coal Liquefaction Product
~40,000 Resolution

Intensity	Meas. m/e	Error, MMU	Formula	(Series)
293	292.0924	0.2	$C_{19}H_{16}SO$	$(_{22}SO)$
758	292.1237	-1.5	$C_{23}H_{16}$	$(_{30})$
589	292.1470	-0.7	$C_{20}H_{20}O_2$	$(_{20}/O_2)$
349	292.1812	-1.5	$C_{21}H_{24}O$	$(_{18}/O)$
245	292.2206	1.5	$C_{22}H_{28}$	$(_{16})$
121	293.0962	0.5	Isotope from	$(_{22}/SO)$
392	293.1223	1.9	$C_{22}H_{15}N$	$(_{29}/N)$
237	293.1490	1.3	Isotope from	$(_{20}/O_2)$
173	293.1764	-1.5	$C_{20}H_{23}NO$	$(_{17}/NO)$
246	293.2151	0.8	$C_{21}H_{27}N$	$(_{15}/N)$

TABLE VIII.
High Sulfur Petroleum Aromatics
~80,000 Resolution

Intensity	Meas. m/e	Error, MMU	Formula	(Series)
423	274.0776	-0.5	$C_{22}H_{10}$	$(_{34})$
1,904	274.0820	0.5	$C_{19}H_{14}S$	$(_{24}/S)$
601	274.0842	-0.5	$C_{16}H_{18}S_2$	$(_{14}/S_2)$
1,608	274.1348	-0.7	$C_{20}H_{18}O$	$(_{22}/O)$
438	274.1362	-2.7	$C_{17}H_{22}SO$	$(_{12}/SO)$
4,206	274.1719	-0.2	$C_{21}H_{22}$	$(_{20})$
5,104	274.1746	-0.7	$C_{18}H_{26}S$	$(_{10}/S)$
3,148	274.2656	-0.2	$C_{20}H_{34}$	$(_{6})$

types from the same series shown in Table VIII. These components are real; they form a regular series appearing at every 14 mass units beginning at the right nuclear molecular weight and additional compound types appear with increasing mass. But even the ultra high resolution MS50 mass spectrometer runs out of resolving power for these close-lying multiplets.

TABLE IX.

High Sulfur Petroleum Aromatics
~80,000 Resolution

		Intensities At Series Shown							
Mass	CN	-6	-10/S	-20	$-14/S_2$	-24/S	-34	$-28/S_2$	-38/S
162	12	10256	8332						
190	14	6880	8700	1851					
246	18	3622	6124	5263	695	410			
288	21	3230	5120	4202	-2081	UR-	903		
372	27	772	-2668	UR-	---1562	UR---		1108	540

CN - Carbon Number; UR - Unresolved; Avg. Error 0.8 MMU

A list of the compound types identified in various petroleum and coal liquid samples is given in Table X. A total of 173 C, H, N, O and S containing types are included, not counting the miscellaneous list of types with the same empirical formulae but different nuclear structures. Fortunately, all of these types are not observed in a single sample.

Quantitative Analysis and Data Presentation.

Our quantitative analysis programs yield arrays of weight percent of compound type by carbon number as the prime data and covers many computer output pages. From such a wealth of detailed data one can calculate many additional paramenters. Some of these are given in Table XI and include quite straightforward items such as average molecular weight, elemental analysis, and carbon number distribution. With some data on boiling points, less obvious information such as distillation curves and compositions of any selected distillate fractions are calculable (2). One seldom needs the detailed compound type by carbon number data. Therefore, we prepare summary tables presenting different aspects of the analysis. Example data from such summary tables are

TABLE X.
Compound Types in Various Samples

Class	Range in General Formulas
Hydrocarbons	C_nH_{2n} to C_nH_{2n-50}
Sulfur Compounds	$C_nH_{2n-2}S$ to $C_nH_{2n-44}S$
Sulfur Compounds	$C_nH_{2n-10}S_2$ to $C_nH_{2n-30}S_2$
Oxygen Compounds	$C_nH_{2n-2}O$ to $C_nH_{2n-48}O$
Oxygen Compounds	$C_nH_{2n-6}O_2$ to $C_nH_{2n-40}O_2$
Oxygen Compounds	$C_nH_{2n-6}O_3$ to $C_nH_{2n-32}O_3$
Oxygen Compounds	$C_nH_{2n-10}O_4$ to $C_nH_{2n-18}O_4$
Nitrogen Compounds	$C_nH_{2n-3}N$ to $C_nH_{2n-49}N$
Sulfur-Oxygen Compounds	$C_nH_{2n-10}SO$ to $C_nH_{2n-30}SO$
Nitrogen-Oxygen Compounds	$C_nH_{2n-7}NO$ to $C_nH_{2n-41}NO$
Miscellaneous N_2, SO_2, SO_3, and N_2O Compounds	---

TABLE XI.
Factors Calculated in Quantitative Analysis.

Data calculated for up to 3000 components, 100 homologous series.[a] Output includes parameters listed below:

1. Weight percent for each component and series.
2. Average MW, carbon atoms, carbon atoms in sidechains for each series and for total aromatics.
3. Elemental analysis, molecular weight distribution, carbon number distribution, distillation, prediction of composition of narrow fractions.

[a]Calculation on IBM 370 computer yields 18 tables, takes 20 seconds.

shown in Tables XII, XIII, and XIV. The compound type summary table, excerpted in Table XII, also gives the average molecular weight and average carbon number for each type. The aromatic ring distribution in a heavy coal liquid (Table XIII) shows that both hydrocarbons and oxygenated types attain maxima at 4 rings/molecule. This sample contained an unusually high amount of oxygenated types, which are primarily hydroxy aromatics. Those from petroleum are mostly aromatic furans. A partial carbon number distribution in a crude oil and in a coal liquid (Table XIV) shows that while the petroleum has a smooth distribution, the coal liquid displays maxima at C_{10}, C_{12}, C_{14}, and C_{16}. These carbon number maxima are due to the naphthalene, acenaphthene, phenanthrene, and pyrene aromatic and hydroaromatic nuclei. The petroleum tends to have long alkyl substituents on the aromatic nuclei while much of the coal liquid is concentrated at the nuclear molecular weight.

TABLE XII.
Partial Compound Type Summary (Crude Oil)

Type	Wt. Pct.	Avg. MW	Avg. C. No.
C_nH_{2n-6}	11.6	158.2	11.7
C_nH_{2n-12}	14.9	192.3	14.6
C_nH_{2n-18}	5.5	256.4	19.6
C_nH_{2n-22}	2.7	307.0	23.5
C_nH_{2n-28}	0.5	327.7	25.4
$C_nH_{2n-10}S$	4.1	215.4	13.8
$C_nH_{2n-16}S$	7.3	250.0	16.7
$C_nH_{2n-16}O$	0.2	179.6	12.8

TABLE XIII.
Aromatic Distribution in a Heavy Coal Liquid

Aromatic Rings	Hydrocarbons	S Compounds	O Compounds
1	2.5	0.4	0.5
2	15.7	1.4	4.3
3	19.5	0.8	6.7
4	26.8	0.4	10.2
5	5.0	<0.1	2.4
6	2.6		0.2
7+	0.5		
	72.6	3.1	24.3

TABLE XIV. (16)
Partial Carbon Number Distribution as Mole %
in a Crude and in a Coal Liquid

C no.	Crude	Coal Liquid
9	0.74	2.18
10	1.00	20.76
11	2.51	9.55
12	6.28	12.15
13	9.18	8.52
14	9.65	15.41
15	10.64	5.36
16	8.06	8.61
17	7.45	3.78

Conclusions.

The methods we have developed are applicable to samples from a variety of sources and which constitute many types. Examples include coal liquefaction products, syncrudes, shale oils, and conventional petroleum streams. The sample characterization is fast and detailed. Minor changes in calibration data and in the presentation of the summary tables as required by different sample sources is accomplished by coded options built into the computer programs. Although prior separation into saturate, aromatic, and polar fractions is desirable for petroleum fractions, it is not required for the mostly aromatic coal liquids. Sample quantity is often limited in some bench-scale exploratory research. With this technique many important characteristics such as elemental analysis, density, and distillation curves can be calculated which would not otherwise be obtainable because of the small sample size.

Literature Cited

1. Aczel, T., Foster, J.Q. and Karchmer, J.H., presented at Division of Fuel Chemistry, 157th National Meeting of ACS, Minneapolis, Minnesota, April (1969).
2. Aczel, T. and Lumpkin, H.E., presented at 18th Annual Conference on Mass Spectrometry and Allied Topics, San Francisco, California, June 14-19, 1970.
3. Aczel, T., Allan, D.E., Harding, J.H. and Knipp, E.A., Anal. Chem., (1970), 42, 341.
4. Aczel, T. and Lumpkin, H.E., presented at 19th Annual Conference on Mass Spectrometry and Allied Topics, Atlanta, Georgia, May 2-7, 1971.
5. Beynon, J.H., Mass Spectrometry and Its Application to Organic Chemistry, Elsevier, Amsterdam, (1960).
6. Brown, R.A., Anal. Chem., (1951) 23, 430.
7. Field, F.H. and Hastings, S.H., Anal. Chem., (1956), 28, 1248.
8. Johnson, B.H. and Aczel, T., Anal. Chem., (1967), 39, 682.
9. Lumpkin, H.E. and Johnson, B.H., Anal. Chem., (1954), 26, 1719.
10. Lumpkin, H.E., Anal. Chem., (1958), 30, 321.
11. Lumpkin, H.E. and Taylor, G.R., Anal. Chem., (1961), 33, 476.
12. Lumpkin, H.E. and Aczel, T., Anal. Chem., (1964), 36, 181.
13. Lumpkin, H.E., Anal. Chem., (1964), 36, 2399.
14. Lumpkin, H.E., Wolstenholme, W.A., Elliott, R.M., Evans, S., and Hazelby, D., presented at 23rd Annual Conference on Mass Spectrometry and Allied Topics, Houston, Texas, May 25-30, (1975).
15. O'Neal, M.J., Jr. and Weir, T.P., Jr., Anal. Chem. (1951), 23, 830.
16. Reprinted with permission from Reviews of Anal. Chem., (1971), 1, 226.

RECEIVED December 30, 1977

15

Organic Mixture Analysis by Metastable Ion Methods Using a Double-Focusing Mass Spectrometer

E. J. GALLEGOS

Chevron Research Co., Richmond, CA 94802

This paper describes two ways in which metastable transition spectra may be used in the analysis of organic compounds or mixtures. The first section describes a technique from which single ion detection GC-MS type information is obtained from the metastable transition spectra without the benefit of a gas chromatograph. The second section describes a technique for obtaining mass ratio and intensity information without the use of magnet, quadrupole, or time of flight techniques.

Theory

Metastable transitions in mass spectrometry were first reported by Hipple and Condon (1) in 1945. The theories of metastable transitions and the techniques of obtaining these kinds of data have been summarized recently. (2,3)

A brief review of how and where metastable ions occur in a double focusing mass spectrometer pertinent to this work will be given here.

The composition of ions reaching the collector of a mass spectrometer is determined by the rate constants, k_n, of many competing consecutive, unimolecular decompositions of excited molecular and fragment ions. The recorded ion current intensity versus the mass-to-charge ratio gives the mass spectrum. The residence time of an ion in source is about 10^{-6} seconds. There is an additional $\sim 10^{-5}$ seconds elapsed time after acceleration to collection. Decompositions requiring less than 10^{-6} seconds occur in the source and will be analyzed and collected to give a normal mass spectrum.

© 0-8412-0422-5/78/47-070-274$10.00/0

Metastable transitions are due to those decompositions which take place after the ion leaves the source but before it reaches the detector. These decompositions can take place in Regions 1-6 of Figure 1. These regions are defined by their different ion residence time and field or field-free configuration. The only metastable transitions which are detected and used in this work are those which occur in Region 2.

Ions which decompose in Region 2 have received full acceleration but have not been energy analyzed in the electrostatic sector. They, having lost mass, will not have the required translational energy to pass through the monitor slit and will impinge on the bottom electrostatic plate at some point $(M_1)^+{}_a$ or $(M_1)^+{}_b$, etc.; see Figure 1.

They can be made to pass through the monitor slit either by increasing the accelerating voltage or by lowering the electrostatic plate voltage.

There are two ways to use this fact. If the magnet is set to focus on some mass, say fragment M_2^+, and the electrostatic voltage is held constant while the accelerating voltage is increased, all metastable peaks observed at the multiplier detector will be due to precursor ions $M_1\ (a,b,\ldots,n)^+$, which decompose in Region 2 to give M_2. The mass of the precursors $M_1^+\ (a,b,\ldots,n)$ can be calculated from the following equation

$$M_1 = M_2 \frac{V_m}{V_o} \qquad (1)$$

where M_2 = mass of daughter or fragment ion
M_1 = mass of the precursor ion
V_m = accelerating voltage at which a metastable is observed
V_o = accelerating voltage at which the main beam is in focus

This, in essence, provides the opportunity to observe and identify all participants in a specific reaction process

$$(M_1)^+a,b,c\ldots \quad M_2^+ - [(M_1)a,b,c\ldots - M_2] \qquad (2)$$

without interference from any other ions or processes which are occurring simultaneously. In effect, this provides an isolation technique without physical separation but with a resulting capability somewhat similar to GC-MS.

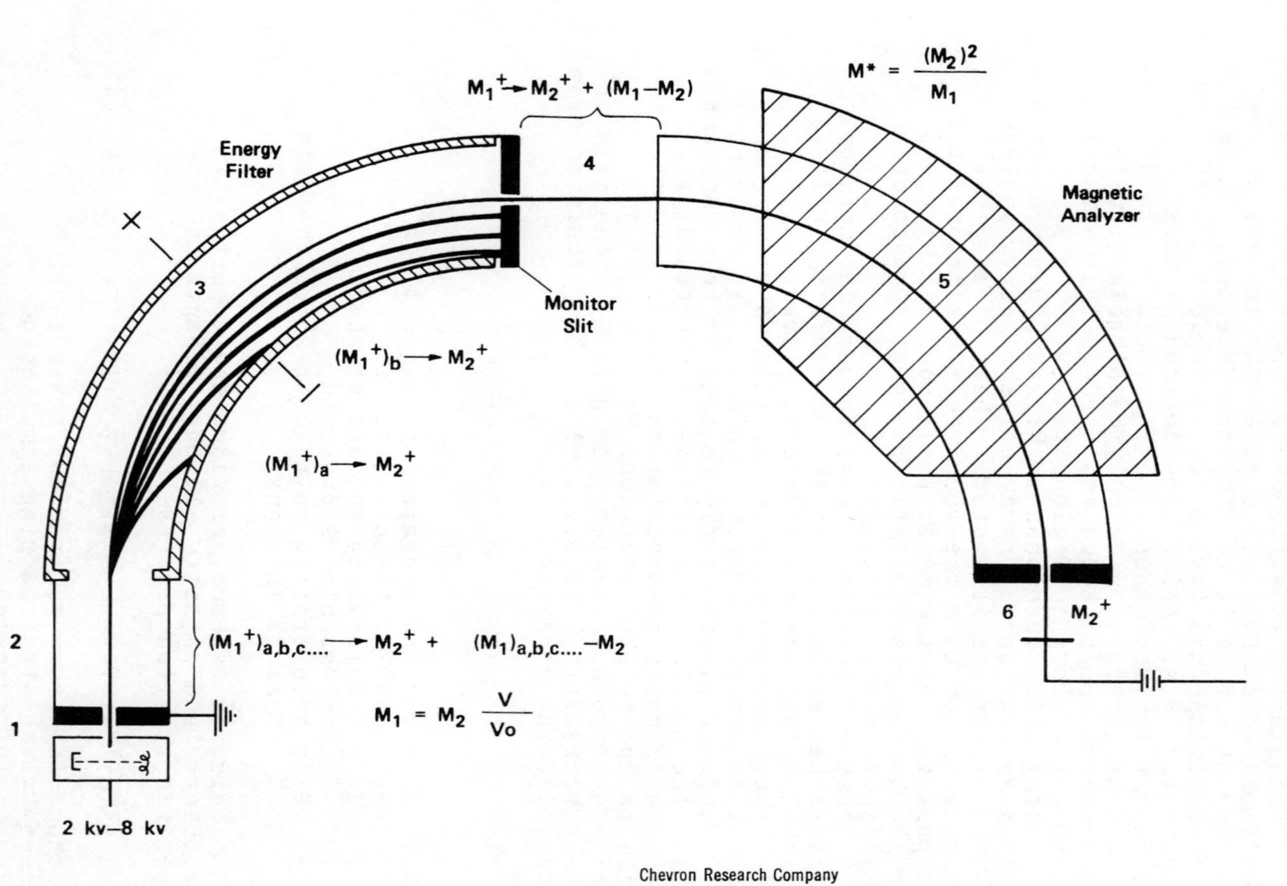

Chevron Research Company

Figure 1. Metastable transition schematic

This method of recording and interpreting metastable transition spectra is called the accelerating voltage scan method and was used to obtain the data described in Section I under results.

The alternate method is to use the total ion monitor of a double focusing instrument as a detector. If the accelerating voltage is scanned in a manner described above all of the metastable transitions which occur in Region 2 will be detected. The spectra obtained carry mass ratio, M_1/M_2, and intensity information. The accelerating voltages V_o and V_m and the mass ratio M_1/M_2 are related according to Equation 1. Only the electrostatic sector is used in this method.

This second method of recording and interpreting metastable transition spectra, called the IKES method, was used to obtain the data described in Section II in the Results Section.

Instrumentation

This work was done on an AEI MS-9 double focusing high resolution mass spectrometer. The instrument was modified so that the accelerating voltage can be scanned from 2-8 kV. The electrostatic sector voltage is held constant at 135 volts which, with an accelerating voltage of 2 kV, will produce normal mass spectra. The sample is introduced through an all-glass, hot inlet (4) or by using a direct insertion probe. In the first method of recording metastable transitions, accelerating voltage scans, the magnet is adjusted to a current which will bring into focus a daughter ion M_2^+ of interest. In the second method of recording metastable transition spectra, the total ion monitor is used as the detector which means the magnet is not used. For both the accelerating voltage is then scanned from 2-8 kV to produce the metastable transition spectra shown in this study.

This arrangement will allow metastable transitions to be observed due to precursor ions M_1^+ with a mass up to four times that of the daughter ion M_2^+.

Results

Section I. Accelerating voltage scan analysis of metastable ions is possible only if a decomposition process can be found that is uniquely characteristic of a compound or compound type. There are many systems which obey the requirements for this technique of analysis. Three examples of this are given.

Terpanes and Steranes

Terpanes and steranes are two types of compounds that fall well inside these requirements. Many terpanes, either pentacyclic- or tricyclic-saturated hydrocarbons found in nature, fragment upon electron impact to give a base peak at m/e 191 fragment ion. (5,6) Figure 2 shows the metastable spectrum of authentic gammacerane, a C_{30} terpane and a proposed scheme for fragmentation of the parent ion to the m/e 191 fragment ion. Similarly, steranes which are saturated tetracyclic hydrocarbons fragment to give a base peak at m/e 217 fragment ion. Figure 3 shows the metastable spectrum of authentic cholestane, a C_{27} sterane and a proposed scheme for fragmentation of the parent ion to the m/e 217 fragment ion.

Green River Shale

The well-analyzed, saturated fraction of the extractable portion of Green River Shale (7) is used to demonstrate the use of the method. Figure 4 shows the metastable scan of m/e 191 of this fraction. The most important peaks are those due to C_{31}, C_{30}, and C_{29}, followed by the C_{20} and C_{21} terpanes. Similarly, Figure 5 presents the metastable results for a scan of m/e 217. These results show that the largest metastable peak is due to a C_{29} sterane followed by the C_{28} and, finally, C_{27} steranes.

Figure 6 shows the mass chromatograms of m/e 191 and 217 for this fraction.

The normal mass spectrum was used to calculate the total concentration of terpanes and steranes in this sample. This is done in the following manner. The m/e 191 fragment ion measured 2.33% of the total ionization of the saturate fraction of Green River Shale. This fragment ion represents one-eighth of the total ionization in the mass spectrum of the average authentic terpane in this sample so that the total terpane contribution to the total ionization of the sample is 18.6%.

The m/e 217 fragment ion measured 1.25% of the total ionization of the saturate fraction of Green River Shale. This fragment ion represents about one-fifteenth of the total ionization in the mass spectrum of the average sterane in this sample so that the total sterane contribution to the total ionization of the sample is about 18.8%. The values of contribution to total ionization are equivalent to concentrations. (8)

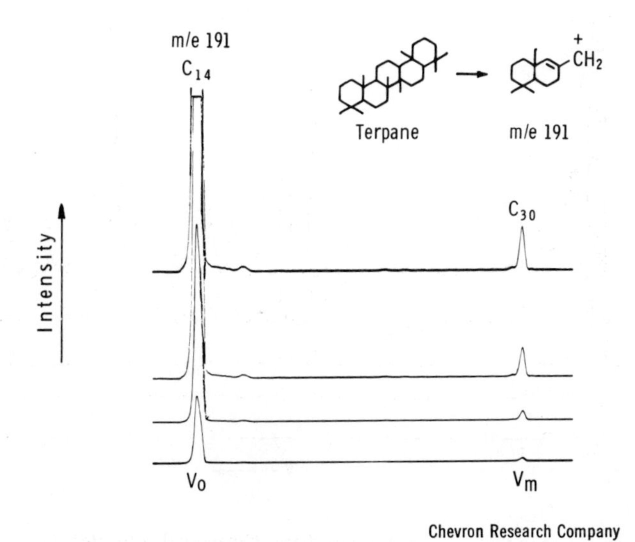

Figure 2. Metastable spectrum of m/e *191 of gammacerane*

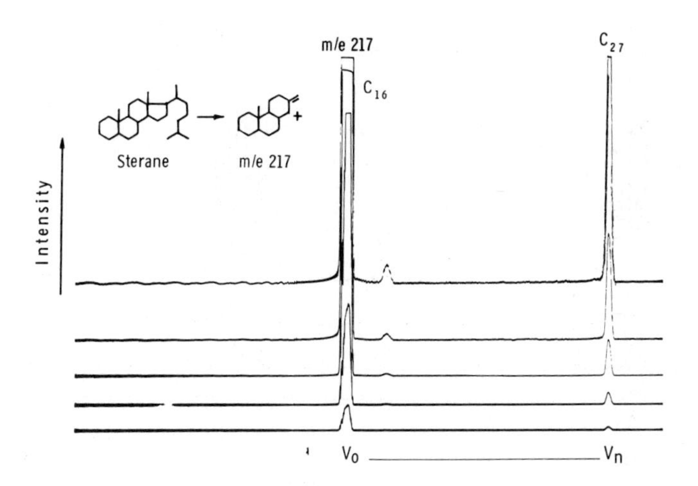

Figure 3. Metastable spectrum of m/e *217 of cholestane*

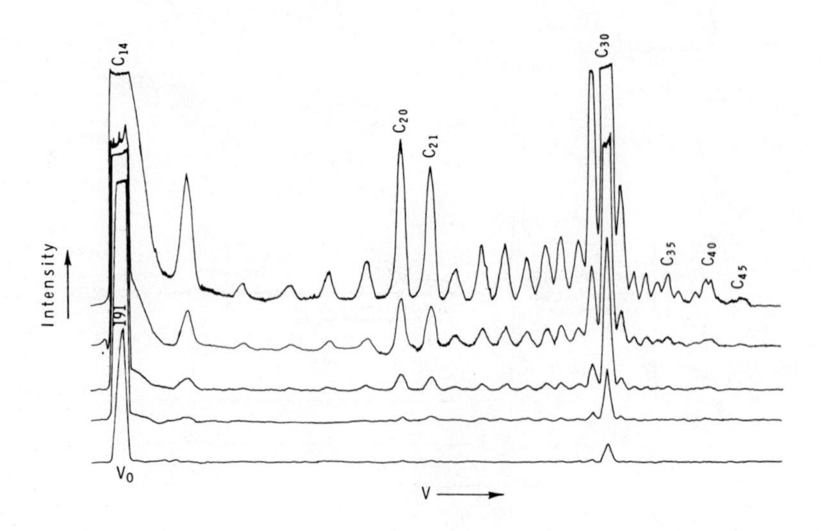

Figure 4. m/e *191 Metastable spectrum (terpanes) saturates, Green River, Colorado; eocene, outcrop*

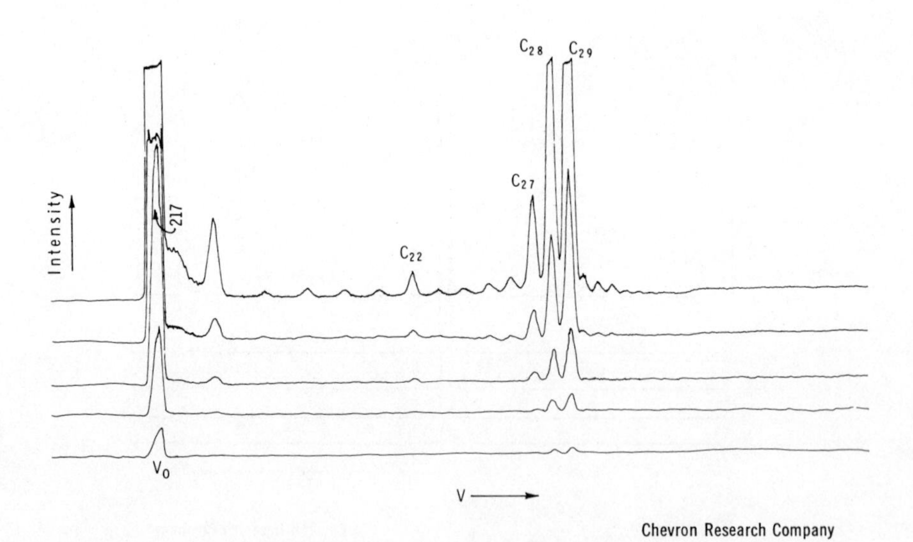

Chevron Research Company

Figure 5. m/e *217 Metastable spectrum saturate fraction of Green River shale*

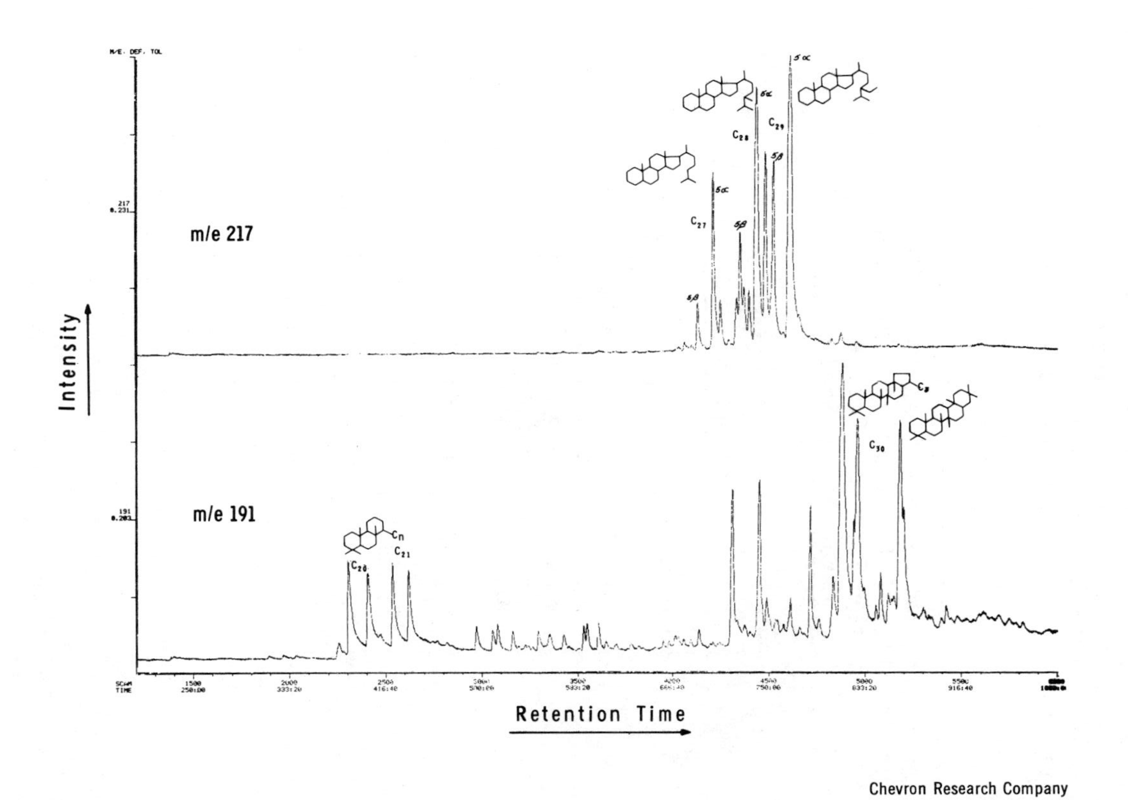

Figure 6. Green River shale saturate cut m/e 191 *terpane and* m/e 217 *sterane mass chromatogram*

The accelerating voltage scan results were obtained by measuring the metastable peak height and then normalizing all components to give 18.6% terpanes and 18.8% steranes.

A comparison of the relative concentrations of terpanes and steranes in the saturate fraction of Green River Shale as determined by GC-MS and the accelerating voltage scan method are given in Table I.

Table I

Comparison of GC-MS and Accelerating Voltage Scan (AVS) Method Results of Terpanes and Steranes in the Saturate Portion of Green River Shale

	GC-MS	AVS Method
Terpanes		
C_{31}	1.0	1.1
C_{30}	13.6	13.2
C_{29}	2.3	2.2
C_{21}	0.4	0.4
C_{20}	1.4	1.7
Total	18.7	18.6
Steranes		
C_{29}	10.3	10.1
C_{28}	7.6	6.3
C_{27}	1.9	1.8
C_{22}		0.6
Total	19.8	18.8

The accelerating voltage scan techniques gives the sum of all terpanes or steranes at a given molecular weight; whereas, isomers can be distinguished by GC-MS. The total terpane and sterane concentrations calculated are within 1% of those obtained by GC-MS.

Pyrolyzed Green River Shale

A similar analysis was made on kerogen pyrolyzate from Green River Shale. The comparison of the relative concentrations of the C_{27} to C_{31} terpanes determined by the accelerated voltage scan method and GC-MS are given in Table II.

Table II

Comparison of GC-MS and Accelerating Voltage Scan (AVS) Method Results of Terpanes and Steranes in the Saturate Cut from the Pyrolysis Oil from Green River Shale

	Percent	
	GC-MS	AVS Method
Terpanes		
C_{31}	0.5	1.3
C_{30}	2.0	2.0
C_{29}	2.0	2.0
C_{28}	0.2	0.9
C_{27}	2.4	1.5
Total	7.1	7.7
Sterane		
C_{29}	0.8	0.7
C_{28}	0.6	0.5
C_{27}	0.3	0.2
Total	1.7	1.4

Total terpanes for accelerated voltage scan results were calculated from the normal mass spectrum as before. Here the accelerating voltage scan method gives a low value for the C_{27} terpanes. This is because one of the C_{27} terpanes does not fragment exclusively to give m/e 191 as the base peak. (9)

Alkyl Phthalates

Alkyl phthalates, used generally as plasticizers in plastic tubing fabrication, have become ubiquitous contaminants in the environment. An organic concentrate from a refinery effluent water sample consisting mainly of naphthenic acids, was found to contain low concentrations of di-n-butyl phthalate. The mass spectra of n-dibutyl phthalate shows a characterizing base peak at m/e 149, which results from the decomposition routes given in Figure 7.

The metastable spectrum is shown in Figure 8, top, for pure di-n-butyl phthalate. Figure 8, bottom, shows the search for phthalates in the organic concentrate from the refinery water sample. The presence of phthalate is clearly shown by the black peaks.

Metastable spectral Peaks A and B are due to Transitions A and B shown in the scheme in Figure 7.

Even in the presence of other components which form an m/e 149 ion on electron impact, the progenitors are sufficiently different in mass so that the phthalate metastable peaks are clearly resolved.

Summary

The accelerating voltage scan technique of analysis may be applied only if the decomposition processes of the molecular types being searched are well understood. Once the magnet is set to accept only one preselected mass, say M_2^+, the only time the multiplier will produce a signal is when the correct parent ion, M_1^+, and accelerating voltage, V_m, are found. All this happens in the presence of many other components and transitions which are ignored. This technique has many of the benefits of GC-MS but doesn't require an actual separation of molecules.

Section II

Mass ratio of precursor mass to final mass, M_1/M_2, spectra as described earlier carry mass ratio and intensity information only. The work to be described here was done to assess the value of these kinds of data for analysis of volatile components. We used alkane, alkene, and cycloalkane isomers, and hydrocarbon mixtures as examples.

IKES spectra are compared to mass spectra in some instances. All of the IKES spectra were taken at low resolution, i.e., the monitor slit assembly of the MS-9 mass spectrometer was used as the detector. Since the monitor slit assembly has an $\sim$0.5 cm^2 hole

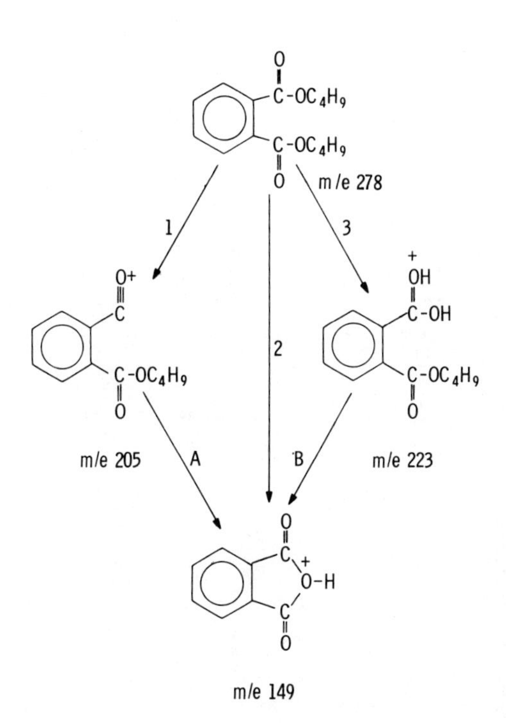

Chevron Research Company

Figure 7. Main fragmentation paths of phthalate

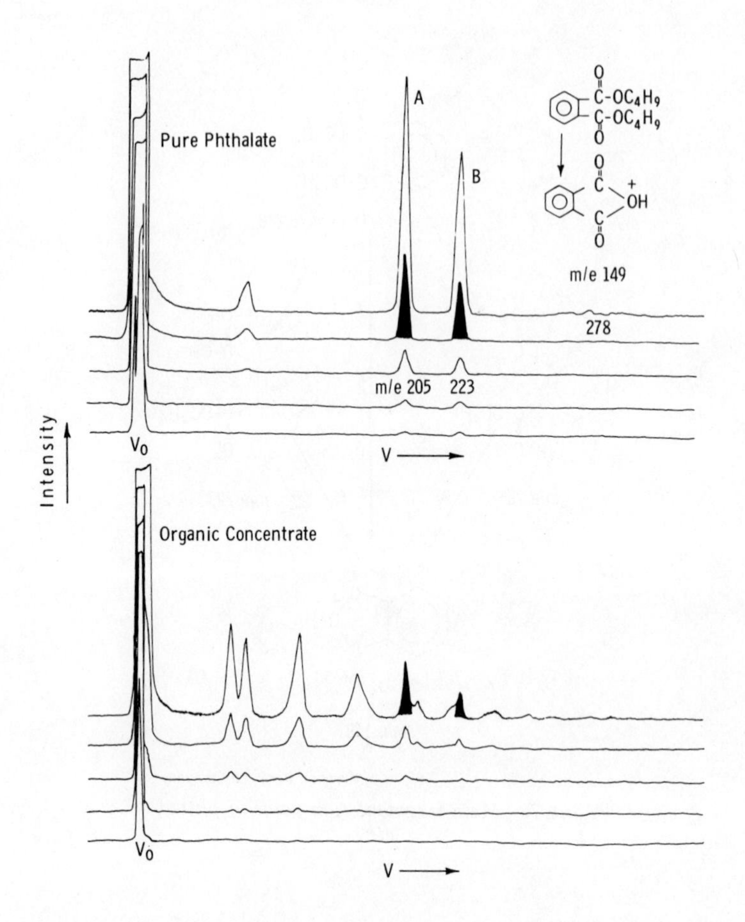

Figure 8. Metastable scan of m/e *149 of pure* n-*dibutyl phthalate and phthalates in organic from a water sample*

in it, all of the IKES spectra peaks will appear as partially resolved doublets or as flat-topped peaks, which reflect the hole in the detector. The masses or carbon number of the molecules involved in any particular transition can be determined from the calculable mass ratio.

Pure Hydrocarbon

Figure 9 shows the IKES and mass spectra of five C_6 alkene isomers. Note that all of the IKES spectra of the branched alkenes have a base peak corresponding to the $C_5^+ \rightarrow C_3^+$ transition. The mass spectra does not show this phenomenon. However, Figure 10 gives the IKES spectra of cis-3-hexene which shows a base peak for the $C_5^+ \rightarrow C_3^+$ transition. In fact, the IKES spectra match very closely that of 2-methyl-1-pentene. Their mass spectra are also practically identical.

Figure 10, in addition, contrasts the IKES spectra of cyclohexane and three hexene isomers. The mass spectra for all four are quite similar. The IKES spectra show much greater contrast.

Figure 11 shows the IKES and mass spectra of three C_5 alkenes. All show a IKES spectra base peak for the $C_5^+ \rightarrow C_2^+$ transition. The base peak in their mass spectra is the (M^+ - 15) fragment ion. This would be due to $C_5^+ \rightarrow C_4^+$ transition. This implies a basic difference in the decomposition kinetics at times of 10^{-5} to 10^{-6} seconds.

Figure 12 shows IKES and mass spectra of some C_8-C_6 alkanes. Note that all the nonsymmetrical branched alkanes have a IKES spectra base peak corresponding to the $C_5^+ \rightarrow C_3^+$ transition.

Figure 13 shows the IKES spectra of mixture of C_8 alkanes, C_8 cycloalkanes, and C_8 alkylbenzenes. There is a definite contrast between the alkanes and the aromatic alkylbenzenes IKES spectra.

The centroid of the IKES spectra is greatest for the alkanes followed by the cyclic alkanes and, finally, the alkylbenzenes.

This just means that the alkanes on the average just break about in half and, if not, the positive charge can stay with the smaller fragment. The cyclics cleave about the same with some preference to leave the charge on the ring system. The alkylbenzenes almost exclusively leave the positive charge on the π bonded system which, in this case, is always of greater mass.

The same may be said of the data shown in Figure 14. These samples are SiO_2-separated cuts from

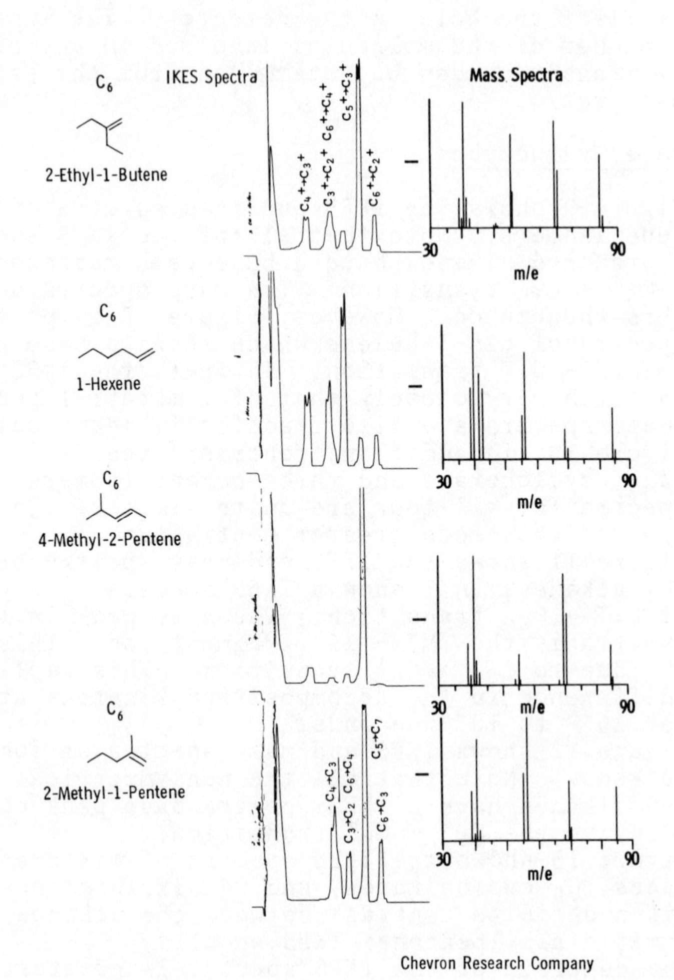

Figure 9. Mass and ikes spectra of some C_6 alkanes

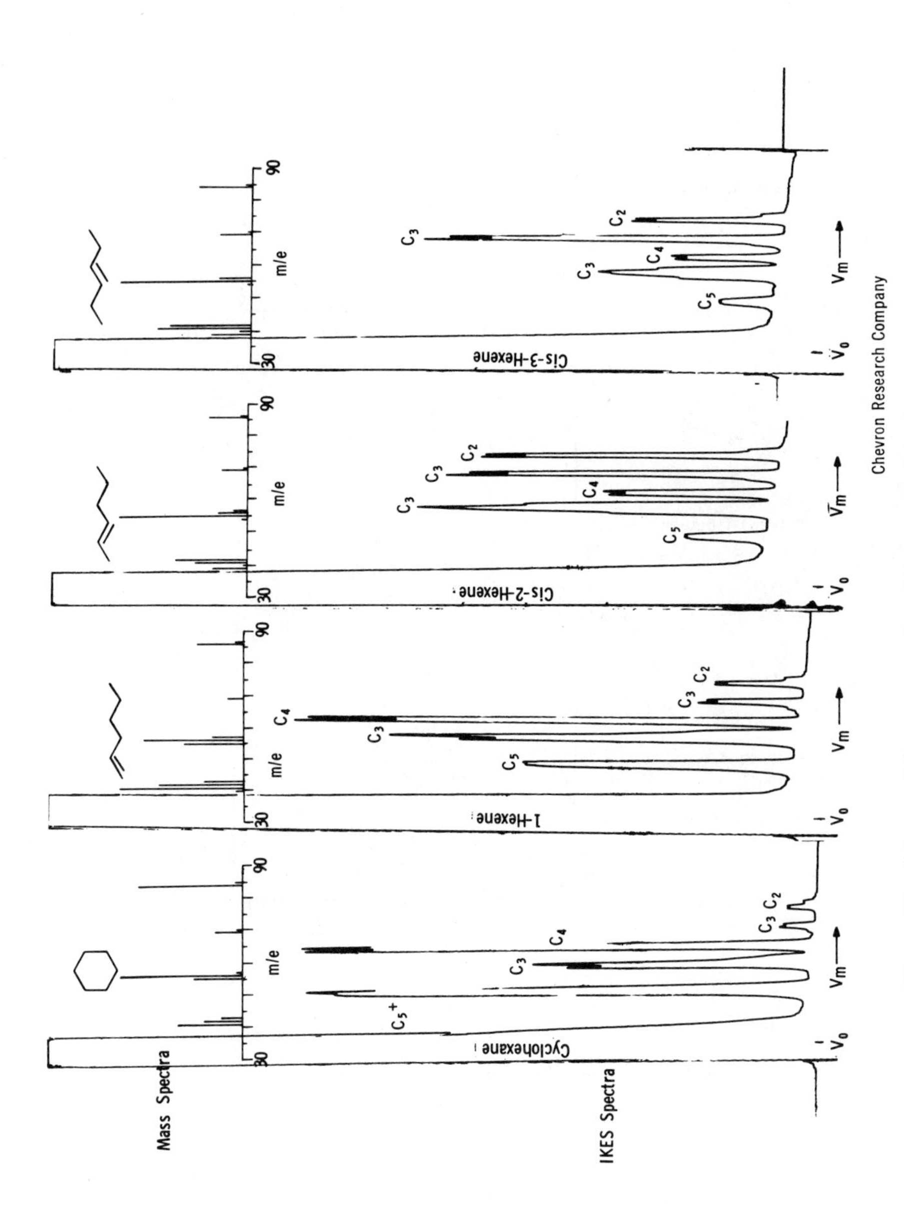

Figure 10. Mass and ikes spectra of cyclohexane and the hexane isomers

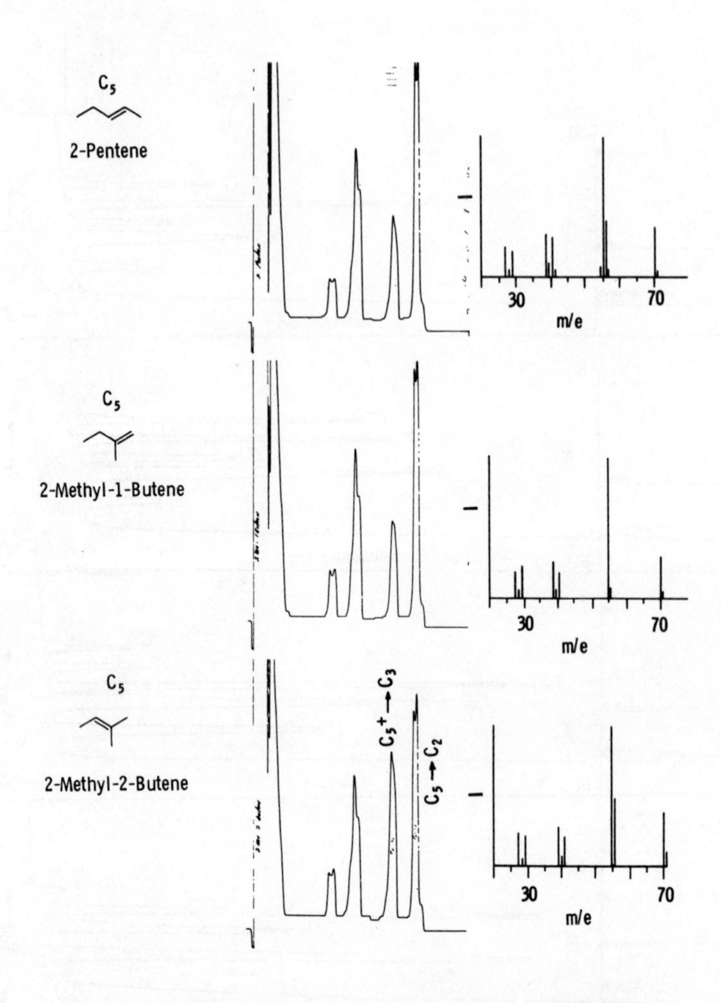

Figure 11. Mass and ikes spectra of some C_5 alkenes

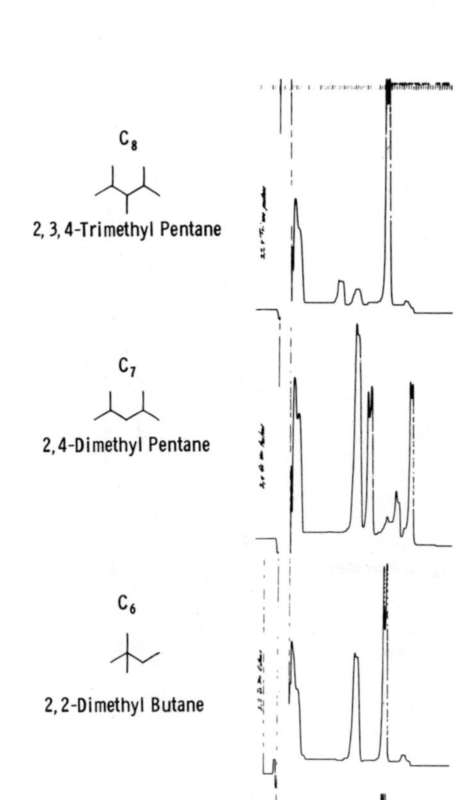

Chevron Research Company

Figure 12. Ikes spectra of some alkanes

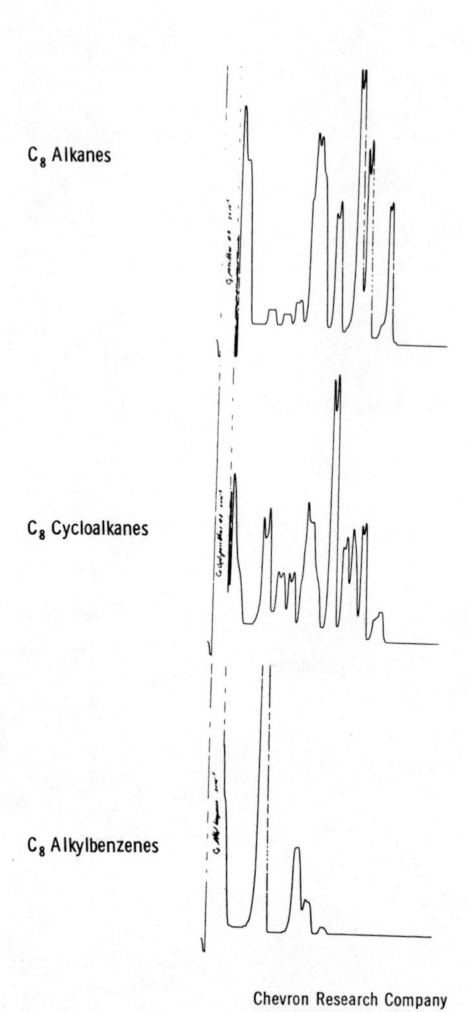

Figure 13. Ikes spectra of C_8 alkanes, cycloalkanes, and alkylbenzenes

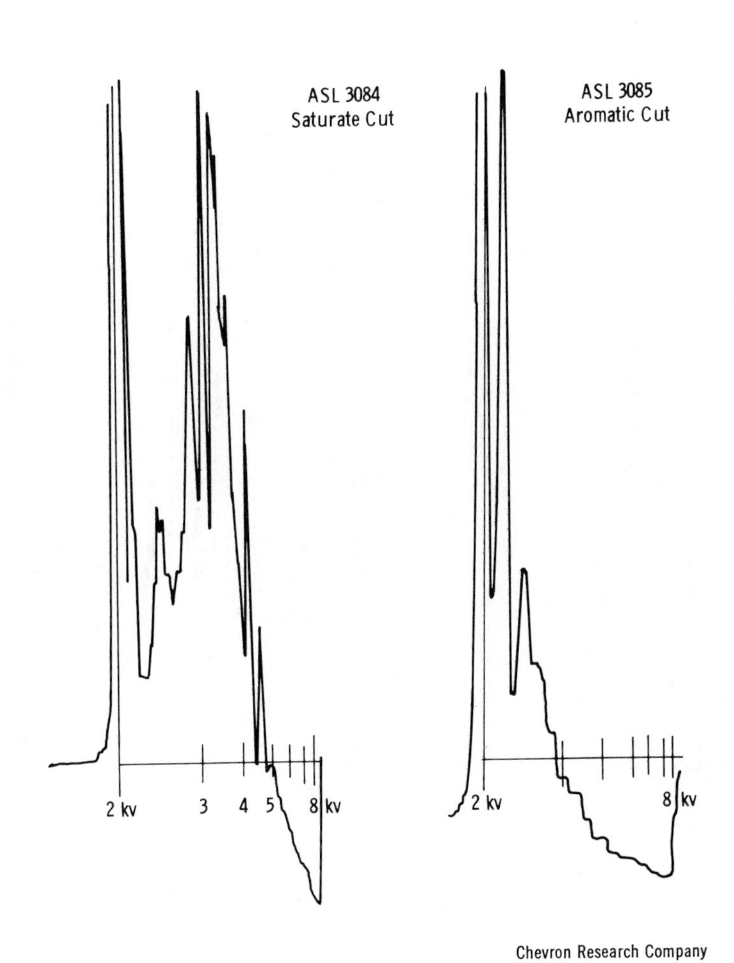

Figure 14. Ikes spectra of the saturate and aromatic fraction of a petroleum sample ASL 3084

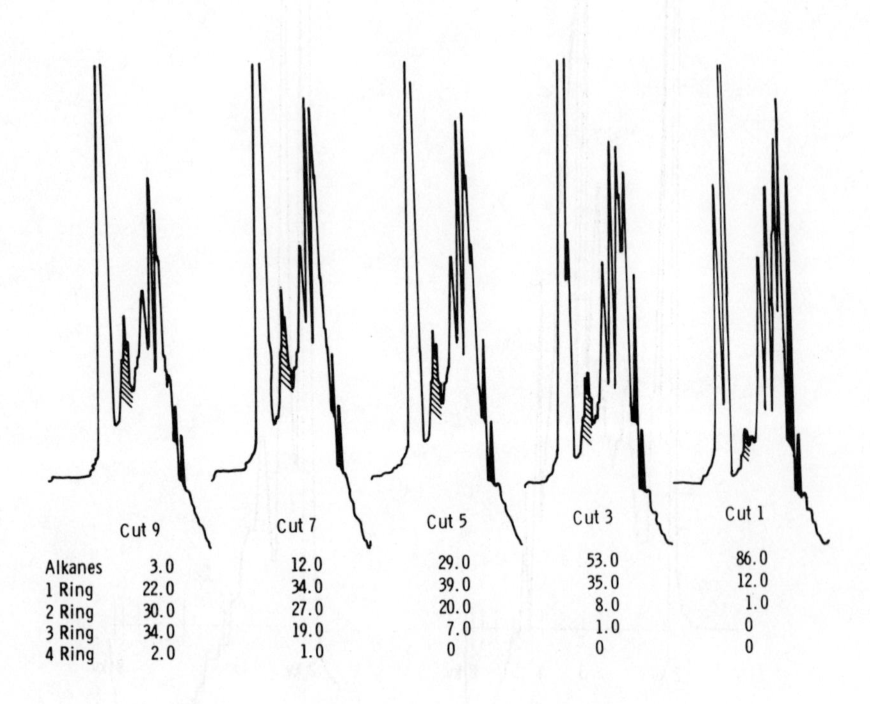

Figure 15. Ikes spectra and Grou-type results of five thermal diffusion cuts of the saturate fraction of a petroleum sample

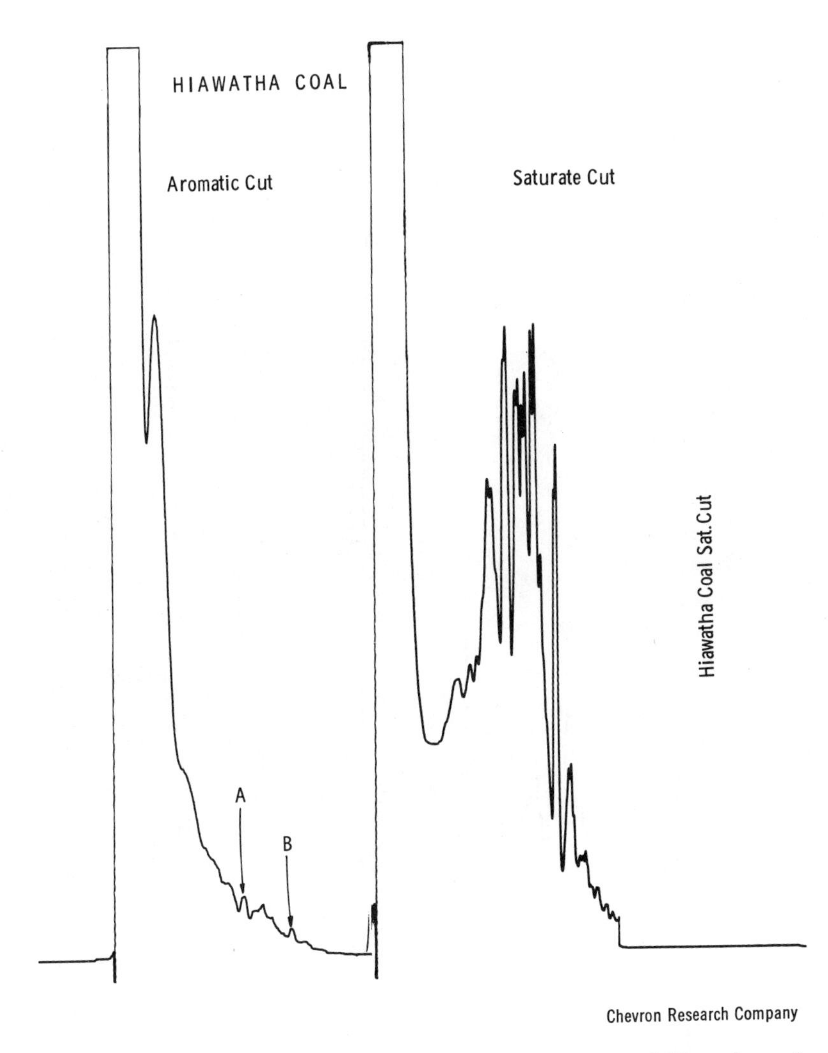

Figure 16. Ikes spectra of the aromatic and saturate fractions of Hiawatha coal

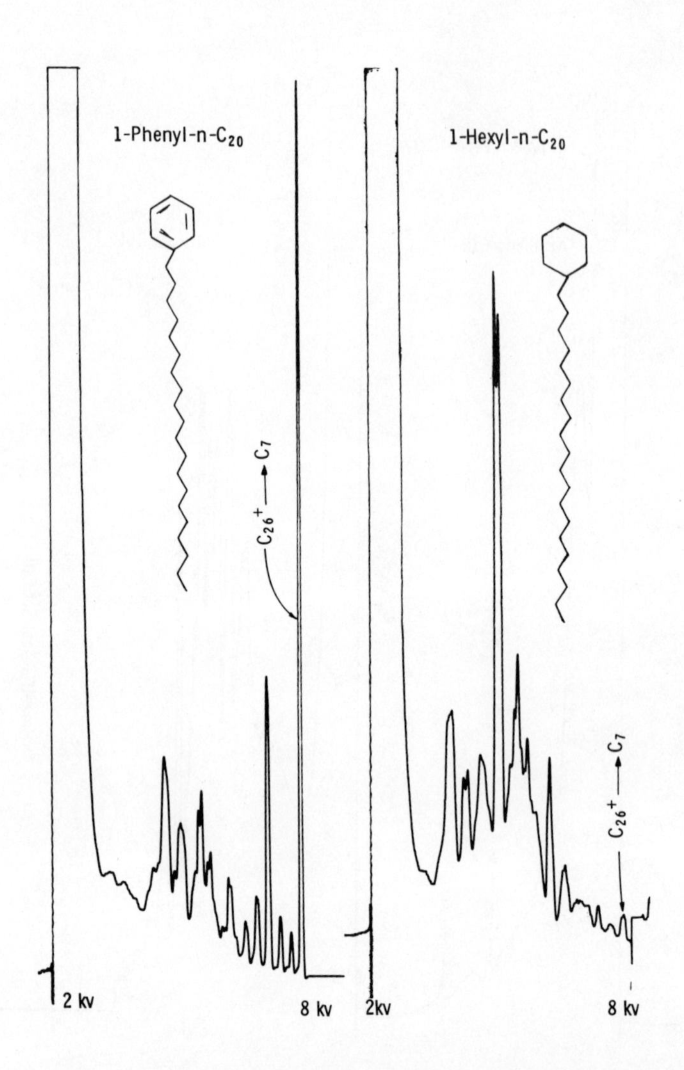

*Figure 17. Ikes spectra of 1-Phenyl-*n-C_{20} *and 1-Hexyl-*n-C_{20}

a petroleum crude oil. The main peaks of the aromatic IKES spectra appear at accelerating voltages less than 3 kV, which means M_1/M_2 ratios are all less than 3/2. This implies two things: first, the charge prefers to stay on the larger piece suggesting a π-bonded or aromatic nucleus on most of the molecules and, second, there is no long-chain branching on any of the molecules of this fraction; otherwise, we would see peaks beyond 3 kV even if an aromatic system is involved.

The saturate cut shows most of its peaks between about 3 and 5 kV, which is a ratio $M_1/M_2 = 3/2$ to $5/2$ averaged at about 4/2. This means that on the average the molecules in this fraction break in half. This is consistent with what is known of saturate-type molecules.

Figure 15 shows the IKES spectra of five thermal diffusion cuts of the ASL 3084 saturate fraction discussed above. The group-type analysis of these cuts is shown below each spectra. The peaks to note are the high ratio $M_1/M_2 > 4/2$ filled peaks and the low ratio $M_1/M_2 < 3/2$ cross-hatched peaks.

The high ratio peaks diminish in intensity along with the paraffins and one-ring components; whereas, the low ratio peak increases with increased concentration of the multiring components.

Miscellaneous Samples

Figure 16 shows the IKES spectra of the SiO_2 saturate and aromatic cut from a Utah Hiawatha coal. The aromatic cut shows two peaks marked A and B which have M_1/M_2 ratios > 3/2 suggesting some long straight chain attachment to a π-bonded nucleus. GC-MS analysis of this fraction reveals the presence of a homologous series of n-alkylbenzenes.

Figure 17 shows the IKES spectra of 1-phenyl-n-C_{20} and 1-hexyl-n-C_{20}. The base peak in the IKES spectra of 1-phenyl-n-C_{20} is due to the $C_{26}^+ \rightarrow C_7$ transition, i.e., the M_1/M_2 ratio is high. The 1-hexyl-n-C_{20} shows only a small peak representing the equivalent transition. The base peak is due to the molecule breaking in half.

Conclusions

This section has shown how IKES spectra may be used, somewhat like IR or UV, in the identification of gaseous materials in both the qualitative and quantitative sense.

Certain isomers not distinguishable by mass spectra alone may well be unraveled using IKES spectra. This sort of approach may well be useful in GC-IKES analysis similar to GC-MS.

Literature Cited

1. Hipple, J. A. and Condon, E. U., Phys. Rev. 68, 54 (1945).
2. Cooks, R. G., Beynon, J. H., Copriol, R. M., and Lester, G. R., "Metastable Ions," Elsevier Scientific Publishing Company, New York, New York, 1973.
3. Jennings, K. R., "Some Aspects of Metastable Transitions, Mass Spectrometry Techniques and Application," G. W. H. Milne, Ed., Wiley-Interscience, New York, New York, 1971.
4. Teeter, R. M. and Doty, W. R., Res. Sci. Inst. 37, p 792 (1966).
5. Budzikiewicz, H., Wilson, J. M., and Djerassi, C, J. Am. Chem. Soc. 85, p 3688, 1963.
6. Tökes, L., Jones, G., and Djerassi, C., J. Am. Chem. Sci. 90, p 5465, 1968.
7. Gallegos, E. J., Anal. Chem. 43, p 1151, 1971.
8. Otvos, J. W. and Stevens, D. P., J. Am. Chem. Soc. 78, p 546, 1956.
9. Gallegos, E. J., Anal. Chem. 47, p 1524, 1975.

Received December 30, 1977

16

Multielement Isotope Dilution Techniques for Trace Analysis*

J. A. CARTER, J. C. FRANKLIN, and D. L. DONOHUE

Oak Ridge National Laboratory, Oak Ridge, TN 37830

Instrumental mass spectrometry includes a variety of techniques for measuring the mass to charge (m/e) ratio of gaseous ions. For analytical purposes, positive ions produced by a suitable source in a vacuum are subsequently separated according to mass-to-charge ratio. The most often used source for trace analysis is the pulsed radiofrequency (RF) ion source.

The early pioneering work in spark source mass spectrometry was conducted prior to the late 1950's (1-3), while commercial instrumentation became available in the early 1960's. The commercial designs employed the Mattauch-Herzog double focusing geometry (4) in which ions of all masses are simultaneously focused in one plane. The field of application of the technique has widened considerably, and with the advent of multielement isotope dilution techniques (5-7), spark source mass spectrometry (SSMS) should produce even better quantitative data.

Basic SSMS Review.

Figure 1 is a schematic diagram of a typical spark source mass spectrometer. The source and analyzer portions are normally operated in the low 10^{-8} torr region while the photographic plate region for prepumping is at a higher pressure, approximately 10^{-6} torr. Ions representative of the sample material are produced by the rf spark source. The spark between the conducting sample electrodes is produced by applying a pulsed radiofrequency potential of 30

* Research sponsored by the Energy Research and Development Administration under contract with the Union Carbide Corporation.

to 50 kV.

Sample electrodes may be made directly from the sample material if it is a conductor or semiconductor, but non-conducting materials must be mixed with pure conductors such as Ag or graphite and compressed.

Ions produced in the source are accelerated into the analyzer region of the mass spectrometer by a potential of about 25 kV. Since the ions have a wide kinetic energy spread, the electrostatic analyzer provides the necessary energy focusing prior to the mass separation in the magnetic stage. Since the Mattauch-Herzog geometry brings ions of different mass to charge ratio to focus at different points in a single plane, a photographic plate placed on the focal plane of the magnetic analyzer simultaneously detects all isotopes with m/e between 7 and 250. A typical photographic plate has 15 graded exposures for a single specimen so that the concentration range from 0.01 to over 1000 ppm is covered.

Figure 2 is a photograph of an AEI MS702 spark source instrument used at ORNL. The source glove box was installed to permit use of an inert atmosphere and to serve as containment for certain radioactive samples.

The technique of SSMS, besides possessing high sensitivity and comprehensive elemental coverage, shows linear response for many elements; i.e., the ion intensity of any single impurity species is proportional to the impurity concentration in the bulk sample, providing the impurity is homogeneously distributed. This property is demonstrated by the linearity plot in Figure 3 for boron concentrations in NBS steel standards in which the linear response extends over several orders of magnitude.

Isotope Dilution Spark Source Mass Spectrometry.

Analyses by SSMS in areas in which standards are not available (as is the case of environmental samples and related energy source samples) require use of the isotope dilution technique. Isotope dilution, described by Hintenberger (8), involves adding to the sample an enriched minor isotope of the element of interest, followed by mass spectrometric measurement of the altered isotope ratios.

The concentration of the element being measured is obtained from equation (1).

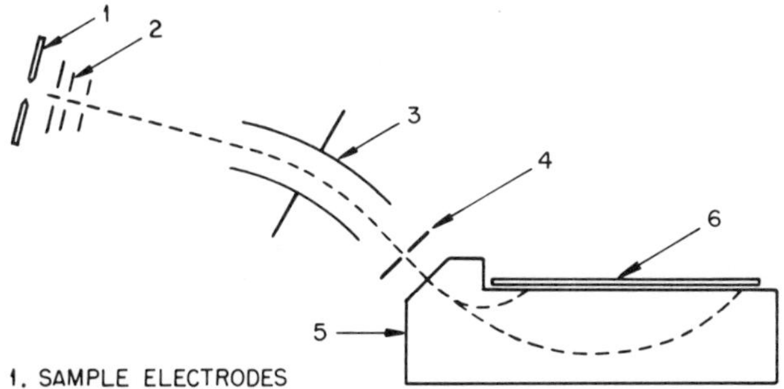

AEI Scientific Apparatus

Figure 1. Double-focusing spark source mass spectrometer

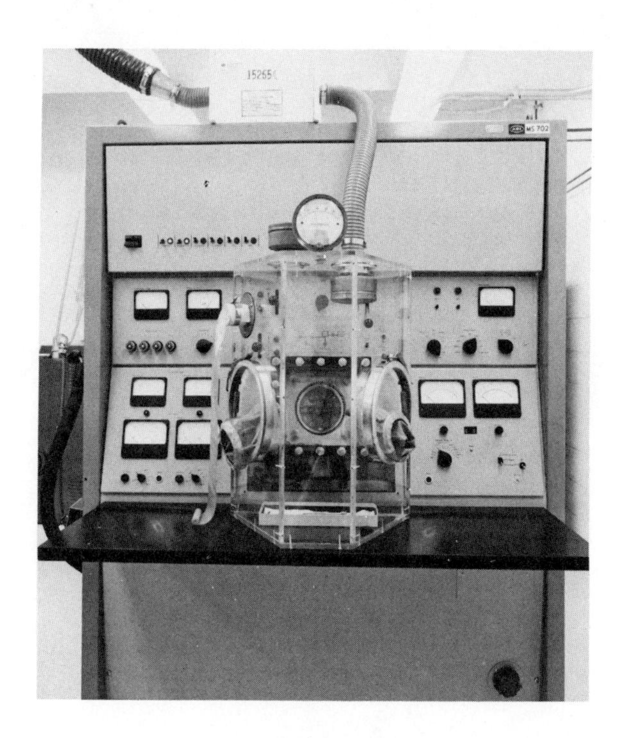

Figure 2. MS-702 mass spectrometer with the special glove box arrangement

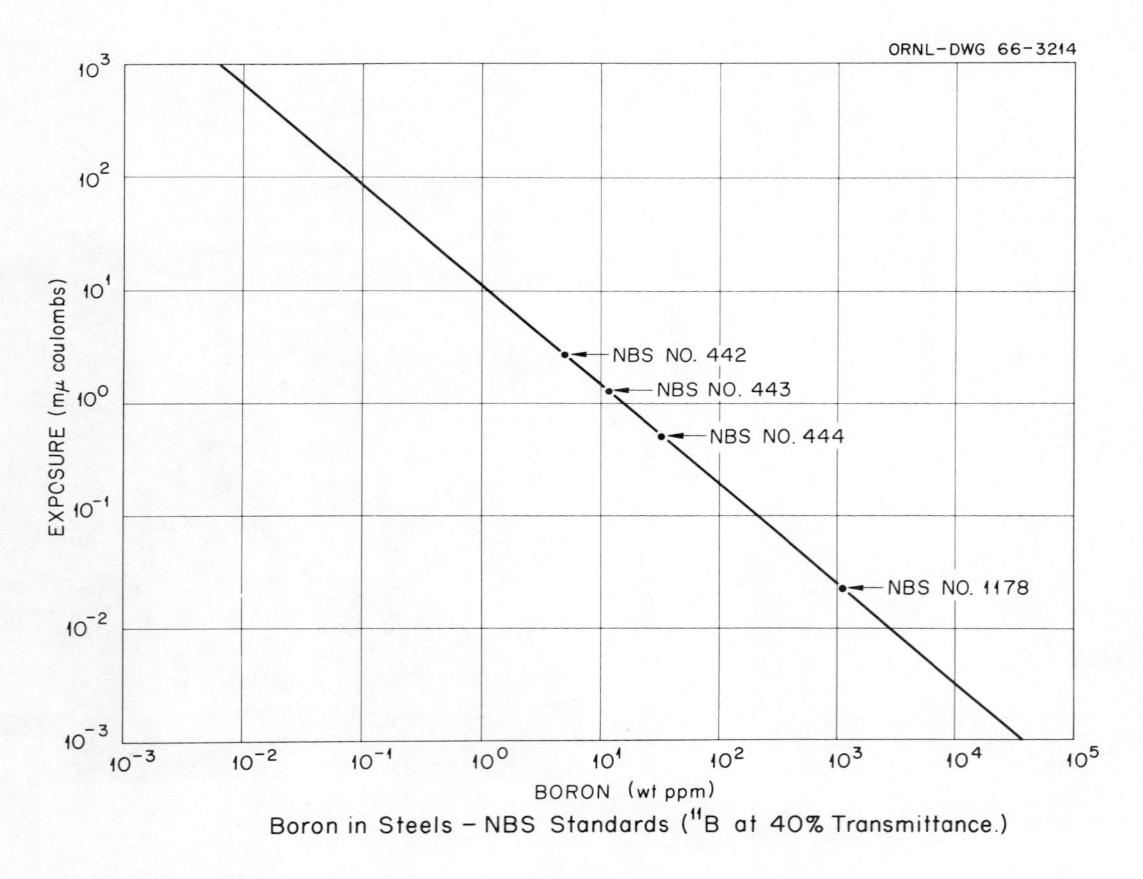

Figure 3. Analytical calibration curve for boron in steel over six orders of magnitude (0.01–10,000 ppm)

$$x = y \left(\frac{R_t - R_m}{R_m - R_s}\right) \cdot \frac{A_s}{A_t} \cdot \frac{T}{S} \quad (1)$$

where x = weight of element in the sample.

y = weight of element in the tracer spike.

R_t = isotope ratio of the spike, $\frac{\text{reference isotope}}{\text{tracer isotope}}$.

R_m = isotope ratio of the mixture, $\frac{\text{reference isotope}}{\text{tracer isotope}}$.

R_s = isotope ratio of the sample, $\frac{\text{reference isotope}}{\text{tracer isotope}}$.

A_s = atomic weight of the sample element.

A_t = atomic weight of the tracer.

T = atom % of the spike isotope in the tracer.

S = atom % of the spike isotope in the sample.

The terms: A_s, A_t, T, S, R_s, and y are usually known in advance, and since R_m can be very accurately measured, it is not unusual to obtain precision and accuracy on the order of 1% for trace element determination. After equilibrium between spike isotope and sample reference isotope is obtained, quantitative results are assured by isotope dilution even when a low yielding chemical purification is required.

Thermal ionization sources are used for solids and electron bombardment sources for gases, giving accurate results with small samples. However, isotope dilution has not been used until recently for analyzing energy-related samples with SSMS.

Even though the method is limited to elements having two or more naturally occurring or long-lived isotopes, it is sensitive and can be very accurate. Since spark source mass spectrometers have similar sensitivities for all elements, SSMS can be used without deleterious effects for many matrices. For example, in order to obtain the cadmium concentration in a sample by isotope dilution SSMS, cadmium enriched in ^{106}Cd is equilibrated with the normal cadmium so that chemical equilibrium is established between the enriched and natural isotopes. Thereafter, any technique may be used that permits the transfer of 0.1-3 ng of cadmium from the solution to the surface of a suitable electrode substrate such as machined graphite rods.

For trace metal concentrations, any suitable extractant may be used to concentrate the metal after equilibration with the enriched spike. High extraction efficiencies are not required since at this point the analysis depends on establishing a ratio between the enriched spike isotope and one of the major isotopes of the metal being sought.

The mass spectra of cadmium and molybdenum spiked with enriched ^{106}Cd and ^{97}Mo are shown in Figure 4. Concentrations of the individual elements can be calculated from the altered isotope ratios.

The results of a typical isotope dilution of cadmium are shown in Table I, which is a reproduction of the format produced by the computer. The data reduction computer-densitometer system which we use has been thoroughly described elsewhere (9). The first entry on the table shows the enrichment of the Cd spike (88.4% ^{106}Cd). The second entry is a report of the abundances of the cadmium in the sample, usually the natural abundance. Sample 318R was analyzed by mixing 1.00 ml of the spike (containing 1.00 ppm enriched cadmium) with a 1.00 ml aliquot of the unknown. Two different photoplate ion exposures were measured to yield the % transmittance value at m/e 106 and 114 as listed in the table. From the emulsion calibration and the observed %T-values, the quantity of cadmium in the unknown is calculated by the computer system and reported in the "Nano-gm" column. Even though the percent transmittances are at opposite ends of the emulsion calibration curve, the agreement is very good. The expected precision for these analyses is ±3-5%.

The data reduction programs for this isotope dilution method are sufficiently flexible to accomodate a variety of experimental approaches. For example, any combination of volume or weight, concentration, isotopic abundance, and % transmittance for both the spike and sample can be accommodated by the data system.

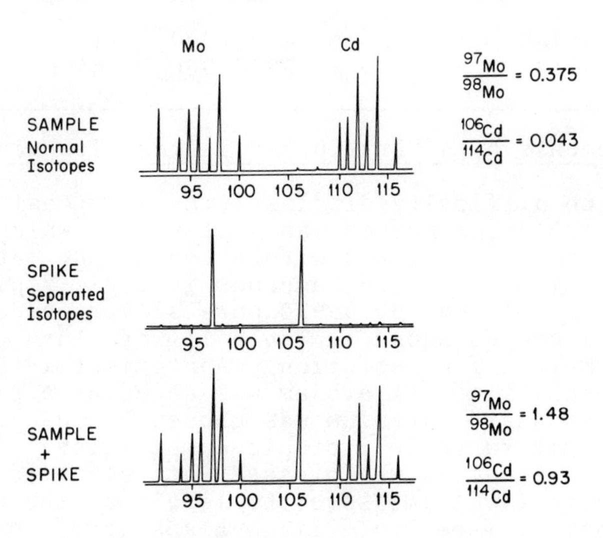

Figure 4. Isotope dilution spark source mass spectra for cadmium and molybdenum

TABLE I.
Computer Printout for Cadmium by Isotope Dilution Spark Source Mass Spectrometry

Cd	106	114	Vol	Conck
Spike	88.400	2.500	1.000	1.000
Sample	1.220	28.900	1.000	

Sample	Nano-gm	106	114
318R	4005.000	19.3	16.2
318R	3933.000	64.6	60.0

Trace Elements by a "Dry Spike" Isotope Technique.

Due to difficulty of dissolving many coal and fly ash samples, a method was employed in which dry sample powders were mixed with a conducting matrix material which contained numerous isotopic "spikes". (10) The matrix was 99.9999% pure silver powder. The enriched isotopes shown in Table II were then deposited onto the matrix from solution. Approximately 50 μg/g of isotopically normal erbium was added as a single internal standard. Erbium was chosen because of the large dynamic range in isotopic composition. The spiked silver matrix was pressed into electrodes and sparked directly to measure the levels of the elements to be analyzed (see Table II). Blank levels of each element present in the silver were determined in a previous experiment.

In order to standardize the amounts of each isotopic spike present in the matrix material, it was necessary to deposit a solution containing known amounts of the natural elements onto a weighed aliquot of the dry matrix. This mixture was dried and homogenized, followed by SSMS analysis and isotope dilution measurements to yield the apparent concentrations of the natural elements. It was possible to back-calculate the actual concentrations of the spike isotopes in the matrix. The accuracy of any measurement should be checked by the use of standards. For the analysis of coal, fly ash and related geological samples, NBS Standard Reference Materials such as NBS SRM 1632 (Coal) and 1633 (Fly Ash) are useful. From the analysis of these standards, relative sensitivity factors or instrumental biases were determined.

Homogenization of the deposited isotopes was ensured by extensive shaking and grinind in a mixer-mill using a plastic container and plastic ball pestles. With 30 minutes of such shaking, the precision obtained for duplicate analyses was 5-10%, indicating that no gross inhomogeneity existed.

TABLE II
Separated Isotopes Deposited Onto Pure Ag Powder

Element	Isotope	% Enrichment	Approx. Conc. in Ag (μg/g)
Pb	204	99.73	10
Tl	203	94.96	5
Ba	135	93.6	50
Te	125	91.23	10
In	113	96.36	5
Cd	111	91.33	5
Mo	97	94.25	5
Sr	86	95.73	100
Se	77	94.38	10
Zn	67	89.68	30
Cu	65	99.70	30
Ni	61	92.11	30
Fe	57	93.63	100
Cr	53	96.4	30
K	41	99.35	100

In the analysis of trace elements in coal by this method, the coal was first ashed (550°C) and an exact amount of ash mixed with a known quantity of silver spike. The mixture was homogenized, pressed, and sparked in the spark source mass spectrometer with the usual ion detection by Ilford Q-2 photoplates. Data were read on a microdensitometer-computer system described elsewhere (9). For the elements listed in Table II, a ratio was measured between the separated isotope and a natural isotope of each element. These ratios were used to calculate the concentrations of each element in the sample according to the internal standard approach to SSMS analyses (10). This technique is similar to, but not the same as, isotope dilution, since the isotopes have not reached chemical equilibrium in solution. Apparently isotopic equilibrium is approached in the ion plasma.

Once the amounts of the elements in Table II are

measured a similar procedure allowed the other elements of interest such as Co, Mn, As, etc., to be measured versus the appropriate isotopic spikes. Again, a sensitivity factor is needed and can be obtained from NBS or other standards. Table III shows average results for NBS SRM 1632 (Coal using SRM 1633 (Fly Ash) as a standard. These data indicate that good precision and accuracy are available at levels less than one part per million. The technique described above for coal ash analysis has also been successfully applied to other ash samples and some geological materials which are difficult to dissolve. As for other analytical endeavors, the availability of standards is important to ensure accuracy of the results by the "dry isotope spike" technique.

TABLE III.
Analysis of SRM 1632 Using SRM 1633 as Standard

Element	Conc. + RSD	NBS Certified
Tl	0.6 ± 0.2	0.59 ± 0.03
Pb	33 ± 3	30 ± 9
Cd	0.4 ± 0.2	0.19 ± 0.03
Zn	32 ± 8	37 ± 4
Cu	15 ± 3	18 ± 2
Ni	15 ± 3	15 ± 1
Cr	19 ± 3	20.2 ± 0.5

Conclusions.

Multielement isotope dilution has added a capability for precision and accuracy to the high sensitivity and comprehensive elemental range of spark source mass spectrometry. Enriched isotopes are available in high purity and high enrichment for a large number of elements so that the number of elements that may be determined by IDSSMS is quite large. We have shown that these isotopes can be used as internal standards for mono-nuclidic elements. We have also shown that dried mixtures of isotopes on conducting powders can be added to solid samples as internal standards with acceptable precision and accuracy. The precision and accuracy of the dry standard method is between 10 and 20%, and for the solution technique the precision and accuracy can approach 5-10% with 1 x 10^{-9} g of an element present on the electrodes.

Literature Cited.

1. Dempster, A.J., Nature, (1935) 135, 452.
2. Gordon, J.G., Jones E.J., and Hipple, J.A., Anal. Chem., (1951) 23, 438.
3. Hannay, N.B., and Ahearn, A.J., Anal. Chem., (1954) 26, 1056.
4. Mattauch, J. and Herzog, R., Z. Physik., (1934) 89, 786.
5. Leipziger, F.D., Anal. Chem., (1965) 37, 171.
6. Alvarez, R., Paulsen, P.J., and Kelleher, D.E., Anal. Chem., (1969) 41, 955.
7. Carter, J.A., Donohue, D.L., Franklin, J.C., and Stelzner, R.W., Trace Substances in Environmental Health-IX, Univ. of Missouri, 1975.
8. Hintenberger, H., "A Survey of the Use of Stable Isotopes in Dilution Analysis", Electromagnetically Enriched Isotopes and Mass Spectrometry, ed. by M. L. Smith, Academic Press, Chap. 21, pp. 177-189. (1956).
9. Stelzner, R.W., McKown, H.S., and Sites, J.R., ACS Spring Meeting, 1975.
10. Jaworski, J.F., and Morrison, G.H., Anal. Chem., (1975) 47, 1173.
11. Ahearn, A.J., ed. "Mass Spectrometric Analysis of Solids", Elsevier, New York, 98 (1966).

RECEIVED December 30, 1977

17

Computer Applications in Mass Spectrometry

F. W. MC LAFFERTY and R. VENKATARAGHAVAN

Department of Chemistry, Cornell University, Ithaca, NY 14853

The information content of the mass spectrum of an average organic compound is unusually high, containing at least 50 bits of information (1); further, with modern instrumentation a mass spectrum can be obtained on a nanogram of compound in one second. Obviously, to utilize this tremendous quantity of data it must be acquired, reduced, and interpreted in a very rapid and efficient manner. The modern digital computer (COM) has proved to be ideal for all three of these duties, and its rapid increase in capabilities, and the concomitant reduction in cost, in the last decade has thus led to a similar revolution in mass spectrometry (MS).

Acquisition and Reduction of MS Data

The book of Waller (2) was very timely in showing the surprising advances in MS/COM systems that had occurred in the few years before 1971. This book contained reports from many leading MS research laboratories concerning the automated data acquisition and reduction systems then in use. There were a wide variety of these, the majority of which had been developed to a substantial extent in the reporting laboratory. In the intervening six years the field has changed dramatically; literally hundreds of mass spectrometry laboratories now have computerized data acquisition and reduction systems, and most of these are manufacturer supplied. Only a small fraction of these acquire data on, for example, magnetic tape for later processing on a central computer; the on-line minicomputer is the rule, but it has increasing competition from microprocessors and even sophisticated calculators. A computer-controlled GC/MS is now available for less than $50,000 (3), and a microprocessor-driven MS is available which gives the confidence of identification as well as quantitative analyses for 16 preselected compounds every second (4).

There are still challenging areas of high potential for further applications of on-line computers in MS data acquisition and reduction, such as high-resolution MS, simplification of MS operation, maintenance and record keeping, collisional activation MS, and continuous analyzers (e.g., patient monitoring, process control) (5). However, just as mass spectrometry evolved so that most laboratories no longer constructed their own instruments, the field has suddenly progressed to where computer data acquisition and reduction equipment which is superior for most applications is now available commercially. This is not true, however, for a highly promising and challenging area, that of the computer identification of unknown mass spectra (6), and this will be emphasized in the remainder of our discussion. If efficient mass spectrometer systems with totally automated data acquisition, reduction, and identification could be made available at reasonable cost, this would surely open many important new areas of application for MS.

Computer Identification of Unknown Mass Spectra

The identification of unknown compounds from their mass spectra, whether as pure compounds or in mixtures, is a problem for which we feel it is better to use two approaches, retrieval and interpretation, with these applied sequentially. The unknown mass spectrum is first matched against a library of all available reference spectra; if no spectrum is retrieved which matches sufficiently well, the unknown spectrum is then interpreted to obtain as much structural information as possible. A variety of systems for both retrieval and interpretation have been proposed (6); interpretation using pattern recognition systems have been described at this meeting by Isenhour (7) and by Wilkins (8) and the Artificial Intelligence system (9) by Smith; thus this report will emphasize the "Probability Based Matching" (PBM) (10, 11) and the "Self-Training Interpretive and Retrieval System" (STIRS) (12 - 15) developed at Cornell for these purposes. Both of these systems are actually available internationally to outside users over the TYMNET computer networking system (16).

Probability Based Matching

For document retrieval in libraries (17), it is well known that optimization requires "weighting" of the descriptors or requirements sought; some are more important than others to the person conducting the search. Low resolution mass spectra contain two principal types of data, masses and abundances; the PBM system (10, 11) employs a probability weighting of these. As first proposed by Grotch (18), the probability of occurrence of particular abundances (based on 100% for the most abundant peak) should follow a log normal distribution. This was shown to be true for a data base of 18,806 different compounds (19); abundance ranges differing by a factor of two in their occurrence probability are ≥0.24%, ≥1.0%, ≥3.4%, ≥9.0%, ≥19%, ≥38%, and ≥73%. The probability of occurrence of the different mass values also varies widely. Because larger molecular fragments tend to decompose to give smaller fragments, higher *m/e* values are less common in mass spectra, the probability decreasing by a factor of 2 approximately every 130 mass units. Although this is a surprisingly smooth function at high *m/e* values, lower values show rather large differences in their probability of occurrence, with these variations repeated every 14 mass units. Thus, although *m/e* 39, 41, and 43 ions of 1% or greater abundance are each found in more than two-thirds of all reference spectra, peaks *m/e* 33, 34, and 35 each occur less than 10% as frequently (19). Thus in an unknown mass spectrum if abundant ions at *m/e* 34, 241, and 343 are matched by comparably abundant ions in a reference spectrum, this is far more significant in indicating that a correct retrieval has been made than if the unknown's *m/e* 39, 41, and 43 peaks were matched in a reference spectrum. These abundance and mass uniqueness weightings are used to calculate the probability that the match occurred by chance and is thus a "false positive"; the reciprocal of this probability (log base 2 "confidence index, K") is used to rate the degree of match.

A second unique feature of PBM, which has been proposed independently by Abramson (20), is "reverse searching", and is valuable for the identification of components in mixtures. In this, PBM ascertains whether the peaks of the reference spectrum are present in the unknown spectrum, not whether the unknown's peaks are in the reference. Thus the reverse search in effect ignores peaks in the unknown which are not in the reference spectrum, as they could be due to other components of the mixture. Although reverse searching should thus reduce the capabilities of PBM for matching unknown spectra of pure compounds, this apparently is more than offset by the increased capabilities resulting from the data weighting. The system has been tested with over 800 "unknown" mass spectra taken from a large collection of mass spectra from diverse sources (21), and its performance has shown to be generally superior (22) to the widely accepted Biemann-MIT system (23) which matches the two largest peaks in each 14 mass unit region.

Performance has been measured utilizing "recall/reliability" plots, methodology developed for library retrieval systems (17) in which *recall* (RC) is defined as the proportion of relevant spectra actually retrieved, and *reliability* (RL, or "% correct" X 0.01) is the proportion of retrieved spectra which are actually

$$RC = I_c/P_c \quad (1)$$

$$RL = I_c/(I_c + I_f) \quad (2)$$

$$FP = I_f/P_f \quad (3)$$

relevant; FP = proportion of false positives, I_c = number of correct identifications, I_f = number of false identifications, P_c = number of possible correct identifications, and P_f = number of possible false identifications ($P_c + P_f$ = total unknowns examined). The "recall/reliability" plot is then constructed by determining the pair of these values achieved for particular K (or "ΔK") value thresholds (11); the higher the K value demanded by the user for an identification, the higher the probability that it is correct (RL), but the lower the chance of making an identification for a particular unknown (RC). Note that if the evaluation is made considering only the highest K selection as the match, the number of compounds tested will be equal to the number of possible correct identifications (P_c), to P_f, and also to ($I_c + I_f$), so that for this evaluation RC = RL = (1 - FP).

Two sets of "unknown" spectra of over 400 each were selected at random from the molecular weight (MW) ranges 144 - 160 and 232 - 312; at 15% recall these showed reliability values of 83% and 85%, respectively; at 25% RC, RL = 76% and 62%; and at 50% RC, RL = 65% and 42%. As discussed later, the 65% RL value corresponds to a false positives of ~1/50,000. The lower results for the high MW set are mainly due to the use of only 15 peaks per spectrum, and this has been increased, depending on MW, to as many as 26. Unknowns present as 30% and 10% components in mixtures gave reliabilities of 60% and 20%, respectively, at the 25% recall level; however, these results were far superior to those from the forward search system on the same unknowns. To repeat, for unknowns giving matches of unsatisfactory reliability, STIRS should be used also.

Because retrieval of the spectrum of a compound whose structure is similar to, but not exactly the same as, that of the unknown can be helpful to the user, four "classes of match" have been defined: I, identical compound or stereoisomer; II, class I or ring position isomer; III, class II or homolog; and IV, class III or an isomer of class III compound formed by moving only one carbon atom. The recall/reliability performance of PBM was evaluated separately for these four matching criteria; at the 50% recall level 95% of the compounds (low MW set) selected matched within the class IV criteria.

As has been suggested previously (24), the subtraction of the reference spectrum of an identified compound from the mixture spectrum should produce a residual unknown spectrum which is easier to identify. This approach has been implemented so that at the end of the PBM run the computer automatically subtracts the best matching reference spectrum (or, at user command, some other reference spectrum) from the unknown spectrum and notifies the user of any significant residual peaks. Although PBM is a reverse search system, its recall is lower for components in lower proportion of the mixture; thus subtracting out the contribution of an abundant component improves the PBM recall for other mixture components. Recently In Ki Mun in our laboratory has applied the PBM principles to the identification of the number of chlorine and bromine atoms present in a fragment ion from the abundances of the isotopic peaks. In a test of more than 1,000 "unknown" spectra taken from our large data base, the predictions of this PBM program were correct 90% of the time, and most of these incorrect predictions were due to inaccurate data.

Self-Training Interpretive and Retrieval System

In some interpretive methods such as Artificial Intelligence (9) the computer is programmed to recognize the mass spectral behavior of particular types of compounds or substructures. In pattern recognition or learning machine approaches, the computer trains itself for such recognition based on reference spectra which do and do not contain a particular structural feature. In contrast, there is no pretraining of STIRS (12) for the mass spectral behavior of particular structural moieties; rather, 15 classes of mass spectral data (Table I) have been selected which are indicative of different compound types or substructures, but without designating what these types or substructures are. STIRS then, in effect, trains itself to interpret the unknown mass spectrum by matching the unknown data for each of these classes against the corresponding data for all of the reference spectra; particular substructures or types of compounds which are found in a substantial proportion of the best matching reference compounds have a correspondingly high probability of being present in the unknown. Thus STIRS does not have to be pretrained to recognize the presence of any particular structural feature; however, it cannot identify any feature which is not present in some compounds of the reference file. In further contrast to some applications of pattern recognition methods, we recommend that STIRS be used for positive information only; the presence of a structural feature in many of the selected reference compounds indicates the presence of that substructure in the unknown, but the absence of a substructure in the selections is not a reliable indication of its absence in the unknown.

The present Cornell PBM system employs a data base of 41,429 different spectra of 32,403 different compounds. To increase the selectivity of STIRS, it uses a data base of only the best spectrum (25) of each compound, limiting this further to the 29,468 compounds which contain only the common elements H, C, N, O, F, Si, P, S, Cl, Br, and/or I. A number of improvements made recently to the system will be described below.

Table I. Mass Spectral Data Classes Used in STIRS

Data Class	Description, maximum number of peaks	Range of mass or mass loss
1	Ion Series (14 amu separation)	<100
2-4	Characteristic ions	
2A	Four even-mass, four odd-mass	6-88
2B	Eight	47-102
3A	Seven	61-116
3B	Seven	89-158
4A	Six	117-200
4B	Six	$159-(M-1)^+$
5-6	Primary neutral losses	
5A	Five	0-2, 15-20, 26-53
5B	Five	34-75
6A	Five	59-109 (MW ≥175)
6B	Five	76-149 (MW ≥250)
5C	Five	16-20, 30-38, 44-51, 59-65, 72-76
6C	Five	26-28, 39-42, 52-56, 66-70, 80-84
7, 8	Secondary neutral losses from most abundant odd-mass (MF7) and even-mass (MF8) loss	<65
11	Overall match factors	
11.0	MF11.1 + MF11.2	
11.1	2A + 2B + 3A + 3B + 4A + 4B	
11.2	5A + 5B + 6A	

It is important to emphasize that STIRS has been designed as an aid to, not as a replacement for, the human interpreter, its use making the interpretation more efficient and possibly more complete. The basic way in which an interpreter uses STIRS is to examine the best matching compounds found for each data class and for the overall match factors (Table I) in an attempt to find common structural features. For example, if one data class showed a substantial proportion of compounds containing an isoxazole ring and another data class found mainly acetoxy compounds, these would be important postulates for the interpreter to test in trying to assemble a molecular structure for the unknown. All of the reference compounds are coded in Wiswesser Line Notation (WLN), so that the interpreter often finds it more convenient to scan these by eye to detect common structural features. There are two ways in which the computer can help in this task. Currently Dr. Michael M. Cone in our laboratory is developing a program for computer identification of "super substructures" which involves the comparison of decoded WLNs of the selected compounds to find the largest common structural features.

A special system which has already been implemented by Dr. H. E. Dayringer in our laboratory (13) predicts the presence of 179 common substructural fragments from the STIRS results. For this, the 15 compounds selected in each data class as providing the best match to the unknown's mass spectrum are examined by the computer for the presence of each of the preselected substructural fragments, and the statistical significance of the number found is evaluated by a "random drawing" model. For example, because 28% of the compounds in the reference file contain phenyl, even if the unknown does not contain phenyl an average of ~4 phenyl-containing compounds would still be expected in the 15 compounds of highest match factor values. However, if for a particular unknown there are 10 phenyl compounds in the top 15, statistically this will occur by coincidence only once in 113 times; it follows that in >99% of cases this will be due to the presence of phenyl in the unknown compound, i.e., in <1% of predictions based on such data will the result be a false positive. To check this, using the overall match factor (MF11) two runs were made of 373 "unknown" mass spectra (every 50th spectrum in the Registry data base) (21); the substructures predicted at the 2.0% or less false positive level actually contained an average of 1.0% false positives, in excellent agreement. The MF11 results (13) at predicted levels of ≤0.5%, ≤0.2%, and ≤0.1% gave average false positives of 0.7%, 0.55%, and 0.45%, respectively; these are not as low as expected statistically, but this appears to be mainly due to the fact that most of the falsely selected substructures have similar mass spectral behavior (e.g., pyridyl identified as phenyl). Typical results are shown in Table II. In practice, the substructures selected are ranked in order of their predicted values, reporting only those of ≤2% false positives. Not surprisingly, more reliable

Table II. Substructure Identification by STIRS and K-Nearest Neighbors[a]

Substructure	% in file	STIRS ≤2% FP[b]		≤0.5% FP		KNN 3/5[c]		3/3	
		RC[b]	FP	RC	FP	RC	FP	RC	FP
Ester, anhydride	19	55	2.1	50	1.4	29	4.8	19	0.8
Phenyl, sub. phenyl	28	75	5.0	67	3.4	71	11.7	51	3.7
Chlorine	8	74	0.3	74	0.0	54	0.7	47	0.5
Doubly-branched C	11	44	3.4	42	1.5	41	4.1	25	1.3
20 Substr.[d] average	14	42	1.9	36	1.2	43	7.9	27	2.1
179 Substr. average[a]	2.9	49	1.9	44	0.7				

[a]References 13 and 26. [b]RC, % recall; FP, % false positives.
[c]Three of the five nearest neighbors contain the substructure.
[d]Others: alkyl $\geq C_1$, $\geq C_2$, $\geq C_3$, carbonyl, -OH, $-NH_2$, -NH-, trisubst. N, S, $-CH_2O-$, F, Cl, -O-, double bond, single-branch C, double-branch C, carbocyclic rings, and heterocyclic rings.

predictions are generally found for the overall match factors MF11.0, 11.1, and 11.2, which are derived from combinations of the values for the individual data classes, as shown in Table I.

The 202 substructural features tested varied widely in their occurrence in the reference file; almost half of the compounds contained the substructures carbonyl and methyl/methylene, with only 6 each of compounds containing azulene and pyrene in a total of 18,806 compounds. The substructures also varied widely in their recall; trimethylsilyl was identified correctly in all but one of the 30 compounds tested which contained it, while 24 of the substructures gave zero recall. As the experienced mass spectrometrist knows, functional group information varies widely in its "visibility" in the mass spectrum, not only between different groups, but also according to the environment of the group in the molecule. In an infrared spectrum, a carbonyl group almost always gives a characteristic peak in the carbonyl-stretching region. However, from a mass spectrum the presence of a carbonyl group often must be inferred from a combination of peaks: the acetyl or benzoyl groups can give characteristic peaks, but not usually the uncombined carbonyl group. Further, in the mass spectrum of acetone the acetyl group is very visible as the base peak at $\underline{m}/\underline{e}$ 43, but in the mass spectrum of dimethylaminoacetone this peak is small (5%) and "hidden" next to a larger peak; the competing fragmentation directed by the amino group is dominant. For the carbonyl group, STIRS-MF11 achieved only a 31% recall at the 98% confidence level. However this does not necessarily mean that STIRS has not been optimized properly for the carbonyl group; unless some other algorithm can show a superior performance, it is more likely that mass spectrometry is not particularly sensitive for the carbonyl group *per se*.

Despite this variation in recall, even the list of 179 substructures with non-zero recall on average gives a substantial amount of information concerning an unknown structure. Multiplying the recall for each substructure by its proportion in the data base and summing these values gives a figure of 2.55, and improvements to STIRS (14, 15) have raised this to approximately 3; this is the number of correct substructure identifications which STIRS should give for the average unknown spectrum. Because 1% of the 179 substructures should be indicated as false positives, this means that on average there will also be 2 incorrect substructure identifications. However, if the interpreter recognizes that a few false positives are possible, he can often discard these as unlikely candidates due to other mass spectral or separate evidence.

We have recently put a substantial effort into the improvement of recall values for the substructure list by optimization of the STIRS data classes, in particular, the "characteristic ions" (14) and the "primary neutral losses" (15); those changes were adopted which showed improvements in the average recall values for 373 "unknown" spectra, without degradation of the reliability performance. Not surprisingly, the best recall values are found for those substructures which are known to have a strong directing influence on mass spectral fragmentations. We are now engaged in a revision of "Mass Spectral Correlations" (27) based on the reference mass spectral file of 29,468 compounds for STIRS; the substructures found to give the best correlations here will be tested with STIRS, and it is expected that this will give a much larger list of substructures with generally higher recall values. We have also made substantial progress on an algorithm to determine the molecular weight from the unknown mass spectrum, even if a molecular ion is not present.

System Evaluation; Effect of File Composition of Tested and Reference Spectra

It is obviously necessary for interested workers to reach a common understanding of evaluation terms such as reliability (used here as synonomous with "% correct" X .01), false positives, and recall (equations 1 - 3), and their applicability for comparison of alternative systems (6 - 15, 26, 29). In testing the efficiency of the STIRS system we based this on false positives and recall values; yet for PBM, as done by others for other retrieval systems, we used reliability as well as recall values. We feel that past reports, including our own, have not fully recognized the effect of the proportion of the component sought in both the tested and reference spectral files (28).

For retrieval systems such as PBM the proportion of false positives is also characteristic of the system's performance; the number of incorrect matches produced should be proportional to the size of the reference file (assuming these spectra are of random composition, and that the proportion of correct answers in the file is small). In fact, PBM is designed to make the K value a direct measure of the probability of a false positive if the mass and abundance data of each reference spectrum are distributed in the same manner as those of the whole file. However, at a predicted FP value of 2^{-50} (K = 50), the observed FP = ~1/50,000; such values are greater than the prediction in large part because mass values tend to be clustered (such as "ion series"); supporting this, class IV matches at K = 50 give FP = ~1/500,000. Because the recall and false positives values are independent of occurrence probabilities (equation 4), it is better to use RC and FP in evaluating the performance of retrieval as well as interpretive systems.

$$RL = P_c \cdot RC/(P_c \cdot RC + P_f \cdot FP) \qquad (4)$$

Retrieval systems are commonly tested using a data base containing one reference spectrum of each unknown; equation 4 shows for such cases that using a larger data base must reduce the reliability value found, as the number of correct retrievals will remain the same while the number incorrect increases. If a reference spectrum is retrieved for a true unknown, the user would like to know the probability that this answer is correct; however, this reliability depends not only on the system's recall and false positives performance, but also on the probability that the unknown compound is represented in the reference file (thus even if it gave "perfect" RC and FP values, PBM could not match an unknown not represented in the data base). Doubling the number of compounds in the data base should increase the probability that the unknown will be represented, but if the new reference spectra are of "rarer" compounds the reliability $[I_c/(I_c + I_f)]$ of answers will decrease because the number of false positives (I_f) must double. This is the basic reason for the use of specialized data bases for unknowns from restricted areas such as drugs or flavors. Thus the reliability of retrieval results depends on the proportion of the compounds tested which are represented in the data base and on the proportion of compounds in the data base which correspond to the compound tested. On the latter, the reliability of PBM results is improved by including in the file multiple reference spectra of many of the more common compounds.

It may be helpful to examine the application of such performance results to evaluate the reliability of the system's prediction for a real unknown. To illustrate, at the 50% recall level (average $\Delta K = 30$) PBM shows ~1/50,000 false positives* (11); if for an unknown PBM finds one match ($\Delta K = 30$) in a 25,000 spectrum file there will be a 50% chance (25,000/50,000) that this is a false positive. However, the offsetting chance that this is the right answer will only correspond to the full 50% recall value if there is a high probability that one of the reference spectra is of the same compound. Thus the user must combine (equation 4) the RC and FP values with his own estimates of the probability that the unknown is present in the file.

In making a similar examination of the use of such evaluation terms for STIRS, it should be noted that in effect STIRS is matching the unknown against the spectral data of each substructure sought. Thus the present STIRS "substructure matching" system has only 179 entries in its reference file, and 1.0% false positives corresponds to an average of two incorrect identifications per search; for our present PBM spectral file of 42,000 compounds, a false positives of only 1/21,000 would yield this number incorrect per search. Allowing this higher false positives level for STIRS makes it able to achieve a higher recall; again, however, the utility to the user will also depend on the probability that one or more of the 179 substructures will be present in the unknown. In our data base collected from a wide variety of sources (21) the occurrence probabilities range from 46% (carbonyl) down to 0.03% (azulene), with an average of 2.9% for each of the 179 selected substructures. At the 1.0% false positives level STIRS (MF11) gave an average recall of 49% (13); thus (as explained above) for an unknown taken at random from this data base on average it should predict 2.55 (2.9% X 179 X 49%) correct (and 2 incorrect) substructures of 179 per unknown [more recent (14, 15) STIRS improvements have raised the recall to give ~3 correct]. Consider the 20 common substructures of Table II, which actually show poorer than average RC and FP performance; on average STIRS should predict 1.01 (14% X 20 X 36%) of these 20 substructures correctly for only 0.24 (1.2% X 20) incorrect. Although the carbonyl and azulene substructure have equivalent 1% chances of being found as false positives, the carbonyl will

---

*The testing of PBM utilized a data base of ~24,000 spectra in which an average of ~2 were of the same compound as the unknown tested. Thus at 50% recall on average one correct answer is found ($I_c = 1$). For this recall (low molecular weight trial set) the reliability = 65% = $1/(1 + I_f)$, so I_f ~0.5, and FP = 0.5/24,000 = ~1/50,000.

be identified on average in 14% (46% X 31% recall) of such unknowns, but the azulene will only be found in 1/1,000 of this frequency (0.03% X 50% recall). Obviously a STIRS identification of such low-frequency substructures as azulene in most cases is a false positive, but usually the interpreter can recognize this from other information on the sample as well as from interpretation of the spectrum. Re-emphasizing, this low frequency refers only to the compounds in our comprehensive data base; the true reliability of the results is determined by the actual frequency in the environment from which the unknown has been taken. If this frequency for azulene should be high, STIRS will show a high reliability for the identification of this substructure.

To summarize, an important difference between tests of compound identification systems such as PBM and substructure identification systems such as STIRS is that generally in the former there is no attempt to study the system's ability to identify different types of compounds. A particular compound is used only once in the retrieval test, and so its occurrence frequency is the same as all other compounds tested; this would also be true of the reference set if it is of all different compounds. Thus if the reference files used in different tests are approximately the same with respect to size and average number of spectra per compound, the "reliability" values obtained will be directly related to the "false positives" values at a particular recall level (equation 4). For evaluations of retrieval systems such as PBM either RL or FP values can be reported, but data necessary for their interconversion should be given. It is better to report "false positives" for the evaluation of substructural identification systems such as STIRS, as the substructure occurrence frequency can vary by orders of magnitude. Also, following the common practice for studies of document retrieval systems (17) [and of PBM (11)], recall should be evaluated as a function of the false positive level. In reports of particular improvements to STIRS (13 - 15) we used the value of RC/(RC + FP) as a modified "reliability" term, really a measure of the utility of the system; this combined recall and false positives in a single term that would be independent of the substructure occurrence frequency. However, there has been considerable confusion over the use of this term, and it now appears preferable to report RC and FP separately.

Comparison of STIRS With Other Available Mass Spectral Systems. One advantage of STIRS is that it is the only interpretive system suitable for most types of organic compounds which is generally available, although interpretive information can be obtained from the results of retrieval systems. The Artificial Intelligence system is available for selected compound types through Stanford University, as discussed in another paper in this program. Very extensive research studies on pattern recognition or learning machine approaches to mass spectral interpretation have been reported over the last half-dozen years, as summarized in Professor Isenhour's talk, although such systems apparently have not seen any appreciable use for real unknowns. Justice and Isenhour have shown that the K-Nearest Neighbor algorithm is superior to 5 other common pattern recognition methods in its ability to extract information from mass spectra (7, 26); to compare KNN and STIRS, Professor Isenhour and his coworkers undertook an evaluation of the ability of KNN to identify 20 of the substructures examined in our STIRS study, using the same data base and substructure identifications. The results of 4 representative substructures and the averages for the 20 are shown in Table II along with the STIRS results at the ≤2% and ≤0.5% false positives levels. Comparing identification requirements giving equivalent average recall values (3/5 KNN vs. >2% FP STIRS) shows a factor of 4 reduction in false positives for STIRS with an increase from 64% to 84% correct. Thus the incorporation of mass spectral knowledge has aided the performance of STIRS, and a similar input into KNN should improve the classifier performance. Note, however, that KNN can only make predictions for substructures for which it has been pretrained, while the self-training feature of STIRS makes possible predictions on any structural feature which is present in the reference data base.

PBM and STIRS as Aids to the Interpreter

For a particular unknown it is possible that a retrieval system such as PBM can identify an unknown mass spectrum with sufficient confidence that no human intervention will be necessary. However, in general, systems such as PBM and STIRS should be employed as an aid to the interpreter; if the system cannot give an identification with sufficient confidence or propose a complete structure, it is still very possible that the interpreter may be able to complete the identification. Processing an unknown spectrum through PBM or STIRS requires less than 1 minute on either our laboratory DEC PDP-11/45 computer or the Cornell/TYMNET computer system (16). Thus for all but the simplest unknowns the information provided by these systems should at least yield a substantial time saving to the interpreter. In the case of total unknowns, such as compounds commonly encountered in pollution samples, insect pheromones, and forensic analyses, even the most experienced interpreter can hardly be familiar with the mass spectral behavior of all compound classes, and here STIRS and PBM often provide the structural information that the interpreter had not been able to elucidate. In the future we expect to see such matching and interpretive systems implemented on the same dedicated computers which now acquire and reduce the data from GC/MS and high-resolution systems. In fact, the UMC microprocessor-driven quadrupole system automatically applies PBM to mass spectral data for routine analyses, giving quantities and confidence levels for the presence of 16 preselected compounds every second (4). Because the further application of GC/MS and LC/MS to important problems is still seriously limited by the ability to identify and interpret the resulting mass spectra, there is still a real need for research on optimization of such programs.

Literature Cited

1. Morrison, J. D. and Dromey, R. G., to be published.
2. Waller, G. R., "Biochemical Applications of Mass Spectrometry", Wiley-Interscience, New York, 1972.
3. Hewlett-Packard Corp., Palo Alto, California 94304.
4. Universal Monitor Corp., a division of McDonnell-Douglas Corp., 430 North Halstead Street, Pasadena, California 91107.
5. Burlingame, A. L., Kimble, B. J., and Derrick, P. J., Anal. Chem., (1976), 48, 368R.
6. Pesyna, G. M. and McLafferty, F. W., in "Determination of Organic Structures by Physical Methods", Nachod, F. C., Zuckerman, J. J., and Randall, E. W., Eds., Vol. 6, pp 91 - 155, Academic Press, New York, 1976.
7. Isenhour, T. L., Kowalski, B. R., and Jurs, P. C., Critical Review Anal. Chem., (1974), 4, 1.
8. Soltzberg, L. J., Wilkins, C. L., Kaberline, S. L., Lam, T. F., and Brunner, T. R., J. Am. Chem. Soc., (1976), 98, 7139.
9. Smith, D. H., Buchanan, B. G., Engelmore, R. S., Adlercreutz, H., and Djerassi, C., J. Am. Chem. Soc., (1973), 95, 6087.
10. McLafferty, F. W., Hertel, R. H., and Villwock, R. D., Org. Mass Spectrom., (1974), 9, 690.
11. Pesyna, G. M., Venkataraghavan, R., Dayringer, H. E., and McLafferty, F. W., Anal. Chem., (1976), 48, 1362.
12. Kwok, K.-S., Venkataraghavan, R., and McLafferty, F. W., J. Am. Chem. Soc., (1973), 95, 4185.
13. Dayringer, H. E., Pesyna, G. M., Venkataraghavan, R., and McLafferty, F. W., Org. Mass Spectrom., (1976), 11, 529.
14. Dayringer, H. E. and McLafferty, F. W., Org. Mass Spectrom., (1976), 11, 543.
15. Dayringer, H. E., McLafferty, F. W., and Venkataraghavan, R., Org. Mass Spectrom., (1976), 11, 895.
16. Office of Computer Services, Cornell University, Ithaca, New York 14853.
17. Salton, G., "Automatic Information Organization and Retrieval", McGraw-Hill, New York, 1968.
18. Grotch, S. L., 17th Annual Conference on Mass Spectrometry, Dallas, May, 1969, p 459.
19. Pesyna, G. M., McLafferty, F. W., Venkataraghavan, R., and Dayringer, H. E., Anal. Chem., (1975), 47, 1161.

20. Abramson, F. P., Anal. Chem., (1975), 47, 45.
21. Stenhagen, E., Abrahamsson, S., and McLafferty, F. W., "Registry of Mass Spectral Data", Wiley-Interscience, New York, 1974.
22. Pesyna, G. M., Ph.D. Thesis, Cornell University, 1975.
23. Hertz, H. S., Hites, R. A., and Biemann, K., Anal. Chem., (1971), 43, 681.
24. Hites, R. A. and Biemann, K., Adv. Mass Spectrom., (1968), 4, 37.
25. Speck, D. D., McLafferty, F. W., and Venkataraghavan, R., in preparation.
26. Isenhour, T. L., Lowry, S. R., Justice, J. B., Jr., McLafferty, F. W., Dayringer, H. E., and Venkataraghavan, R., Anal. Chem., submitted.
27. McLafferty, F. W., "Mass Spectral Correlations", Advances in Chemistry Series No. 40, American Chemical Society, Washington, 1963.
28. McLafferty, F. W., Anal. Chem., submitted.
29. Rotter, H. and Varmuza, K., Org. Mass Spectrom., (1975), 10, 874.

Acknowledgment

We are grateful to the National Institutes of Health (grant 16609) and to the Environmental Protection Agency (grant R804509) for financial support of this work, and to the National Science Foundation for partial funding of the DEC PDP-11/45 computer used in this research.

RECEIVED December 30, 1977

18

Structure Elucidation Based on Computer Analysis of High and Low Resolution Mass Spectral Data

DENNIS H. SMITH and RAYMOND E. CARHART

Departments of Chemistry, Genetics, and Computer Science,
Stanford University, Stanford, CA 94305

A tremendous effort has been directed toward development of advanced instrumentation for mass spectrometric analysis. Advancements include ever-increasing sensitivities and resolving powers, new ionization techniques, metastable ion probes of ion decomposition and structure and computer systems for rapid acquisition and reduction of data. We sometimes lose sight of the fact that these developments are designed to provide information about chemical and biochemical structures at greater depth and in greater detail than previously available. The ultimate goal in most research in mass spectrometry is to provide powerful tools for molecular structure elucidation, either directly, by exploitation of existing techniques, or indirectly by development of new techniques.

Concurrently, several computer-based techniques designed to assist chemists in the analysis and interpretation of mass spectral data have been developed. Analytical procedures for treatment of combined gas chromatographic/mass spectrometric data obtained at low resolving powers (1) (gc/lrms)

© 0-8412-0422-5/78/47-070-325$10.00/0

provide mass spectra of high quality for subsequent examination by manual or computer methods. Library search procedures (2) and their extensions (3) or pattern recognition programs (4) may provide clues to the identity of the structure or be used to determine the structure uniquely. A computer program for analysis of spectra based on class-specific fragmentation rules is available (5). These techniques have obvious limitations (3,4,5); in fact, the ability to interpret mass spectral data in terms of molecular structure lags far behind the capabilities of modern spectrometers to produce high quality data. There are several reasons for this lag: 1. There is no formal theory relating molecular structures to their respective mass spectra which has predictive power of use to the structural chemist. 2. (a corollary of 1) Mass spectrometry rarely provides detailed substructural information to assist in elucidating a structure except in known chemical contexts where previously developed rules may be applied retrospectively. 3. Current methods do not make adequate use of other knowledge about a particular compound, and; 4. The the combinatorial complexity of dealing with the actual information content of a mass spectrum has not, until now, been addressed. Points (3) and (4) will be discussed in some detail subsequently.

In our laboratories we have been trying to bring newly-developed computational tools to bear on general approaches to assisting structural chemists in interpretation of mass spectra. In this paper we will discuss the strengths and limitations of these new tools while assuming that the requisite mass spectral data are available. We are engaged in research which involves the gc/ms analysis of complex mixtures together with subsequent analysis of these data to extract spectra of individual components (1b) and search for the spectra in libraries of mass spectral data. The gc/lrms analyses provide an important pre-screening of mixtures. Combined gc/ms data obtained at high mass spectrometer resolving powers (gc/hrms) (6) yield elemental composition data for novel components. In any case the computer programs described in subsequent sections accept either low or high resolving power data (nominal masses or elemental compositions, respectively).

Computer-Assisted Structure Elucidation Based

Primarily on Mass Spectral Data.

Assume that an important unknown component has been observed in data from a gc/ms analysis of a mixture (*e.g.*, Figure 1). Assume further that the spectrum of the component (Figure 2) was not found in existing libraries. This problem becomes a classic problem of structure elucidation. One, for example, might attempt to isolate larger quantities and obtain additional spectral data. For many problems this is time-consuming and difficult. Realizing that high resolving power data are less ambiguous than data provided by the low resolution spectrum (Figure 2), one might obtain a gc/hrms spectrum and determine elemental compositions for the observed ions. The data obtained in the example are presented in Table I for major ions in the spectrum. These data may be the only data one can

TABLE I. Elemental Compositions for Significant Fragment Ions in the Mass Spectrum Given in Figure 2.

m/e	Composition
293	$C_{15}H_{19}NO_5$
261	$C_{14}H_{15}NO_4$
234	$C_{13}H_{16}NO_3$
202	$C_8H_{12}NO_5$
174	$C_7H_{12}NO_4$
142	$C_6H_8NO_3$
116	$C_5H_{10}NO_2$
91	C_7H_7

easily obtain to determine structural information about the unknown. Many current programs operate under the assumption that these are the only data. But, of course, this is not true. In almost every instance a great deal of additional information about the unknown is available. Some of this information is factual, for example, the physical and chemical properties and the source of the materials,

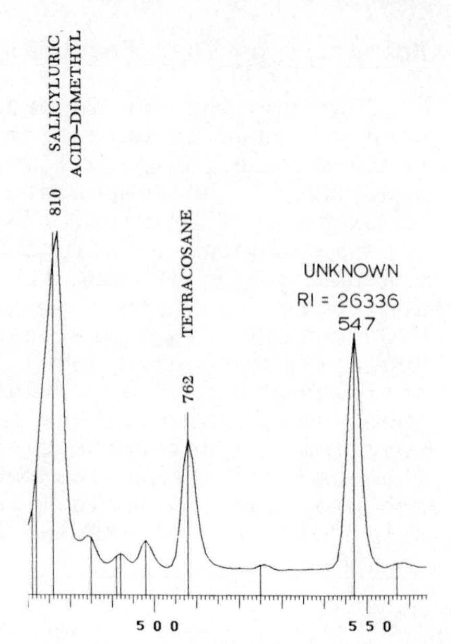

Figure 1. Total ion current plot, scans 470–565, of the GC/lrms analysis of the ether/ethyl acetate extracts of an acidified human urine subsequent to a 1 hr 1N NaOH hydrolysis. Numbers and names associated with component spectra detected by the CLEANUP program (1b) refer to the match scores and names of close library matches. Tetracosane, scan 508, is an internal standard for relative retention index calculations.

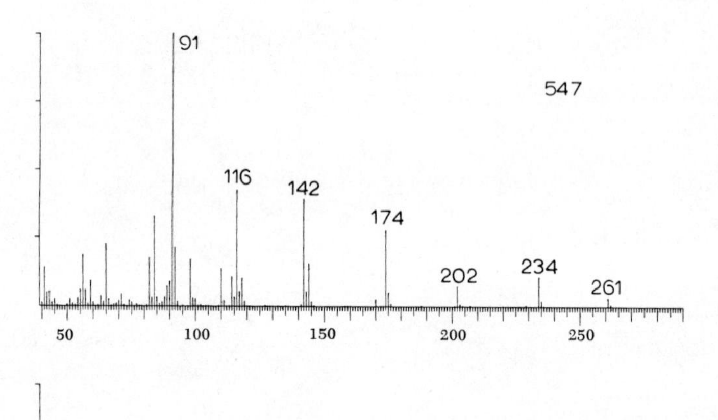

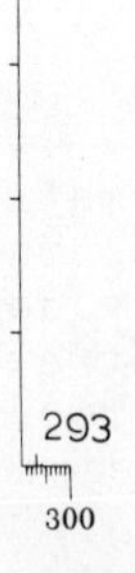

Figure 2. 70 eV low resolution mass spectrum obtained for the component eluting at scan 547 (see total ion current plot, Figure 1)

and isolation and derivatization procedures. Other information is judgmental; for example, knowledge of other compounds present in the same mixture, plausibility of chemical or biochemical processes which may have yielded the compound, and good intuitions. If this information can be brought to bear on the problem, it should be easier to solve.

We have developed the CONGEN program for computer-assisted structure elucidation. (7) This program has a flexible mechanism for expression and use of constraints on the plausibility of certain chemical features (substructures, ring systems). (8) Although designed for the general problem of incorporating structural inferences from different spectroscopic techniques or other sources of chemical information, we are introducing capabilities for more thorough use of mass spectral data. The capabilities were developed initially in earlier DENDRAL work. (5,9) The importance of CONGEN in problems such as that outlined above is that it gives the chemist a mechanism for exploring structural possibilities under constraints expressing his/her factual or judgmental knowledge. This knowledge can be defined, applied to a current problem, and saved for future use in related problems, as illustrated below.

Returning to the example, it is foolhardy to consider structural possibilites in the light of the presumed molecular ion, $C_{15}H_{19}NO_5$. Without constraints, the number of possible structures is huge. However, knowledge that the compound is a component of a mixture of organic compounds isolated from human urine, and that the urine was subjected to basic hydrolysis prior to extraction provides additional information which can to some extent be expressed as structural constraints (see below). More specific structural information is available from the fact that the fraction consists of ether/ ethyl acetate extractable organic acids, which were subsequently esterified using diazomethane prior to gc/ms analysis.

This "history" of the sample provides a tremendous reduction in the scope of possible structures. Chemists use this reasoning automatically in manual examination of structural possibilities. To be truly effective, a program must somehow provide the same capabilities when confronted with a mass spectrum. To assist chemists in making use of such reasoning, we are extending CONGEN to allow exploration of structural possibilities for an unknown within constraints provided by the mass spectrum and by factual and judgmental knowledge supplied by the chemist. We are developing two general approaches to the use of mass spectral data. These two approaches, MSPRUNE and Mass Distribution Graphs (MDG's), mirror the classic interplay between maximum use of information in retrospective testing vs. prospective guidance (planning) toward hypothetical solutions in the problem-solving paradigm of heuristic search. (5,7) In principle, the approches are complementary. They will yield the same answers by working on a problem from two different directions. In practice we have made more progress to date on the former (see below). We will illustrate both with examples.

I. MSPRUNE - Retrospective Testing of Structural Candidates.

If it is possible to arrive at a set of structural candidates for an unknown based on constraints derived from chemical considerations, other spectroscopic data and/or characteristic ions in a mass spectrum, it should be possible to test each candidate in some detail to determine if it is capable of producing the observed spectrum. MSPRUNE makes such tests and will "prune", or reject those structures which could not have yielded observed ions.

In many problems, including the example (Figure 2, Table I), it is possible to arrive at a reasonable set of candidate structures for an unknown using available data and the following general procedure.

A. Determine the molecular weight and formula. In the example, a candidate molecular ion is found at m/e 293, of composition $C_{15}H_{19}NO_5$. This ion is selected by MOLION (10), our molecular ion determin-

TABLE II. Molecular Ion Candidates and Rankings for the Spectrum Given in Figure 2.

Candidate	Ranking
m/e 293	100
294	71
325	50
352	46
308	43

ation program, and makes sense chemically. The five best candidates and their rankings are given in Table II.

B. Derive superatoms (7) and constraints from available data. In the example, knowledge of the source of the sample tells us that it is an organic acid from human urine and was esterified to form methyl esters prior to gc/ms analysis. This fraction contains aromatic and aliphatic acids in addition to conjugates of these acids with basic nitrogens, such as in amino acids. We can define a set of superatoms, or building blocks, which can be used to construct structures and can be saved on a computer file for future use in related problems. Such a set of superatoms with their associated names is shown in Figure 3, where the bonds to unspecified atoms are free valences which will subsequently be bonded to other atoms including hydrogen. In the example, the abundant m/e 91 ion suggests the superatom BZ (Figure 3), with no other substituents attached to the aromatic ring. The number of oxygen atoms and degree of unsaturation suggest two methyl ester functionalities (EST, Figure 3). The single nitrogen suggests at least the part structure AMI, arising from an acid conjugate with a basic nitrogen. There are perhaps other ways to phrase this problem, and alternative assumptions, but these assumptions will suffice for illustrative purposes.

C. Generate structures under appropriate constraints from the composition of superatoms and remaining atoms. In our example, the composition is $BZ_1EST_2AIM_1C_3H_5$. Without constraints there are 78 structural possibilities. This list contains many implausible structures. For example, if we assume that the compound is an amino acid conjugate, then a part structure similar to ACI must be present, in fact $-NHCH_{(1-2)}-COOCH_3$. Implementation of this constraint leaves 16 structures.

D. Use MSPRUNE to test the remaining structural candidates to determine which could yield key ions in the observed spectrum. MSPRUNE is an extension to CONGEN which allows interaction with the program to carry out the tests. MSPRUNE operates using the following sequence of steps:

D.1 Obtain fragmentation rules: A series of questions to the user of MSPRUNE/CONGEN elicits the mass spectrometric fragmentation rules to be used in interpretation of the data. We are currently restricted to rules used previously in the Meta-DENDRAL program INTSUM (9). These rules include constraints on cleavage of aromatic rings, multiple bonds, more than one bond to the same atom, number of steps in a fragmentation process, hydrogen transfers and loss (or transfer) of other neutral species such as water or carbon monoxide. For the example, the constraints summarized in Table III were used. This set of constraints is particularly restrictive. The only danger in this level of restriction is that an incorrect structure may yield a simpler explanation of the spectrum than the correct structure. The correct structure in such a case may be missed.

D.2 Input mass spectral ions to be explained. The user is then asked to input the ions he/she wishes to be explained. These ions may be entered

TABLE III. Fragmentation Process Constraints Used in the Analysis of the Mass Spectrum Given in Figure 2.

Constraints	User Response
Allow Adjacent Breaks:[a]	No
Allow Aromatic Breaks:	No[b]
Allow Breaks of Double or Triple Bonds:	No
Max Bonds to Break in a Single Step:[c]	1
Max Steps Per Process:	2
Max Bonds to Break in a Process	2
Allowed H Transfers:	-2 -1 0 1 2
Allowed Neutral Transfer:	-

[a] Cleavage of more than one, non-hydrogen bond to the same atom.

[b] I.e., do not cleave the aromatic ring.

[c] There are no aromatic rings or multiple bonds allowed to cleave; any number >1 is meaningless because there are no other degrees of unsaturation (rings); thus every cleavage yields a fragment ion. (9)

as either nominal masses or elemental compositions. Obviously the latter form is much more effective than nominal masses of possible compositional ambiguity. The method of selecting the ions to be used in the analysis is up to the user. In the example, we chose ions on the basis of a) high mass (intuitively of greater structural utility), and, b) high abundance. Ions of low mass and low abundance have a greater chance of resulting from either several different places in the molecule or from complex processes beyond the ability of the simple rules (Table III) to explain.

D.3 Test each candidate structure to determine if it could yield ions input. All possible fragmentations allowed by the constraints are determined for each structure, using an algorithm similar to that developed for INTSUM (9). For each fragmentation the mass and composition of the resulting ion is determined, including allowed hydrogen and/or neutral transfers. A simple comparison of these ions with the ions input reveals whether or not the structure could yield all ions input. If not, the structure is rejected. If so, the structure is retained. This experimental version of MSPRUNE takes no cognizance of ion abundances in this comparison. It makes a simple existence test only. Given the ions of Table I and the constraints of Table II, the list of 16 best candidates is trimmed to five 1 -5, (see Figure 4). If the original set of 78 possibilities is tested under the above conditions for MSPRUNE, 15 structures remain; the spectrum itself is a powerful constraint on possible structures. If only nominal masses are input, the set of 16 structures is not reduced by MSPRUNE; 16 structures remain. This kind of comparison can quantitate the information content of low *vs.* high resolution spectra.

E. Evaluate Remaining Structures. The structures which result can be examined with the help of CONGEN to determine additional constraints or to design experiments to differentiate among the possibilities. With knowledge of human metabolic processes and the chemistry of the isolation procedure, it is easy to assign 1, (see Figure 4) phenylacetylglutamic acid dimethyl ester (6), as the correct structure. Phenylacetic acid is normally conjugated with glutamine and excreted as phenylacetylglutamine. The base catalyzed hydrolysis converted the primary amide functionality into the observed carboxylic acid (6, see Figure 4): the dimethyl ester was formed on subsequent derivatization.

In larger problems, it is possible to use other features of CONGEN to test intuitions on possible structures. For example, in this problem where an amino acid conjugate was suspected, it was possible to test automatically every structure for the presence of one of the known amino acid skeletons. Of the five final structures, (1-5, see Figure 4) only one, 1, possesses a known amino acid skeleton. Of the 16 assumed conjugates, four formally possess a glycine, two a phenylalanine, one an aspartic and one 1, a glutamic skeleton. Possible origins of important fragmentations in the spectrum of 1 are illustrated in Scheme 1. We have no isotopic labelling data to support these suggestions.

An Application of MSPRUNE.

The procedure outlined above proved extremely helpful in analysis of unknown compounds observed in a gc low resolution ms experiment. A patient exhibiting signs of mental retardation was referred to Stanford. We examined organic compounds in this patient's urine using procedures described above for the example of structure 1. In fact, these compounds were observed in the same organic acid fraction, but this time prior to any alkaline hydrolysis.

The portion of the total ion current *vs.* scan number plot where the unknowns were observed is shown in Figure 5. The components in question, A-C, were detected by the program CLEANUP (1b) at scans 382, 402, and 406 (Figure 4). Resolved spectra (1b) are shown in Figures 6a and 6b. The spectra bear obvious similarities; in fact the spectra at scans 402 and 406 are nearly superimposable. Gc/hrms (6) analysis of this fraction lent further evidence to support the relationship of the unknown structures; all significant ions common to the spectra possess the same elemental compositions (shown in Figure 6a for the significant, higher mass ions). The MOLION program finds *m/e* 207, $C_{11}H_{13}NO_3$, the highest ranking molecular ion candidate for all three components.

The unknowns are apparently structural isomers. Thus, they can be investigated by CONGEN and MSPRUNE in a single run. For this application, we assumed the presence of an aromatic ring, a methyl ester functionality and an amide (conjugate)

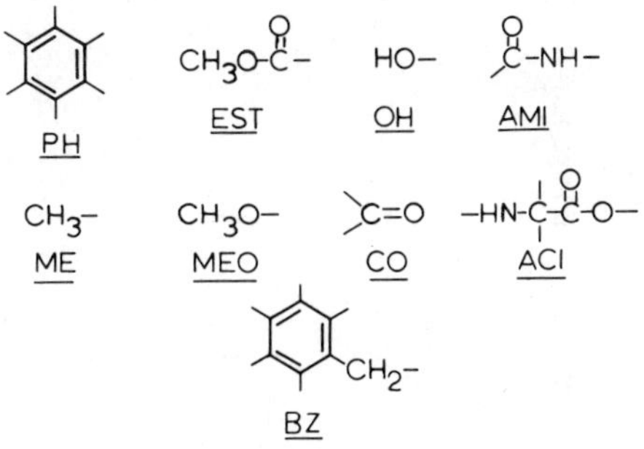

CH_3O-C(=O)– HO– –C(=O)–NH–

PH EST OH AMI

CH_3– CH_3O– >C=O –HN-C-C(=O)-O–

ME MEO CO ACI

CH_2–

BZ

Figure 3. A set of superatoms useful for considering structural candidates in the context of solvent extractable organic acids from human urine (derivatized to methyl esters)

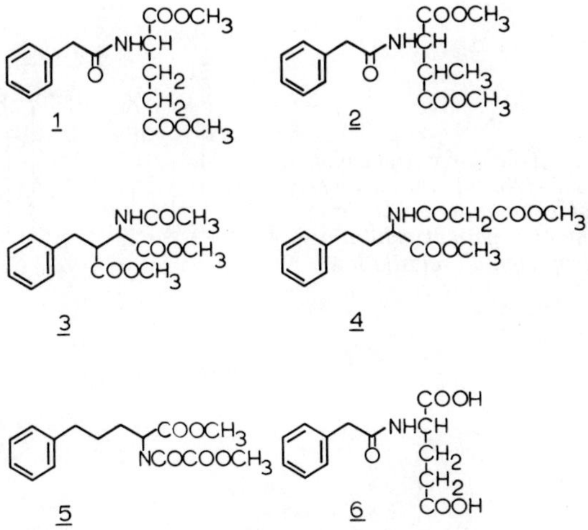

$COOCH_3$ NHCH CH_2 CH_2 $COOCH_3$

1

$COOCH_3$ NHCH $CHCH_3$ $COOCH_3$

2

$NHCOCH_3$ $COOCH_3$ $COOCH_3$

3

$NHCOCH_2COOCH_3$ $COOCH_3$

4

$COOCH_3$ $NCOCOOCH_3$

5

COOH NHCH CH_2 CH_2 COOH

6

Figure 4. Candidate structures for unknown represented by the mass spectrum in Figure 2

Scheme 1

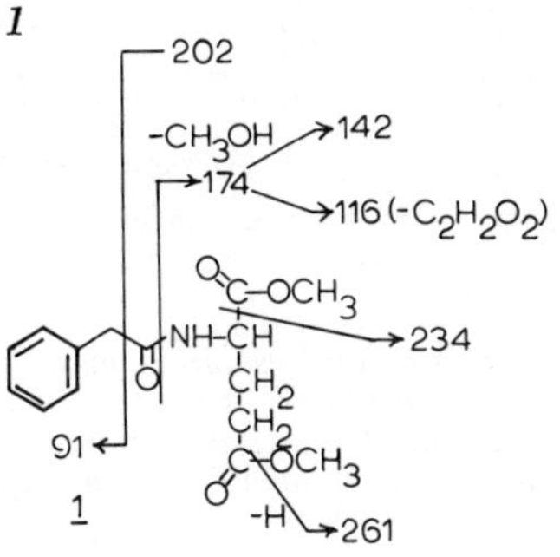

202

$-CH_3OH$ → 142

→ 174 → 116 $(-C_2H_2O_2)$

C(=O)–OCH_3

NH–CH → 234

CH_2

CH_2

C(=O)–OCH_3

91 ←

1

-H → 261

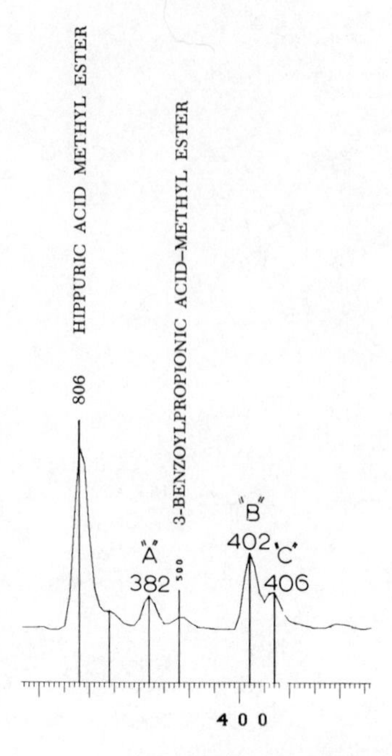

Figure 5. Total ion current plot, scans 357–426, of the solvent extractable organic acids from the urine of a patient exhibiting signs of mental retardation

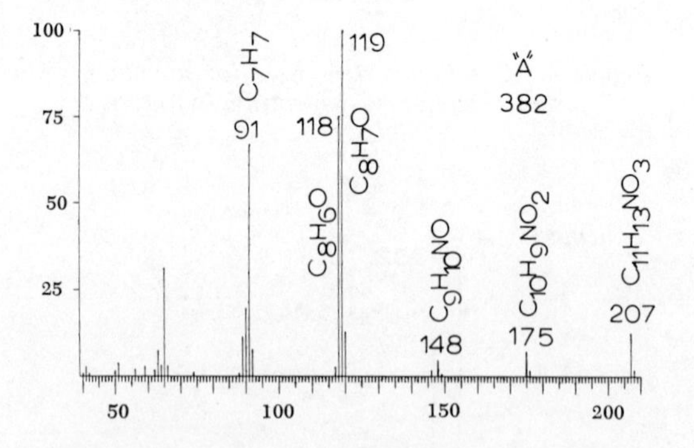

Figure 6a. 70 eV low resolution mass spectrum of component A, scan 382, of the total ion current plot of Figure 5. Elemental compositions were determined by a subsequent GC/hrms experiment (6)

linkage (superatoms PH, EST, and AMI respectively). Constraints forbid direct connection of the amide and ester functions, and alkyl chains of two or more carbons connected to the aromatic ring.

Generation of structures under these constraints yielded 52 possibilities. A number of methods were used to examine these structures. For example, automatic survey of the structural possibilities showed 6 structures which formally possess known amino acid skeletons. Use of MSPRUNE under constraints used for the previous example (Table III), and with the ions shown in Figure 5a was not too helpful; 36 candidates remained after this test. More severe constraints were then used, specifically considering only single rather than two step processes. This effected a dramatic reduction, leaving only four structures, 7 (see Figure 7) and the three isomers represented by 8. A possible source of each ion is shown in Scheme 2 for 8. Literature surveys revealed that 7 has been observed in the dog but never in man. Structure 8 is particularly attractive because there are three substitution isomers possible. These compounds are formally conjugates of toluic acids with glycine. We substantiated this hypothesis and proved the structures by synthesis of the three isomers. The retention indices (Table IV) and mass spectra agree completely. Further investigations indicated that xylenes are excreted by the body by first, oxidation of one of the methyl groups, and second, conjugation with glycine. Further, the relative concentrations of the three compounds closely approximate the relative amount of *ortho*, *meta* and *para* isomers in commercial mixtures of xylene. The patient had somehow been exposed to quantities of xylene.

TABLE IV. Relative Retention Indexes (R.R.I.) for Unknowns *A-C* and Synthetic *Ortho*, *Meta* and *Para*-toluylglycines, as Determined by CLEANUP (1b).

Unknown	Scan#	R.R.I.	Synthetic Compound	R.R.I.
A	382	2060	*ortho*-toluylglycine methyl ester	2058
B	402	2128	*meta*-toluylglycine methyl ester	2128
C	406	2141	*para*-toluylglycine methyl ester	2143

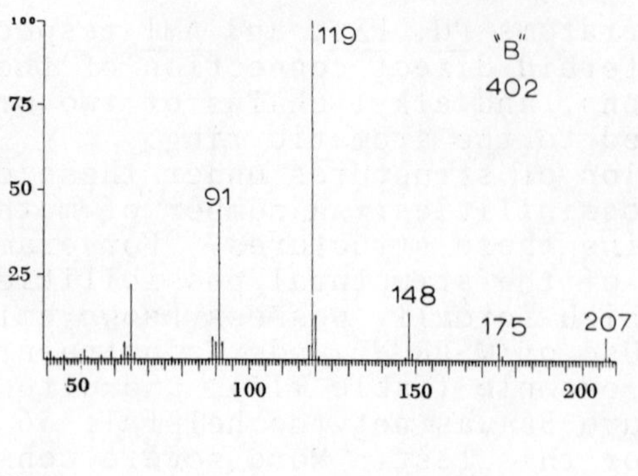

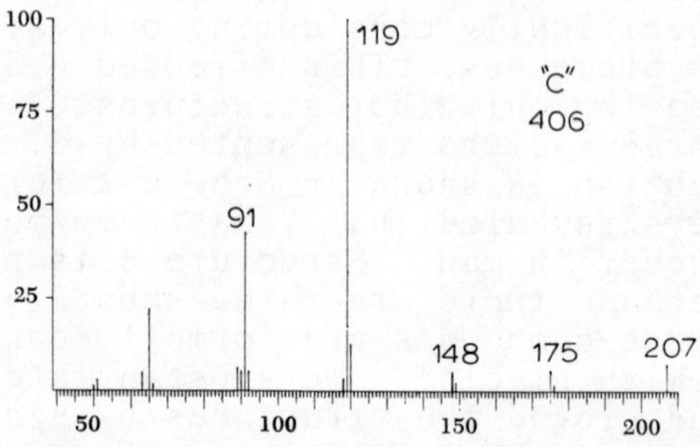

Figure 6b. 70 eV low resolution mass spectra of components B and C (Figure 5). Elemental compositions of major ions are those given in Figure 6a.

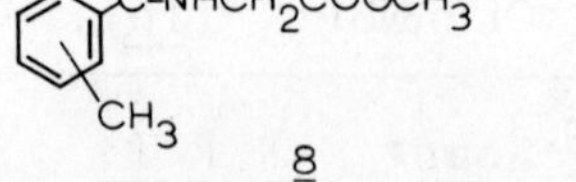

$C_6H_5CH_2C(=O)$-$NHCH_2COOCH_3$

7

Figure 7. Candidate structures for unknowns represented by spectra in Figures 6a and 6b

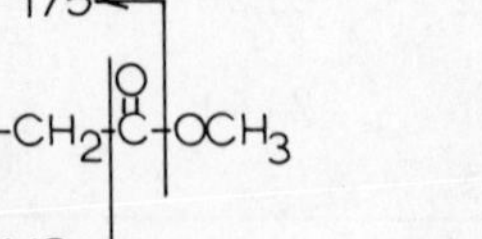

$CH_3C_6H_4C(=O)$-$NHCH_2COOCH_3$

8

Scheme 2

119 ← ; 175 ← −H ; 91 ← ; 148 ←

CH_3-C_6H_4-$C(=O)$-NH-CH_2-$C(=O)$-OCH_3

8

II. Prospective Generation of Structural Possibilities by MDGGEN.

It is a common intuition that mass spectra possess a tremendous amount of structural information, most of which goes unused. Every ion is a piece of the original structure. There are many such ions representing pieces which may overlap to an unknown extent. Ideally, one would like a computational method to perform all possible overlaps of pieces of structure represented by the ions, thereby constructing the set of candidate structures. We are exploring an approach which can in principle carry out this general analysis of mass spectra prospectively, by generation of structural possibilities directly from the mass spectral data. Note that this approach differs from MSPRUNE, which utilizes retrospective testing of previously generated candidate structures.

Our approach introduces the concept of a "mass distribution graph" (an "MDG"). An MDG is a graph whose nodes possess the properties of mass and/or elemental composition. Edges between nodes possess the property of multiplicity, *i.e.* single, double, *etc.* MDG's are incompletely specified chemical structures in that nodes of MDG's contain only mass or numbers of atoms without specification of how such mass or atoms are interconnected within each node. In addition, the edges which connect nodes are not precise chemical bonds in that multiple edges may or may not become multiple bonds in complete structures (see below). We have developed a method ("MDGGEN") to construct MDG's, and subsequently, complete structures, directly from the mass spectral data.

We illustrate the use of the concept of MDG's with a simple example, the mass spectrum (Figure 8) and structure of hexanal 9. The spectrum, Figure 8, shows the elemental compositions of ions used in the subsequent discussion. Although MDG's and the section of CONGEN, MDGGEN, which generates them can use nominal masses, results are more precise using elemental compositions, if known (see also MSPRUNE, above).

At the most general level, the MDG for a structure is a single node possessing all atoms and degrees of unsaturation (rings plus multiple bonds), or $C_6H_{12}O$ (one degree of unsaturation) for this

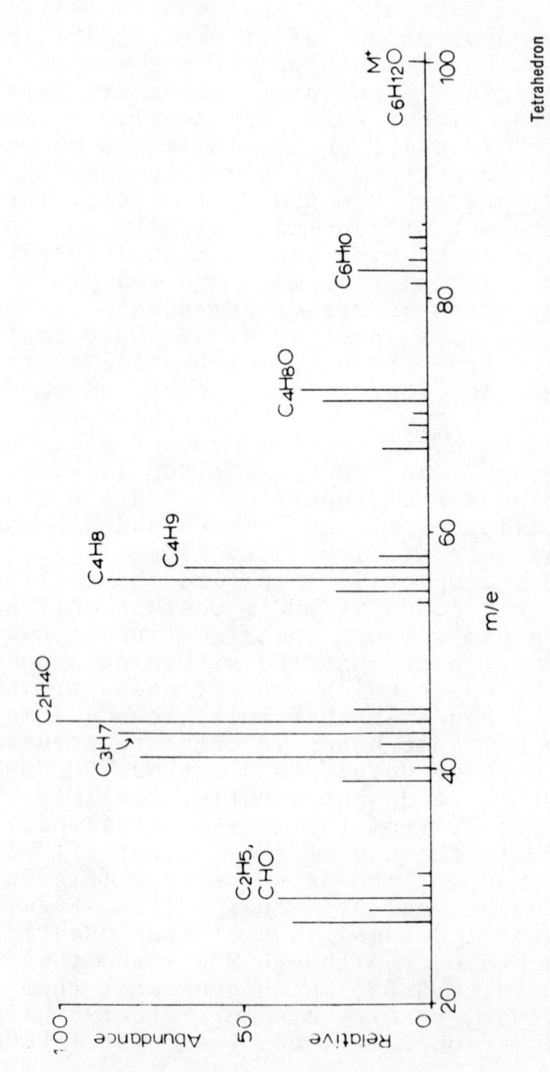

Figure 8. 70 eV low resolution mass spectrum of hexanal. Elemental compositions were determined by accurate mass measurements in a subsequent experiment at high resolving power.

example. MDGGEN requires input similar to that of MSPRUNE. It requires mass spectrometric fragmentation rules. The suite of rules is the same as that mentioned previously (Table III) with an additional parameter for the number of ions which are allowed to go unexplained. It also requires mass spectral data, input by the user, who again selects whatever peaks are deemed relevant.

MDGGEN operates by selecting input ions one at a time. Each ion selected results in an attempt by the program to apportion atoms (or mass) in nodes of the current MDG among nodes of new MDG's such that fragmentation of the new MDG's under input rules would yield the ion. The results for hexanal beginning with the molecular ion MDG ($C_6H_{12}O$) and using the observed ion C_3H_7 as the first selected ion are shown in Figure 9. The number of MDG's resulting after incorporation of a new ion depends on the fragmentation theory in use. Simple bond cleavage of the first MDG (Figure 9) without hydrogen transfer yields C_3H_7 directly. However, if hydrogen transfers (0, 1 or 2) are allowed, the second MDG is also a possibility. Cleavage of the single bond with transfer of two hydrogens into the charged fragment would yield C_3H_7. If two step processes are allowed, together with hydrogen transfers, then there are five other MDG's generated at this level (Figure 9) each of which would yield C_3H_7 with the indicated cleavages and hydrogen transfers.

MDGGEN uses some simple parity considerations to avoid generating nonsense MDG's which have unspecified connections. These considerations are the same as those used by mass spectroscopists. An odd mass ion containing an even number of nitrogens (e.g., 0) can arise from cleavage of an odd number of bonds (e.g., a single bond) accompanied by transfer of an even number of hydrogen atoms (e.g., 0). Or the same ion can arise from cleavage of an even number of bonds (e.g., 2) accompanied by a transfer of an odd number of hydrogens, and so forth. Thus the MDG C_3H_6-C_3H_6O as a single step, one hydrogen transfer process to yield C_3H_7 is nonsense -- only ionic structures or free radicals can be constructed from that MDG. Other constraints are applied to the MDG's if selected by the user, and some examples include forbidding cleavage of more than one bond to the same carbon atom or forbidding cleavage of multiple bonds. Note that the third MDG of Figure 9 cannot be rejected on

$M^+ = C_6H_{12}O$

	No. bonds	No. steps	Hyd trans.
C_3H_7 — C_3H_5O	1	1	0
C_3H_5 — C_3H_7O	1	1	2
C_3H_6 = C_3H_6O	2	1	1
C_3H_5 — C_3H_6 — OH	2	2	1
C_2H_5 — C_3H_6 — CHO	2	2	1
C_2H_3 — C_3H_6 — CH_3O	2	2	1
CH_3 — C_3H_6 — C_2H_3O	2	2	1

Figure 9. Mass distribution graphs illustrating the various ways of obtaining a C_3H_7 ion from a $C_6H_{12}O$ molecular ion under different assumptions concerning numbers of bonds cleaved, number of steps, and numbers of hydrogen atoms transferred in the fragmentation process

the latter constraint. The two interconnecting bonds may or may not be a double bond in final structures depending on how interconnections are made among atoms in various nodes of the MDG.

Each MDG resulting from application of the first ion is then further elaborated by selecting another ion and asking how, under the existing fragmentation theory, the next ion could be obtained from each MDG. This generally results in expansion of the MDG's into more complex graphs, accompanied however by a greater specificity of each node. This expansion is under the control of an operator which determines legal overlaps of existing MDG's with new MDG's implied by the next ion. An example is shown in Figure 10. The first MDG represents one way of obtaining the observed ion (Figure 8) C_4H_8 from hexanal, by cleavage of two bonds with no accompanying hydrogen transfer. Assume that the next ion applied was C_2H_5, which can be obtained from the MDG C_2H_5-C_4H_7O. There is only one way to perform the overlap (the operator is designated $\oplus$, Figure 10), yielding the MDG C_2H_5-C_2H_3=C_2H_4O. The elaborated MDG can yield both C_4H_8 and C_2H_5. However, it is structurally more specific because the assembly of atoms C_4H_8 is apportioned into more precise units, C_2H_5-C_2H_3. Assume that the next ion chosen was C_3H_7. One of the possible MDG's for obtaining C_3H_7, cleavage of the two bonds interconnecting C_3H_6=C_3H_6O with one hydrogen transfer, was shown in Figure 9. The overlap operator expands the previous MDG into five more elaborate, but more specific MDG's (Figure 10), from which all ions to that point (C_4H_8, C_2H_5 and C_3H_7) can be obtained. Note that MDG;s 1, 2, 4 and 5 at this level would be eliminated at this point given the constraint forbidding cleavage of more than one bond to the same carbon atom; there is no other way of obtaining the ion C_3H_7 from these MDG's given the assumed origin of C_3H_7 (above).

This procedure (under the constraints of single step processes, forbidding cleavage of multiple bonds, forbidding cleavage of more than one bond to the same carbon atom, transfer of 0 or 1 hydrogen atoms and allowing at most one ion to go unexplained) results in two possible MDG's for hexanal, using the ions labelled with elemental compositions in Figure 8. These are shown in Scheme 3. The first MDG is structurally very specific and yields only the correct structure of hexanal **9**. The second MDG indicates a more common

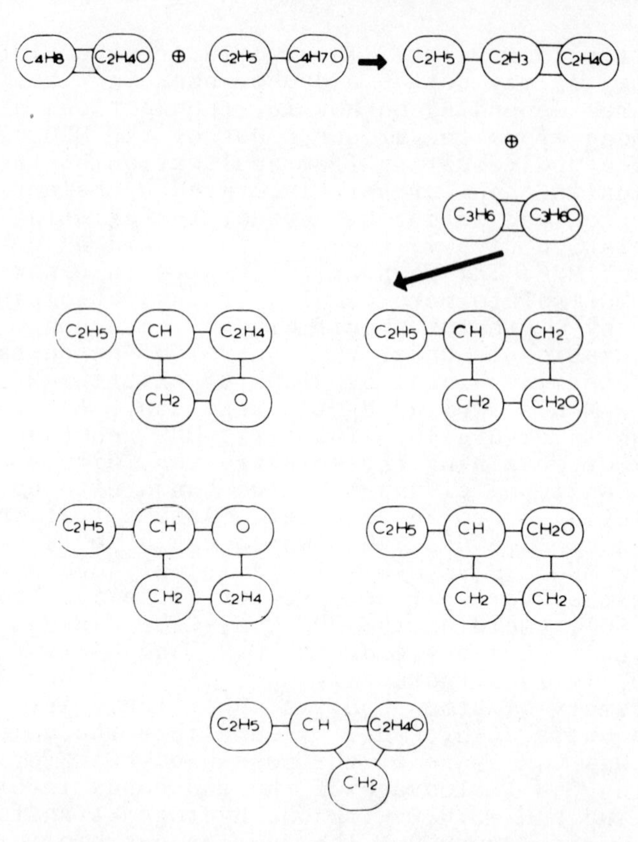

Figure 10. A sequence illustrating the elaboration of MDG's as explanations are determined for new data points (ions of specific elemental composition; see text for details)

Scheme 3

CHO

9

CH3 — CH2 — CH2 — CH2 — CH2 — CHO ⟶ CHO

9

CH3 — CH2 — CH2 — CH2 — CH2 — OH

–CH=CH–   CH2=C<

OH

10   11   OH

occurrence, where one or more nodes of the MDG can yield different part structures. In this example, the node $-C_2H_2-$ yields either the 1,1- or 1,2-disubstituted C=C, thus producing two alternative structures for the second MDG, 10 and 11, Scheme 3, both enols.

As a developing, experimental procedure, MDGGEN has several limitations. The foremost is the inherent combinatorial complexity of even simple problems. Even with severe constraints there are usually several MDG's resulting from the application of a single ion to each MDG at the previous step. The task of determining all the overlaps for several ions in the spectrum of a molecule of significant size is time-consuming. This is due in large part to our current inability to add chemical intelligence to the procedure. As we discussed above, any problem is simplified if there is a way of representing and utilizing constraints based on knowledge of the problem. We are pursuing this problem in MDGGEN by devising ways to incorporate known superatoms in the procedure. As in the problem of structure generation, problems are simplified when collections of atoms are aggregated into known substructures. Another limitation is the extent of duplication inherent in the procedure. The same structure in many cases can be constructed from two or more MDG's. For example, the structure of hexanal can be obtained from MDG's 1, 5 and 7, Figure 9. This results because MDG's are in a sense only explanations of ions. A given structure may account for a given ion in several different ways. We obtain a separate MDG for each of these explanations, each of which will yield the same structure. Another limitation is in the fragmentation theory itself (see Conclusions). If ions arise from processes more complex than those allowed by the theory, then the correct structure may well be missed or no structures obtained at all. For this reason, we currently allow some number of ions to go unexplained. In the hexanal example, this number was one. Under the theory used, structure 9 cannot lose the elements of H_2O observed in the spectrum, and 10 and 11 cannot yield HCO^+, also observed in the spectrum. But with one ion allowed to go unexplained, 9-11 result.

Conclusions.

We have presented some preliminary results of our efforts toward programs to assist in the general analysis of mass spectra. We have pointed out the need to incorporate as much factual and judgmental knowledge as is reasonable. Only in this way can mass spectral data, in conjunction with a generator of chemical structures, provide significant constraints on structural possibilities. Indeed, we feel that successful systems of the future must make use of such knowledge. Chemists use it in their reasoning about molecular structures and mass spectra; program assistants must do the same.

We are progressing along several lines to make the MSPRUNE and MDGGEN extensions to CONGEN more useful. We are currently investigating ways to make much more detailed statements of fragmentation theory to CONGEN, for use either by MSPRUNE or MDGGEN. We are using a subgraph representation of rules for input by a user (the representation used in Meta-DENDRAL [11]). Such rules may refer to alpha-cleavages, allylic cleavages, etc., or be detailed, class-specific rules in instances where that information about a structure is available. We are extending MSPRUNE to allow investigation of a complete spectrum and a ranking of candidate structures based on the agreement of predicted vs. observed spectra. These are all features of earlier DENDRAL work (5) now brought to bear on chemical problems in the general framework of CONGEN to provide a powerful tool for computer-assisted structure elucidation.

Acknowledgment.

We wish to thank the National Aeronautics and Space Administration (NGR-05-020-004), and the National Institutes of Health (RR 00612 and GM 20832) for their support of this research and for the NIH support of the SUMEX computer facility (RR 00785) on which the CONGEN program is developed, maintained and made available to a nationwide community of users.

Literature Cited.

1. a) Biller, J.E. and Biemann, K., Anal. Lett., (1974), 7, 515; b) Dromey, R.G., Stefik, M.J., Rindfleisch, T.C. and Duffield, A.M., Anal. Chem., (1976), 48, 1368.
2. See review by R.G. Ridley in "Biochemical Applications of Mass Spectrometry", G.R.

Waller, Ed., Wiley-Interscience, New York, NY, (1972), p. 177.
3. Kwok, K.-S., Venkataraghavan, R. and McLafferty, F.W., J. Amer. Chem. Soc., (1973), 95, 4185.
4. Jurs, P.C. and Isenhour, T.L., "Chemical Applications of Pattern Recognition", Wiley-Interscience, New York, NY (1975).
5. Smith, D.H., Buchanan, B.G., Engelmore, R.S., Duffield, A.M., Yeo, A., Feigenbaum, E.A., Lederberg, J. and Djerassi, C., J. Amer. Chem. Soc., (1972), 94, 5962.
6. Gc/hrms data were obtained using a Varian-Mat 711 mass spectrometer connected to a Digital Equipment Corp. PDP 11/45 computer system of our own design. 70 ev mass spectra were obtained at a resolving power of 5000 (extended by software doublet resolution routines) at 8 or 10 sec/decade scans.
7. Carhart, R.E., Smith, D.H., Brown, H. and Djerassi, C., J. Amer. Chem. Soc., (1975), 97, 5755.
8. Carhart, R.E., and Smith, D.H., Computers in Chemistry, 1, 79 (1976).
9. Smith, D.H., Buchanan, B.G., White, W.C., Feigenbaum, E.A., Lederberg, J. and Djerassi, C., Tetrahedron, (1973), 29, 3117.
10. Dromey, R.G., Buchanan, B.G., Smith, D.H., Lederbert, J. and Djerassi, C., J. Org. Chem., (1975), 40, 770.
11. Buchanan, B.G., Smith, D.H., White, W.C., Gritter, R.J., Feigenbaum, E.A., Lederbert, J., and Djerassi, C., J. Amer. Chem. Soc., (1976), 98, 6168.

RECEIVED December 30, 1977

INDEX

D

N

O

P

Q

R

New methods and techniques in mass spectral fragmentation have boosted the field of mass spectrometry into a major scientific discipline. Ongoing research in chemical analysis problems through identification and quantification of trace amounts of material with high performance mass spectrometry methods is reviewed in these 18 chapters. Special emphasis is placed on structure, property, and energy surface determinations of gas-phase ions. These mass spectrometry methods include chemical and field ionization, ultrahigh resolution, new methods of defocused metastable scans, collisional activation spectrometry, field ionization kinetics, spark source ionization and new techniques in gas chromatography/mass spectrometry.

Thirteen of the chapters describe analytical applications of these methods in biological and environmental systems that require exact mass measurements, GC/MS, and sample vaporization and ionization. Other areas in which the application of mass spectrometry to important problems has proved successful are in the fields of bio-